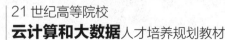

# 大数据技术
## 与应用基础项目教程

李俊杰 谢志明 ◎ 主编
肖政宏 石慧 谢高辉 杨泽强 ◎ 副主编

人民邮电出版社
北京

图书在版编目（CIP）数据

大数据技术与应用基础项目教程 / 李俊杰，谢志明主编. -- 北京：人民邮电出版社，2017.12（2020.8重印）
21世纪高等院校云计算和大数据人才培养规划教材
ISBN 978-7-115-47333-2

Ⅰ．①大… Ⅱ．①李… ②谢… Ⅲ．①数据处理—高等学校—教材 Ⅳ．①TP274

中国版本图书馆CIP数据核字(2017)第287410号

## 内 容 提 要

全书共十个项目，除了项目一介绍大数据基础理论外，其余项目均以实战为主线，内容循序渐进，逐步深入，围绕大数据技术的应用层层展开。内容主要包括大数据的基本概念、Ubuntu及服务安装配置、Hadoop集群部署、MapReduce编程、HBase数据库部署与应用、Hive数据仓库安装与应用、Pig数据分析、Sqoop数据迁移、Spark部署及数据分析等知识，最后以大数据技术的具体应用介绍了MapReduce大数据编程、Mahout的K-Means计算、决策树和随机森林的分类预测、频繁项集运算和关联分析等知识。本书秉承"实践为主、理论够用，注重实用"原则，将实验环节及实操内容融入各个知识点与课程教学中，以便读者能更好地学习和掌握大数据关键技术。

本书可作为院校计算机专业的教材，也可作为培训机构云计算和大数据等相关课程的培训用书，还可作为相关技术人员的参考书。

◆ 主　编　李俊杰　谢志明
　副主编　肖政宏　石　慧　谢高辉　杨泽强
　责任编辑　左仲海
　责任印制　马振武

◆ 人民邮电出版社出版发行　北京市丰台区成寿寺路11号
邮编　100164　电子邮件　315@ptpress.com.cn
网址　http://www.ptpress.com.cn
大厂回族自治县聚鑫印刷有限责任公司印刷

◆ 开本：787×1092　1/16
印张：19　　2017年12月第1版
字数：455千字　2020年8月河北第4次印刷

定价：49.80元

读者服务热线：(010)81055256　印装质量热线：(010)81055316
反盗版热线：(010)81055315
广告经营许可证：京东市监广登字20170147号

# 序 PREFACE

　　大数据产业的发展，人才培养是关键，特别是用得上、留得住的职业技术人才更是稀缺。本书的作者以面向应用、面向实战为指导思想，紧扣职业技术人才培养的特点，在知识点的讲解和实验中避免复杂的理论，使学生能快速上手体验、验证大数据系统的魅力，不断激发学生的学习兴趣。本书覆盖了大数据技术的主要技术要点，目的是让学生在实践的基础上养成在大数据系统环境下工作的习惯。本书可以作为大数据相关专业核心实践课程的教材。

　　在《大数据技术与应用基础项目教程》一书即将出版之际，作者邀请我写几句话，我欣然应允。本书的作者来自于汕尾职业技术学院云计算与大数据教学团队，由于工作的关系，我在汕尾职业技术学院挂职两年，见证了这个团队艰难成长的历程。汕尾职业技术学院地处广东省相对欠发达的粤东地区，发展条件和资源远远落后于珠三角地区的职业院校，但在云计算、大数据这一新的技术概念提出之后，汕尾职业技术学院的领导和老师敏锐地意识到这一新兴技术所带来的对相关职业技术人才需求的爆发，率先设立了云计算、大数据专业，并与广州五舟科技股份有限公司发起成立了广东省高等教育学会高职高专云计算与大数据专业委员会，并编写了国内第一本面向高职高专的云计算与大数据专业的国家级规划教材。在新专业建设的初期，师资问题是困扰专业发展的一个核心问题，本书的作者团队在最为困难的时候承担起了专业建设的重任，相继建设了多门云计算和大数据专业核心课程，所培养的学生在国内云计算和大数据竞赛中屡获佳绩，使汕尾职业技术学院在这一新专业建设上实现了弯道超车，成为广东省云计算、大数据专业建设的示范和标杆。

　　数据科学的春天已经到来，大数据、云计算、人工智能、高性能计算技术已呈现出相互交织和相互促进的局面，本书的出版正是响应了这一新时代的召唤，我想是时候我们共同去拥抱这一新技术时代的到来了。

广东省高等教育学会高职高专云计算与大数据专业委员会理事长

西南民族大学计算机科学与技术学院

并行计算实验室　王鹏

# 前言 FOREWORD

近年来，国家陆续出台了许多与大数据有关的各类专项支持政策，并把大数据列为国家重点支持和发展的战略新兴产业，大数据的迅速发展已然成为当今科技界、企业界甚至世界各国政府关注的热点。一个国家拥有数据的规模和运用数据的能力将成为综合国力的重要组成部分，对数据的占有和控制将成为国家间和企业间新的争夺焦点。全球著名管理咨询公司麦肯锡率先提出"大数据时代"的到来，并声称："数据已经渗透到当今各行各业的职能领域，成为重要的生产因素。"

本书的体系结构及知识点的分布按照学习思维逻辑由浅入深、循序渐进、以操代教的模式编排，编者建议初学者尽可能按照章节编排顺序学习和开展实训，这样有助于较为全面地了解大数据技术及其应用。为了快速提升学习者掌握大数据技术实战的能力，我们在每个项目或任务学习后安排多个精选的具有代表性的应用实例作为同步训练。全书注重实用性，图文并茂，知识点也以精练为主，读者在学习每一任务之时只需花上少许时间学习该任务的知识点即可直接进行实操或实训。

本书由广东省高等教育学会高职高专云计算与大数据专业委员会牵头，并组织专家、教师与企业共同参与编写，是全国高等院校计算机基础教育研究会 2016 年度高职科研规划纵向课题（课题编号：2016GHB02025、2016GHB02005）、广东省高等职业教育质量工程教育教学改革项目（课题编号：GDJG2015245、GDJG2015244）和广东省高职教育信息技术教指委教改项目（课题编号：XXJZW2015002）、广东省高等教育学会高职高专云计算与大数据专业委员会教育科研课题（课题编号：GDYJSKT16-01、GDYJSKT16-03、GDYJSKT16-05）、汕尾职业技术学院教学改革与科研重点立项课题（课题编号：SWKT15-002、SWKT16-002、swjy15-004、swjy16-012）的科研成果。本书编写期间得到了广东技术师范学院、广州五舟科技股份有限公司、汕尾市创新工业设计研究院的鼎力支持，同时还得到了汕尾职业技术学院各处系领导和老师的关怀和支持，正因为有了他们的支持和帮助，我们才能如期完成本书的撰写和编排工作。

大数据技术涉及面很广且更新较快，编者经过大量的重复实验与多年的工作经验积累，参考并引用了无数前辈学者的研究成果、论述，并在与之相关的科学研究上不断扩充和改进，编者在此向这些前辈学者深表敬意。大数据技术作为一门正在高速发展的新兴技术日新月异，新技术、新方法、新架构层出不穷，加之编者的经验和水平有限，本书的结构、内容肯定存在诸多疏漏和不妥之处，诚恳接受读者的批评与指正。如有任何问题和建议，可发送电子邮件至 441814853@qq.com，以便我们及时修正完善。

为方便读者学习和教学需要，本教材配备了大量的电子资源，欢迎读者登录人民邮电出版社教育社区（http://www.ryjiaoyu.com）免费下载使用，同时欢迎相关课程的教师加入云计算大数据 HPC 教育 QQ 群（321168742）讨论交流。读者还可以通过发送邮件给编者以获得更多下载资源（资源配套邮箱：gdswyun@126.com）。

感谢您使用本书，期待本书能成为您的良师益友。

编 者
2017 年 6 月于云计算与数据中心工程设计研究所

# 目录 CONTENTS

## 项目一　走进大数据　1

- 任务 1　概述大数据的内涵　2
- 任务 2　关注大数据的影响　6
- 任务 3　认识常见的大数据计算模式　11
- 任务 4　厘清大数据处理的基本流程　14
- 任务 5　大数据应用大显神通　15
- 任务 6　大数据的发展及面临的挑战　18
- 【同步训练】　22

## 项目二　Ubuntu 及服务安装配置　23

- 任务 1　安装 Ubuntu Server　24
- 任务 2　搭建 FTP 系统　33
- 任务 3　搭建 MySQL 数据库系统　37
- 任务 4　安装 Ubuntu Desktop　41
- 【同步训练】　47

## 项目三　Hadoop 集群部署　48

- 任务 1　构建集群系统　49
- 任务 2　SSH 证书登录　54
- 任务 3　Hadoop 部署与使用　56
- 【同步训练】　76

## 项目四　MapReduce 编程　77

- 任务 1　搭建 MapReduce 开发平台　78
- 任务 2　编写单词计数程序　82
- 任务 3　编写气象数据分析程序　96
- 【同步训练】　111

## 项目五　HBase 数据库部署与应用　112

- 任务 1　HBase 部署　113
- 任务 2　HBase Shell　125
- 任务 3　HBase 编程　136
- 任务 4　MapReduce 与 HBase 集成　144
- 【同步训练】　154

## 项目六　Hive 数据仓库安装与应用　155

- 任务 1　安装 Hive　155
- 任务 2　Hive CLI　168
- 任务 3　Hive 编程　182
- 任务 4　Hive 与 HBase 集成　186
- 【同步训练】　187

## 项目七　Pig 数据分析　188

- 任务 1　Pig 安装及使用　188
- 任务 2　Pig 高级编程　200
- 【同步训练】　209

## 项目八　Sqoop 数据迁移　210

任务1　Sqoop 安装及 MySQL 与 HDFS 数据迁移　210

任务2　MySQL 与 Hive/HBase 数据转移　216

【同步训练】　218

## 项目九　Spark 部署及数据分析　219

任务1　Spark 部署　220

任务2　Spark 数据分析　229

任务3　Spark 编程　241

【同步训练】　252

## 项目十　大数据综合实例编程　253

任务1　MapReduce 大数据处理　254

任务2　Mahout 的 K-Means 计算　266

任务3　决策树和随机森林的分类预测　272

任务4　频繁项集计算与关联分析　287

【同步训练】　297

## 参考文献　298

# PART 1 项目一 走进大数据

【项目介绍】

本项目的主要目标是理解大数据的内涵和定义，学会常用的大数据的计算模式和厘清大数据处理的基本流程，了解大数据的应用情况及对当今社会各界产生的影响，乐观看待大数据发展所带来的一系列问题，积极面对各种挑战等。

本项目分为以下 6 个任务：
- 任务 1  概述大数据的内涵
- 任务 2  关注大数据的影响
- 任务 3  认识常见的大数据计算模式
- 任务 4  厘清大数据处理的基本流程
- 任务 5  大数据应用大显神通
- 任务 6  大数据的发展及面临的挑战

【学习目标】

一、知识目标

- 了解大数据的基本概念、产生的原因及其特性。
- 理解科学研究的 4 种范式。
- 清楚了解大数据对社会发展及思维方式产生的影响。
- 学会常用的几种大数据计算模式。
- 掌握大数据处理的基本流程，如数据采集、数据清洗、数据分析和数据解释等过程。
- 了解大数据在各个领域的应用情况，熟悉大数据推荐系统及搜索引擎系统等。
- 了解大数据的发展历程及发展现状，积极面对各种挑战。

二、能力目标

- 能够从大数据角度理解大数据对生产力的影响。
- 学会按大数据处理流程挖掘分析有价值的信息。
- 能够理解使用大数据计算模式。
- 能够理解使用大数据推荐系统及搜索引擎系统。

● 能够理解大数据的发展及面临的挑战。

# 任务1　概述大数据的内涵

## 【任务概述】

大数据已成为社会各界研究及关注的焦点。本任务着重介绍大数据的内在含义，其中包括大数据的多种定义表述、大数据产生的原因、大数据特性的演进以及在大数据时代才出现的一些数据计量单位。

## 【支撑知识】

近几年，大数据迅速发展成为科技界和企业界甚至世界各国政府关注的热点。人们对于大数据的挖掘和运用，预示着新一波生产力增长和消费盈余浪潮的到来。美国政府认为大数据是"未来的钻石矿和新石油"，一个国家拥有数据的规模和运用数据的能力将成为综合国力的重要组成部分，对数据的占有和控制将成为国家间和企业间新的争夺焦点。全球著名管理咨询公司麦肯锡（McKinsey&Company）首先提出了"大数据时代"的到来并声称："数据已经渗透到当今各行各业的职能领域，成为重要的生产因素。"

数据的产生方式由"人机""机物"的二元世界向着融合社会资源信息系统及物理资源的三元世界转变，数据规模呈膨胀式发展，例如，互联网领域中，谷歌搜索引擎的每秒使用用户量达到200万；科研领域中，仅某大型强子对撞机在一年内积累的新数据量就达到15PB左右；电子商务领域中，eBay的分析平台每天处理的数据量高达100PB，超过了纳斯达克交易所每天的数据处理量；"双十一"大型商业活动中，淘宝商城屡创神话，销售额由2010年的9亿元一路攀升到现今的1200多亿元，支付宝平台平均每秒成功交易12万笔，交易覆盖235个国家和地区；航空航天领域中，仅一架双引擎波音737飞机在横贯大陆飞行的过程中，传感器网络便会产生近240TB的数据。综合各个领域，目前积累的数据量已经从TB量级上升至PB、EB甚至已经达到ZB量级，其数据规模已经远远超出了现有通用计算机所能够处理的量级。

根据全球著名咨询机构互联网数据中心（Internet Data Center，IDC）做出的估测，人类社会产生的数据一直都在以每年50%的速度增长，也就是说，每两年数据量就会增加一倍，即已形成了"大数据摩尔定律"，这意味着人类在最近两年产生的数据量相当于之前产生的全部数据量之和。据IDC统计，2011年全球被创建和复制的数据总量为1.8ZB，到2020年这一数据将攀升到40ZB，是2012年的12倍。而我国的数据量到2020年将超过8ZB，是2012年的22倍。其中80%以上来自于个人（主要是图片、视频和音乐），远远超过人类有史以来所有印刷材料的数据总量（200PB）。目前，全球的数据量正以每18个月翻一番的速度呈膨胀式增长，数据量的飞速增长同时也带来了大数据技术和服务市场的繁荣发展。

### 一、大数据的定义

"大数据"一词由英文"Big Data"翻译而来，是近几年兴起的概念。往前追溯却发现由来已久，早在1980年就已由美国著名未来学家阿尔文·托夫勒在《第三次浪潮》一书中，将大数据赞颂为"第三次浪潮的华彩乐章"。

"大数据"并不等同于"大规模数据",那么何谓"大数据"呢?迄今并没有公认的定义,由于大数据是相对概念,因此,目前的定义都是对大数据的定性描述,并未明确定量指标。维基(Wiki)百科从处理方法角度给出的大数据定义,即大数据是指利用常用软件工具捕获、管理和处理数据所耗时间超过可容忍时间限制的数据集。麦肯锡公司认为将数据规模超出传统数据库管理软件的获取存储管理,以及分析能力的数据集称为大数据;高德纳咨询公司(Gartner)则将大数据归纳为需要新处理模式才能增强决策力、洞察发现力和流程优化能力的海量高增长率和多样化的信息资产;徐宗本院士在第 462 次香山科学会议上的报告中,将大数据定义为不能够集中存储并且难以在可接受时间内分析处理,其中个体或部分数据呈现低价值性而数据整体呈现高价值的海量复杂数据集。虽说这些关于大数据定义的定义方式角度及侧重点不同,但是所传递的信息基本一致,即大数据归根结底是一种数据集,其特性是通过与传统的数据管理及处理技术对比来凸显,并且在不同需求下,其要求的时间处理范围具有差异性,最重要的一点是大数据的价值并非来自数据本身,而是来自由大数据所反映的"大决策""大知识""大问题"等。

从宏观世界角度来看,大数据则是融合物理世界、信息空间和人类社会三元世界的纽带,因为物理世界通过互联网、物联网等技术有了在信息空间中的大数据反映,而人类社会则借助人机界面、脑机界面、移动互联等手段在信息空间中产生自己的大数据映像。从信息产业角度来讲,大数据还是新一代信息技术产业的强劲推动力。所谓新一代信息技术产业,本质上是构建在第三代平台上的信息产业,主要是指云计算、大数据、物联网、移动互联网(社交网络)等。

**二、大数据产生的原因**

"大数据"并不是一个凭空出现的概念,其出现对应了数据产生方式的变革,生产力决定生产关系的道理对于技术领域仍然是有效的,正是由于技术发展到了一定的阶段才导致海量数据被源源不断地生产出来,并使当前的技术面临重大挑战。归纳起来大数据出现的原因有以下几点。

(1)数据生产方式变得自动化

数据的生产方式经历了从结绳计数到现在的完全自动化,人类的数据生产能力已不可同日而语。物联网技术、智能城市、工业控制技术的广泛应用使数据的生产完全实现了自动化,自动数据生产必然会产生大量的数据。甚至当前人们所使用的绝大多数数字设备都可以被认为是一个自动化的数据生产设备:我们的手机会不断与数据中心进行联系,通话记录、位置记录、费用记录都会被服务器记录下来;我们用计算机访问网页时访问历史、访问习惯也会被服务器记录并分析;我们生活的城市、小区遍布的传感器、摄像头会不断产生数据并保证我们的安全;天上的卫星、地面的雷达、空中的飞机也都在不断地自动产生着数据。

(2)数据生产融入每个人的日常生活

在计算机出现的早期,数据的生产往往只是由专业的人员来完成的,能够有机会使用计算机的人员通常都是因为工作的需要,物理学家、数学家是最早一批使用计算机的人员。随着计算机技术的高速发展,计算机得到迅速普及,特别是手机和移动互联网的出现使数据的生产和每个人的日常生活结合起来,每个人都成为数据的生产者:当你发出一条微博时,你在生产数据；当你拍出一张照片时，你在生产数据；当你使用手中的市民卡和银行卡时，你在生产数据；当你在 QQ 上聊天时，你在生产数据；当你在用微信发朋友圈或聊天时，你在生产数据；当你在玩游戏时，你在生产数据。数据的生产已完全融入人们的生活：在地铁上，你在生产数据；在工作单位，你在生产数据；在家里，你也在生产数据。个人数据的生产呈现出随时、随地、移动化的趋势，我们的生活已经是数字化的生活，如图 1-1 所示。

图 1-1　数据生产融入人们的生活

（3）图像和音视频数据所占比例越来越大

人类在过去几千年主要靠文字记录信息，而随着技术的发展，人类越来越多地采用视频、图像和音频这类占用空间更大、更形象的手段来记录和传播信息。从前聊天我们用文字，现在用微信和视频，人们越来越习惯利用多媒体方式进行交流，城市中的摄像头每天都会产生大量视频数据，而且由于技术的进步，图像和视频的分辨率变得越来越高，数据变得越来越大。

（4）网络技术的发展为数据的生产提供了极大的方便

前面说到的几个大数据产生原因中还缺乏一个重要的引子：网络。网络技术的高速发展是大数据出现的重要催化剂：没有网络的发展就没有移动互联网，我们就不能随时随地实现数据生产；没有网络的发展就不可能实现大数据视频数据的传输和存储；没有网络的发展就不会有现在大量数据的自动化生产和传输。网络的发展催生了云计算等网络化应用的出现，使数据的生产触角延伸到网络的各个终端，使任何终端所产生的数据能快速有效地被传输并存储。很难想象在一个网络条件很差的环境下能出现大数据，所以，可以这么认为：大数据的出现依赖于集成电路技术和网络技术的发展，集成电路为大数据的生产和处理提供了计算能力的基础，网络技术为大数据的传输提供了可能。

（5）云计算概念的出现进一步促进了大数据的发展

云计算这一概念是在 2008 年左右进入我国的，而最早可以追溯到 1960 年人工智能之父麦卡锡所预言的"今后计算机将会作为公共设施提供给公众"。2012 年 3 月在国务院政府工作报告中云计算被作为附录给出了一个政府官方的解释，表达了政府对云计算产业的重视，在政府工作报告中云计算的定义是这样的："云计算：是基于互联网的服务的增加、使用和交付模式，通常涉及通过互联网来提供动态易扩展且经常是虚拟化的资源。是传统计算机和网络技术发展融合的产物，它意味着计算能力也可作为一种商品通过互联网进行流通。"云计算的出现使计算和服务都可以通过网络向用户交付，而用户的数据也可以方便地利用网络传递，云计算这一模式网络的作用被进一步凸显出来，数据的生产、处理和传输可以利用网络快速地进行，改变传统的数据生产模式，这一变化大大加快了数据的产生速度，对大数据的出现起到了至关重要的作用。

三、大数据特性

在大数据的定义中，已经包含了大数据的特性，即数据量大、处理速度要求快、价值密度低等，目前对于大数据的特性认可度较高的是 3V 特性：数据的规模性（Volume）、高速性（Velocity）及数据结构多样性（Variety），而在此基础上已经有不同的公司及研究机构对其进行了扩展，大数据特性描述的演化如表 1-1 所示。

表 1-1 大数据特性描述的演化情况

| 特 点 | 提出时间 | 作者或机构 | 内 涵 |
|---|---|---|---|
| 规模性（Volume） | 2001 年 | Doug Laney（高德纳咨询公司） | 体量大，数据量级可达 TB、PB 乃至 EB 以上 |
| 高速性（Velocity） | | | 数据分析和处理速度快，俗称"秒级定律" |
| 多样性（Variety） | | | 数据类型多样 |
| 价值性（Value） | 2012 年 | 咨询机构 IDC | 价值稀疏性，即具有高价值低密度的特点 |
| 真实性（Veracity） | 2012 年 | IBM（国际商业机器公司） | 数据反映客观事实 |
| 易变性（Variability） | 2012 年 | Brian Hopkins&Boris Evelson（弗雷斯特研究公司） | 大数据具有多层结构 |

由表 1-1 可以看出，随着时间的演化，业界对于大数据的认识也更深入、全面。除以上对大数据特性的通用性描述之外，不同应用领域的大数据的具体特性也存在差异性。如互联网领域需要实时处理和分析用户购买行为，以便及时制定推送方案，返回推荐结果来迎合和激发用户的消费行为，精度及可靠性要求较高；医疗领域需要根据用户病例及影像等信息判断病人的病情，由于其与人们的健康息息相关，所以，其精度及可靠性要求非常高。表 1-2 列举了不同领域大数据的具体特点及应用案例。

表 1-2 不同领域大数据的具体特点及应用案例

| 领 域 | 用户数目 | 响应时间 | 数据规模 | 可靠性要求 | 精度要求 | 应用案例 |
|---|---|---|---|---|---|---|
| 科学计算 | 小 | 慢 | TB | 一般 | 非常高 | 大型强子对撞机数据分析 |
| 金融 | 大 | 非常快 | GB | 非常高 | 非常高 | 信用卡营销 |
| 医疗领域 | 大 | 快 | EB | 非常高 | 非常高 | 病历、影像分析 |
| 物联网 | 大 | 快 | TB | 高 | 高 | 迈阿密戴德县的智慧城 |
| 互联网 | 非常大 | 快 | PB | 高 | 高 | 网络点击流入侵检测 |
| 社交网络 | 非常大 | 快 | PB | 高 | 高 | Facebook、QQ 等结构挖掘 |
| 移动设备 | 非常大 | 快 | TB | 高 | 高 | 可穿戴设备数据分析 |
| 多媒体 | 非常大 | 快 | PB | 高 | 一般 | 史上首部大数据制作的电视剧《纸牌屋》 |

由表 1-2 可以看出，不同应用领域的数据规模、用户数目及精度要求等均存在较大的差异，例如，互联网领域与人的正常活动息息相关，其数据量达 PB 级别，用户数目非常大，而且以用户实时性请求为主。与此不同，在科研领域中，其用户数目相对较少，产生的数据量级别在 TB 级。因此，对大数据后续的分析及处理必须因地制宜，才能实现大数据价值的最大化。

### 四、数据的计量

大数据出现后人们对数据的计量单位也逐步变化，常用的 KB、MB 和 GB 已不能有效地描述大数据。在大数据研究和应用时我们经常会接触到数据存储的计量单位。下面对数据存储的计量单位进行介绍。

计算机学科中一般采用 0、1 这样的二进制数来表示数据信息，信息的最小单位是 bit（比特），一个 0 或 1 就是一个比特，而 8bit 就是一字节（Byte），如 10010111 就是一 Byte。习惯上人们用大写的 B 表示 Byte。信息的计量一般以 $2^{10}$ 为一个进制，如 1024Byte=1KB（KiloByte，千字节），更多常用的数据单位换算关系如表 1-3 所示。

表 1-3 数据存储单位之间的换算关系

| 单位名称 | 换算关系 |
| --- | --- |
| Byte（字节） | 1 Byte=8 bit |
| KB（KiloByte，千字节） | 1 KB=1024 Byte |
| MB（MegaByte，兆字节） | 1 MB=1024 KB |
| GB（GigaByte，吉字节） | 1 GB=1024 MB |
| TB（TeraByte，太字节） | 1 TB=1024 GB |
| PB（PetaByte，拍字节） | 1 PB=1024 TB |
| EB（ExaByte，艾字节） | 1 EB=1024 PB |
| ZB（ZettaByte，泽字节） | 1 ZB=1024 EB |
| YB（YottaByte，尧字节） | 1 YB=1024 ZB |
| BB（Brontobyte，珀字节） | 1 BB=1024 YB |
| NB（NonaByte，诺字节） | 1 NB=1024 BB |
| DB（DoggaByte，刀字节） | 1 DB=1024 NB |

目前市面上主流的硬盘容量大都为 TB 级，典型的大数据一般都会用到 PB、EB 和 ZB 这 3 种单位。

## 任务 2　关注大数据的影响

### 【任务概述】

大数据对科学研究、思维方式和社会发展都具有重要而深远的影响。本任务除了重点介绍曾为大数据做出卓越贡献的科学家之外，还着重介绍了大数据所带来的影响，其中影响较深的有大数据对科学研究的影响及大数据对社会发展的影响，主要体现在大数据改变了科学研究的思维方式、大数据改变了人们的生存方式、大数据改变了人类的生产方式。

### 【支撑知识】

大数据对科学研究、思维方式和社会发展都具有重要而深远的影响。在科学研究方面，大数据使得人类科学研究在经历了实验、理论、计算 3 种范式之后，迎来了第四种范式——数据；在思维方式方面，大数据具有"全样而非抽样、效率而非精确、相关而非因果"三大显著特征，完全颠覆了传统的思维方式；在社会发展方面，大数据决策逐渐成为一种新的决策方式，大数据应用有力促进了信息技术与各行业的深度融合，大数据开发大大推动了新技术和新应用的不断涌现；在就业市场方面，大数据的兴起使得数据科学家成为热门职业；在人才培养方面，大数据的兴起，将在很大程度上改变我国高校计算机信息技术相关专业的现有教学和科研体制。

### 一、大数据之父——吉姆·格雷（Jim Gray）

云计算和大数据是密不可分的两个概念，云计算时代网络的高度发展，使每个人都成为数据产生者，物联网的发展更是使数据的产生呈现出随时、随地、自动化、海量化的特征，大数据不可避免地出现在了云计算时代。吉姆·格雷（见图1-2）生于1944年，在著名的加州大学伯克利分校计算机科学系获得博士学位，是声誉卓著的数据库专家、1998年度的图灵奖获得者。2007年1月11日在美国国家研究理事会计算机科学与通信分会上吉姆·格雷明确地阐述了科学研究第四范式——"数据密集型科学"，认为依靠对数据分析挖掘也能发现新的知识，其实质是科学研究将从以计算为中心向以数据为中心转变，即数据思维的到来。这一认识吹响了大数据前进的号角，计算应用于数据的观点在当前的云计算大数据系统中得到了大量的体现。在发表这一演讲后的十几天，2007年1月28日格雷独自驾船出海就再也没有了音信，虽然经多方努力搜寻，也没有发现他的一丝信息，人们再也没能见到这位伟大的天才科学家。

图1-2 吉姆·格雷

### 二、大数据对科学研究的影响

第四范式的命名是与之前的3种科学范式"实验科学""理论科学""计算科学"相呼应和一脉相承的，是人类在科学研究领域上新的发现与突破。这4种范式在不同时代或时期都给人类社会带了巨大的财富与文明，是人类发现世界、探索世界的利器，下面将分别对这4种范式进行简明扼要的表述，如图1-3所示。

（1）第一种范式：观测与实验科学

出于好奇的天性，人类一直都在不断地认识自己所生活的世界。最早人类通过自己的观察来认知这个世界，发现了火能烤熟食物、石头能够凿开坚果，发现月亮有阴晴圆缺。随着知识的不断积累，人类开始将之前通过观察和实验得到的感性认识总结为理论。伽利略在比萨斜塔进行"两个大小不同的铁球同时落地"的实验推翻了持续1900年之久的亚里士多德"物体下落速度和重量成比例"的学说，这就是人类的认识由感性经验上升到理性理论的重要实验之一。

（2）第二种范式：理论科学

随着科学的进步，人类开始采用各种数学、几何、物理等理论，有了理论以后人类可以用理论来分析世界、预测世界和寻求科学的解决方案。我们有了历法，能够预言一年四季，能够指导春耕秋收，能预测尚未被发现的行星，譬如，海王星、冥王星的发现就不是通过观测而是通过理论计算而得到的。我们还能运用各种理论，如牛顿三大定律、麦克斯韦方程组、相对论等去认识世界和改造世界。

（3）第三种范式：计算与仿真科学

随着世界上第一台通用计算机ENIAC在美国宾夕法尼亚大学的诞生，人类社会开始步入计算机时代,科学研究也进入了一个以"计算"为中心的全新时期。理论的逐步完善使人类仅仅通过计算和仿真就能发现和认识新的规律,目前在材料科学研究中物质大量的特性正是利用"第一性原理",通过软件的仿真来完成的,在全面禁止核爆条款下,原子弹的研究也完全依赖计算模拟核爆炸来进行。人类认识世界的方法就这样走过了实验科学、理论科学和计算科学三大阶段。

（4）第四种范式：数据密集型科学

网络技术和计算机技术的发展使人类在近期获得了一种新的认识世界的手段,就是利用大量数据来发现新的规律,这种认识世界的方法被称为"第四范式",是美国著名的科学家图灵奖得主吉姆·格雷在2007年提出的。这标志着数据正式成为大家公认的认识世界的方法。大数据出现后人类认识世界的方法就达到4种：实验、理论、计算和数据,如图1-3所示。现在人类在一年内所产生的数据可能已经超过人类过去几千年产生的数据的总和,即使是复杂度为$O(n)$的数据处理方法,在面对庞大的$n$时也显得力不从心,人类逐步进入大数据的时代。在大数据环境下,一切都将以数据为中心,从数据中发现问题、解决问题,真正体现出数据的宝贵价值。第四范式的出现正说明了可以利用海量数据加上高速计算发现新知识,计算和数据的关系在大数据时代将会变得更加紧密。

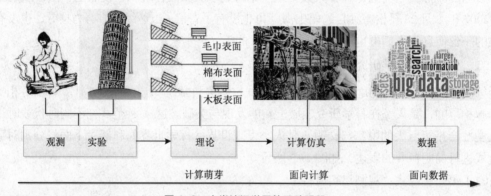

图1-3　人类认识世界的四种手段

### 三、大数据对社会发展的影响

大数据的发展不仅改变了科学思维,也必然会引起企业、政府及个人的思维方式的变革,维克托·迈尔·舍恩伯格在《大数据时代：生活、工作与思维的大变革》一书中指出,对于大数据时代,应放弃对因果关心的渴求,而更关注相关关系,正如其在福布斯·静安南京路论坛上的演讲所述："在大数据时代,人们每天醒来,要想的事情就是面对如此庞大复杂的数据可以用来做什么,其价值可以体现在哪些方面,是否可以找到一个别人从未涉及的事情使得思路及想法成为重要的资产。"由此可见,大数据时代必然会引起思维的转变,而且思维的转变越快,越能在如今竞争激烈的社会中抢占先机。

（1）大数据改变科学研究的思维方式

① 要全体不要抽样。在以往的科学分析中,由于数据存储和处理能力的限制,通常采用抽样的方法,即从全体数据集中抽取一部分样本数据,通过对样本数据的分析,来推断全体数据集的总体特征。通常,样本数据规模要比全集数据小得多,因此,可以在可控的代价内实现数据分析的目的。在大数据时代,其核心技术就是对海量数据进行处理和存储,分布式

文件系统和分布式数据库技术提供了理论上近乎无限的数据存储能力，分布式并行编程框架 MapReduce 提供了强大的海量数据并行处理能力。因此，有了大数据技术的支持，科学分析完全可以直接针对全集数据并可以在短时间内快速得到分析结果，例如，Google 公司的 Dremel 可以在 2～3 秒完成 PB 级的数据查询，其速度之快，超乎想象。

② 要效率不要绝对精确。抽样分析方法是科学研究人员常用的一种科学实验分析方法，一般来说，把采集到的数据进行抽样，并以精确性的分析方法分析样本数据，其样本分析结果通常来说较为精准，但是如将其分析结果应用到全体数据集后，微小误差也将会被放大许多，这就意味着抽样分析的微小误差，被放大到全体数据集后，其误差也有可能会随之放大很多。正是由于这个原因，传统的数据分析方法往往更加注重提高算法的精确性，其次才是提高算法效率。现在，大数据时代采用的是全体数据集分析而非抽样数据分析，其分析结果就不存在误差被放大的问题；因此，算法的高精确性已经不是现在所要追求的首要目标，相反，大数据时代具有"秒级响应"的特征，要求在几秒内就迅速给出针对海量数据的实时分析结果，否则，就会丧失数据的价值，因此，数据分析的效率将会是大数据时代关注的焦点。

③ 要相关不要因果。数据分析的目的主要体现在如下两方面：一方面是解释事物背后的发展规律或机理，例如，找出某地区大型企业某类产品销售业绩大幅下滑的原因，并分析问题所在；另一方面是用于预测未来可能发生的事件，比如，通过实时分析各大媒体最新报道及微博等数据，当发现人们对雾霾的讨论明显增加时，就可建议销售部增加口罩的进货量，因为人们关注雾霾的一个直接结果是在这样的一个环境下如何保障身体损害最小，而简单方便的口罩将会是一个不错的产品。其实，不管是何目的，反映的都是"因果关系"。而在大数据时代，因果关系将不再那么重要，人们转而追求"相关性"而非"因果性"。譬如，当我们在淘宝网购买了一个键盘后，淘宝网还会自动提示你，与你购买相同产品的其他客户还购买了鼠标，也就是说，淘宝网只会告诉你"键盘"和"鼠标"之间存在相关性，但是，它不会告诉你其他客户为什么会在购买键盘之后还会继续购买鼠标。

（2）大数据改变人们的生存方式

在 21 世纪，信息技术突飞猛进的今天，物联网、嵌入式技术、传感技术等的发展，为人类更全面地感知客观存在的物理世界提供了基础；而互联网、云计算等信息技术的发展更是改变了人类通信与管理信息的方式。随着技术的发展及工具的更新换代，人类也提出了更高的生存需求，美国国家科学基金委员会在 2006 年提出了信息物理系统（Cyber Physical Systems，CPS）的概念；2007 年，不同机构及研究学者对其进行了定义，包括 LEE、Baheli、Sastry 及 Krogh 等，强调计算元素及物理元素，实体与虚拟网络的关系，并注重通信计算及控制能力，尽管不同定义的描述不同，但是都明确了 CPS 的内涵：Cyber 与 Physical 的深度融合后形成的智能系统。CPS 的运行状况如图 1-4 所示。

在图 1-4 中，最外面一层是物理实体，其代表我们生活的物理世界；中间一层为感知层，包括传感器等具有采集功能的设备；第三层为计算机等具有计算功能的设备，其负责实现对采集数据的分析及可视化呈现；最里面一层为决策层（具有决策能力的人或者其他事物），其通过感知及分析结果做出决策，并作用于物理实体。CPS 的运行图体现了在感知基础上，人、机、物的深层融合。CPS 的有效工作将改变人类的生存方式，如其可应用于无人机、自主导航的汽车等以实现物理实体的自主工作，医疗领域中可应用于自动手术，物联网领域中可实现生活中的智能家居及智慧城市等。上述 CPS 的成功实现，最重要的基础就是系统中收集的大量数据的有效分析及处理，其是决策支持的重要来源。即如果没有大数据的积累及分析，那么 CPS 也就无从谈起。由此可见，大数据的产生及有效分析是 CPS 的重要资源和基础，结合其他技术的发展，将为改变人类生存方式提供重要动力。

图 1-4　CPS 的运行状况

（3）大数据改变人类的生产方式

目前已经先后经历了三次工业革命，包括 1760—1840 年因为蒸汽动力的发明产生了生产制造的机械化，开创了"蒸汽时代"；1840—1950 年因为电的发明开创了"电气时代"，使得生产得以批量化；1950 年至今，电子技术和计算机等信息技术的发展开创了"信息时代"，使得产品更为丰富，功能性更强；而随着科技的进一步发展，科技的进步也必定引起生产方式的变革。为此，德国提出了"工业 4.0"，即第四次工业革命，以智能制造为主导实现生产制造人机一体化，"工业 4.0"的提出预示着革命性的生产方式的诞生，而实现"工业 4.0"的基础就是大数据的分析及 CPS 的推广，其标志着生产制造业必须转向以数据分析为中心。由此可见，大数据的发展将在生产方式改变中起到关键作用。四次工业革命演化如图 1-5 所示。

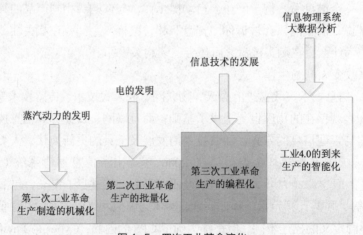

图 1-5　四次工业革命演化

"工业 4.0"是在 2013 年举办的汉诺威（Hannover）工业博览会上正式提出的，虽然第四次工业革命是否到来还存在很大争议，但是目前很多国家已经投入了大量资金及精力来推进"工业 4.0"的进程。成功的典范是特斯拉及西门子，特斯拉将自己的核心定位于大型可移动的智能终端，通过互联网将汽车设计为包含软件、硬件，以及内容和服务的体验工具，将互联网思维引入制造业；西门子的电子车间更是将"工业 4.0"付诸实践的典型代表，其建立了一个紧密结合的技术网络，通过技术整合形成更智能、高效的整体，使生产线的可靠性达到99%，追溯性达到100%。

"工业 4.0"将要达到的目标是通过物联网系统实现智能工厂，即每件产品、零部件都会包含大量的信息，包括何时生产、可以用多久、是否需要替换等，通过非人为干预的智能方式实现自主处理。由此可见，大数据将在改变生产方式中扮演重要角色，由大数据到决策的实现将加速"工业 4.0"时代的到来。

# 任务 3 认识常见的大数据计算模式

## 【任务概述】

大数据处理技术除了使用频率较高的 MapReduce 之外，还有多种大数据计算模式。本任务主要介绍几种常用的大数据计算模式，主要包括查询分析计算（HBase、Hive、Dremel、Cassandra、Impala、Shark、HANA）、批处理计算（Hadoop、Spark）、流计算（Scribe、Flume、Storm、S4、Spark Streaming）、迭代计算（Haloop、iMapReduce、Twister、Spark）、图计算（Pregel、Giraph、Trinity、GraphX、PowerGraph）、内存计算（Spark、HANA、Dremel）。

## 【支撑知识】

当人们提到大数据处理技术时就会自然而然地先想到 MapReduce，而实际上，MapReduce 仅是大数据计算模式中使用频率较高的一种，大数据处理的问题复杂多样，数据源类型也较多，包括结构化数据、半结构化数据、非结构化数据，由此可见，单一的计算模式早已无法满足不同类型的计算需求。例如，有些场合需要对海量已有数据进行批量处理，有些场合需要对大量的实时生成的数据进行实时处理，有些场合需要在数据分析时进行反复迭代计算，有些场合需要对图数据进行分析计算。目前主要的大数据计算模式有查询分析计算、批处理计算、流计算、迭代计算、图计算和内存计算等。

### 一、查询分析计算

大数据时代，查询分析计算系统需要具备对大规模数据实时或准实时查询的能力，数据规模的增长已经超出了传统关系型数据库的承载和处理能力。目前主要的数据查询分析计算主要有 HBase、Hive、Dremel、Cassandra、Impala、Shark、HANA 等。

HBase：开源、分布式、面向列的非关系型数据库模型，是 Apache 的 Hadoop 项目的子项目，源于 Google 论文《Bigtable：一个结构化数据的分布式存储系统》，它实现了其中的压缩算法、内存操作和布隆过滤器。HBase 的编程语言为 Java。HBase 的表能够作为 MapReduce 任务的输入和输出，可以通过 Java API 来存取数据。

Hive：基于 Hadoop 的数据仓库工具，用于查询、管理分布式存储中的大数据集，提供完整的 SQL 查询功能，可以将结构化的数据文件映射为一张数据表。Hive 提供了一种类 SQL 语言（HiveQL），可以将 SQL 语句转换为 MapReduce 任务运行。

Dremel：由谷歌公司开发，是一种可扩展的、交互式的实时查询系统，用于只读嵌套数据的分析。通过结合多级树状执行过程和列式数据结构，它能做到几秒内完成对万亿张表的聚合查询。系统可以扩展到成千上万的 CPU 上，满足谷歌上万用户操作 PB 量级的数据，并且可以在 2~3 秒完成 PB 量级数据的查询。

Cassandra：开源 NoSQL 数据库系统，最早由 Facebook 开发，并于 2008 年开源。由于其良好的可扩展性，Cassandra 被 Facebook、Twitter、Rackspace、Cisco 等公司使用，其数据模型借鉴了 Amazon 的 Dynamo 和 Google BigTable，是一种流行的分布式结构化数据存储方案。

Impala：由 Cloudera 公司参考 Dremel 系统开发，是运行在 Hadoop 平台上的开源大规模并行 SQL 查询引擎。用户可以使用标准 SQL 接口的工具快速查询存储在 Hadoop 的 HDFS 和 HBase 中的 PB 量级大数据。

Shark：Spark 上的数据仓库实现，即 Spark SQL，与 Hive 相兼容，但处理 HiveQL 的性能比 Hive 快 100 倍。

HANA：由 SAP 公司开发的与数据源无关、软硬件结合、基于内存计算的平台。

## 二、批处理计算

批处理计算主要解决针对大规模数据的批量处理，也是日常数据分析工作中常见的一类数据处理需求。MapReduce 是最具有代表性和影响力的大数据批处理技术，可以并行执行大规模数据集（TB 量级以上）的处理任务。MapReduce 对具有简单数据关系、易于划分的海量数据采用"分而治之"的并行处理思想，将数据记录的处理分为 Map 和 Reduce 两个简单的抽象操作，提供了一个统一的并行计算框架，但是，MapReduce 的批处理模式不支持迭代计算。批处理计算系统将并行计算的实现进行封装，大大降低了开发人员的并行程序设计难度。典型的批处理计算系统除了 MapReduce，还有 Hadoop 和 Spark。

Hadoop：目前大数据处理最主流的平台，是 Apache 基金会的开源软件项目，是使用 Java 语言开发实现的。Hadoop 平台使开发人员无须了解底层的分布式细节，即可开发出分布式程序，在集群中对大数据进行存储、分析。

Spark：由加州伯克利大学 AMP 实验室（Algorithms Machines and People Lab）开发，适合用于机器学习、数据挖掘等迭代运算较多的计算任务。由于 Spark 引入了内存计算的概念，运行 Spark 时服务器使用内存替代 HDFS 或本地磁盘来存储中间结果，大大加速数据分析结果的返回速度。Spark 提供比 Hadoop 更高层的 API，同样的算法在 Spark 中的运行速度比 Hadoop 快 10~100 倍。Spark 在技术层面兼容 Hadoop 存储层 API，可访问 HDFS、HBASE、SequenceFile 等。Spark-Shell 可以开启交互式 Spark 命令环境，能够提供交互式查询，因此，可将其运用于需要互动分析的场景。

## 三、流计算

大数据分析中一种重要的数据类型——流数据，是指在时间分布和数量上无限的一系列动态数据集合体，数据的价值随着时间的流逝而降低，因此，必须采用实时计算的方式给出秒级响应。流计算具有很强的实时性，需要对应用不断产生的流数据实时进行处理，使数据不积压、不丢失，经过实时分析处理，给出有价值的分析结果。常用于处理电信、电力等行业应用及互联网行业的访问日志等。Facebook 的 Scribe、Apache 的 Flume、Twitter 的 Storm、Yahoo 的 S4、UCBerkeley 的 Spark Streaming 是常用的流计算系统。

Scribe：Scribe 是由 Facebook 开发的开源系统，用于从海量服务器实时收集日志信息，对日志信息进行实时的统计分析处理，应用在 Facebook 内部。

Flume：Flume 由 Cloudera 公司开发，其功能与 Scribe 相似，主要用于实时收集在海量节点上产生的日志信息，存储到类似于 HDFS 的网络文件系统中，并根据用户的需求进行相应的数据分析。

Storm：基于拓扑的分布式流数据实时计算系统，由 BackType 公司（后被 Twitter 收购）开发，现已经开放源代码，并应用于淘宝、百度、支付宝、Groupon、Facebook 等平台，是主要的流数据计算平台之一。

S4：S4 的全称是 Simple Scalable Streaming System，是由 Yahoo 开发的通用、分布式、可扩展、部分容错、具备可插拔功能的平台。其设计目的是根据用户的搜索内容计算得到相应的推荐广告，现已经开源，是重要的大数据计算平台。

Spark Streaming：构建在 Spark 上的流数据处理框架，将流式计算分解成一系列短小的批处理任务进行处理。网站流量统计是 Spark Streaming 的一种典型的使用场景，这种应用既需要具有实时性，还需要进行聚合、去重、连接等统计计算操作。如果使用 Hadoop MapReduce 框架，则可以很容易地实现统计需求，但无法保证实时性。如果使用 Storm 这种流式框架，则可以保证实时性，但实现难度较大。Spark Streaming 可以以准实时的方式方便地实现复杂的统计需求。

## 四、迭代计算

针对 MapReduce 不支持迭代计算的缺陷，人们对 Hadoop 的 MapReduce 进行了大量改进，Haloop、iMapReduce、Twister、Spark 是典型的迭代计算系统。

HaLoop：HaLoop 是 Hadoop MapReduce 框架的修改版本，用于支持迭代、递归类型的数据分析任务，如 PageRank、K-Means 等。

iMapReduce：一种基于 MapReduce 的迭代模型，实现了 MapReduce 的异步迭代。

Twister：基于 Java 的迭代 MapReduce 模型，上一轮 Reduce 的结果会直接传送到下一轮的 Map。

Spark：Spark 是一种与 Hadoop 相似的开源集群计算环境，但 Spark 启用了内存分布数据集，除了能够提供交互式查询外，它还可以优化迭代工作负载。

## 五、图计算

社交网络、网页链接等包含具有复杂关系的图数据，这些图数据的规模巨大，可包含数十亿顶点和上百亿条边，图数据需要由专门的系统进行存储和计算。常用的图计算系统有 Google 公司的 Pregel、Pregel 的开源版本 Giraph、微软的 Trinity、Berkeley AMPLab 的 GraphX，以及高速图数据处理系统 PowerGraph。

Pregel：Pregel 是由谷歌公司开发的一种基于 BSP（Bulk Synchronous Parallel）模型实现的并行图处理系统，采用迭代的计算模型。为了解决大型图的分布式计算问题，Pregel 搭建了一套可扩展的、有容错机制的平台，该平台提供了一套非常灵活的 API，可以描述各种各样的图计算。在谷歌的数据计算任务中，大约 80% 的任务处理采用 MapReduce 模式，如网页内容索引；图数据的计算任务约占 20%，采用 Pregel 进行处理。

Giraph：Giraph 是一个迭代的图计算系统，最早由雅虎公司借鉴 Pregel 系统开发，后捐赠给 Apache 软件基金会，成为开源的图计算系统。Giraph 是基于 Hadoop 建立的，此外，Apache Giraph 开源项目已被 Facebook 广泛应用在搜索服务中，同时还对 Giraph 做了大量功能性改进和扩展，目前基于 Giraph 开发的平台稳定且易使用。

Trinity：Trinity 是微软公司开发的图数据库系统，该系统是基于内存的数据存储与运算系统，源代码不公开。

GraphX：GraphX 是由 AMPLab 开发的运行在数据并行的 Spark 平台上的图数据计算系统。

PowerGraph：PowerGraph 是高速图数据处理系统，常用于广告推荐计算和自然语言处理。

### 六、内存计算

随着内存价格的不断下降和服务器可配置内存容量的不断增长，使用内存计算完成高速的大数据处理已成为大数据处理的重要发展方向。目前常用的内存计算系统有分布式内存计算系统 Spark、全内存式分布式数据库系统 HANA、谷歌的可扩展交互式查询系统 Dremel。

Spark：是一种基于内存计算的开源集群计算系统，启用了内存分布数据集，由 Scala 语言实现并将其作为应用程序框架。

HANA：SAP 公司开发的基于内存技术、面向企业分析性的产品。

Dremel：谷歌的交互式数据分析系统，可以在数以千计的服务器组成的集群上发起计算，处理 PB 级的数据。Dremel 是 Google MapReduce 的补充，大大缩短了数据的处理时间，成功地应用在谷歌的 BigQuery 中。

# 任务 4　厘清大数据处理的基本流程

## 【任务概述】

进入大数据时代，数据采集来源广泛，且数据类型多以半结构化和非结构化海量数据为主，因此，要想获得有价值的数据信息，需要对这些采集到的海量数据在适合的辅助工具下进行技术处理。本任务介绍的大数据处理基本流程主要包括三个方面：一是数据清洗，二是数据分析，三是数据解释，数据最终以可视化的方式呈现给用户，供用户做决策。

## 【支撑知识】

大数据并非仅指数据本身，而是海量数据和大数据处理技术这两者的综合。通常，大数据的处理流程可以定义为在适合工具的辅助下，对广泛异构的数据源进行抽取和集成，按照一定的标准统一存储，利用合适的数据分析技术对存储的数据进行分析，从中提取有益的知识并利用恰当的方式将结果展示给终端用户。从数据分析全流程的角度来看，大数据处理的基本流程如图 1-6 所示。

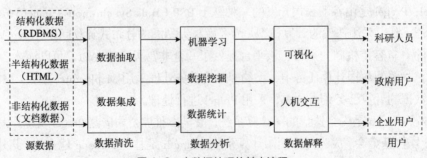

图 1-6　大数据处理的基本流程

### 一、数据清洗

由于大数据处理的数据来源类型丰富,大数据处理的第一步是对数据进行抽取、清洗、转换和集成,从中提取出关系和实体,经过关联和聚合等操作,按照统一定义的格式对数据进行存储。现有的大数据清洗方法有三种:基于物化或数据仓库技术方法的引擎(Materialization or ETL Engine)、基于联邦数据库或中间件方法的引擎(Federation Engine or Mediator)、基于数据流方法的引擎(Stream Engine)。

### 二、数据分析

数据分析是大数据处理流程的核心步骤,通过数据抽取和集成环节,我们已经从异构的数据源中获得了用于大数据处理的原始数据。用户可以根据自己的需求对这些数据进行分析处理,如机器学习、数据挖掘、数据统计等。数据分析可以用于决策支持、商业智能、推荐系统、预测系统等。

### 三、数据解释

大数据处理流程中用户最关心的是数据处理的结果,正确的数据处理结果只有通过合适的展示方式才能被终端用户正确理解,因此,数据处理结果的展示非常重要,可视化和人机交互是数据解释的主要技术。开发调试程序的时候经常通过打印语句的方式来呈现结果,这种方式非常灵活、方便,但只有熟悉程序的人才能很好地理解打印结果。使用可视化技术,可以将处理的结果通过图形的方式直观地呈现给用户,标签云(Tag Cloud)、历史流(History Flow)、空间信息流(Spatial Information Flow)等是常用的可视化技术,用户可以根据自己的需求灵活地使用这些可视化技术。人机交互技术可以引导用户对数据进行逐步分析,使用户参与到数据分析的过程中,深刻地理解数据分析结果。

需要指出的是,大数据技术是许多技术的一个集合体,这些技术也并非全部都是新生事物,诸如关系型数据库、数据仓库、数据挖掘、OLAP、ETL、数据安全和隐私保护、数据可视化等已经是发展多年的技术,在大数据时代得到不断补充、完善、提高后又有了新的升华,因此,也可将这些技术视为大数据技术的一个重要组成部分。

## 任务 5  大数据应用大显神通

### 【任务概述】

大数据无处不在并已融入社会各行各业,其在各个领域的应用也是相当广泛。本任务主要介绍大数据在各个领域应用的基本情况,其中包括电信行业、金融行业、餐饮行业等,并重点介绍了高能物理、推荐系统、搜索引擎系统和百度迁徙方面的应用。

### 【支撑知识】

大数据无处不在,包括电信、金融、餐饮、零售、政务、医疗、能源、娱乐、教育等在内的社会各行各业都已经融入了大数据的印迹,表1-4是大数据在各个领域的应用情况。

了解了大数据在各个领域运用的基本情况,下面再举几个大数据典型应用的示例让读者能更加清楚大数据在各行各业发挥的作用,以及确切知道大数据与人们的日常生活是息息相关的。

表 1-4　大数据在各个领域的应用一览

| 领　域 | 大数据的应用 |
|---|---|
| 电信行业 | 利用大数据技术实现客户离网分析，及时掌握客户离网倾向，出台客户挽留措施 |
| 金融行业 | 大数据在高频交易、社交情绪分析和信贷风险分析三大金融创新领域发挥重要作用 |
| 餐饮行业 | 利用大数据实现餐饮 O2O 模式，彻底改变传统餐饮经营方式 |
| 城市管理 | 可以利用大数据实现智能交通、环保监测、城市规划和智能安防 |
| 生物医学 | 大数据可以帮助人们实现流行病预测、智慧医疗、健康管理，同时还可以帮助人们解读 DNA，了解更多的生命奥秘 |
| 能源行业 | 随着智能电网的发展，电力公司可以掌握海量的用户用电信息，利用大数据技术分析用户用电模式，可以改进电网运行，合理地设计电力需求响应系统，确保电网运行安全 |
| 教育和娱乐 | 大数据可以帮助教学和实训，决定投拍哪种题材的影视作品，以及预测比赛结果 |
| 互联网行业 | 借助大数据技术，可以分析客户行为，进行商品推荐和有针对性的广告投放 |
| 物流行业 | 利用大数据优化物流网络，提高物流效率，降低物流成本 |
| 安全领域 | 政府可以利用大数据技术构建起强大的国家安全保障体系，企业可以利用大数据抵御网络攻击，警察可以借助大数据来预防犯罪 |
| 个人生活 | 大数据还可以应用于个人生活，利用与每个人相关联的"个人数据"，分析个人生活行为习惯，为其提供更加周到的个性化服务 |

**一、大数据在高能物理中的应用**

高能物理学科一直是推动计算技术发展的主要学科之一，万维网技术的出现就是来源于高能物理对数据交换的需求。高能物理是一个天然需要面对大数据的学科，高能物理科学家往往需要从大量的数据中去发现一些小概率的粒子事件，这跟大海捞针一样。目前世界上最大的高能物理实验装置是在日内瓦欧洲核子中心（CERN）的大型强子对撞机（Large Hadron Collider，LHC），如图 1-7 所示，其主要物理目标是寻找希格斯（Higgs）粒子。高能物理中的数据处理较为典型的是采用离线处理方式，由探测器组负责在实验时获取数据，现在最新的 LHC 实验每年采集的数据达到 15PB。高能物理中的数据特点是海量且没有关联性，为了从海量数据中甄别出有用的事件，可以利用并行计算技术对各个数据文件进行较为独立的分析处理。中国科学院高能物理研究所的第三代探测器 BESⅢ产生的数据规模已达到 10PB 量级，在大数据条件下，计算、存储、网络一直考验着高能所的数据中心系统。在实际数据处理时 BESⅢ数据分析甚至需要通过网格系统调用俄罗斯、美国、德国及国内的其他数据中心来协同完成任务。

图 1-7　大型强子对撞机（LHC）

**二、推荐系统**

推荐系统可以利用电子商务网站向客户提供商品信息和建议，帮助用户决定应该购买什么东西，模拟销售人员帮助客户完成购买过程。我们经常在上网时看见网页某个位置出现一些商品推荐或者系统弹出一个商品信息，而且这些商品可能正是我们自己感兴趣或者正希望购买的商品，这就是推荐系统在发挥作用。目前推荐系统已变得无处不在，如商品推荐、新闻推荐、视频推荐，推荐方式包括网页式推荐、邮件推荐、弹出式推荐。例如，在京东商城查找你想购买关于云计算和大数据相关的书籍时，系统会根据你近期搜索的关键词列出人气指数排行较高的书供你参考选择，如图1-8所示。

图1-8 京东商城书籍推荐页面

推荐过程的实现完全依赖于大数据，在进行网络访问时访问行为被各网站所记录并建立模型，有的算法还需要与大量其他人的信息进行融合分析，从而得出每一个用户的行为模型，将这一模型与数据库中的产品进行匹配从而完成推荐过程。为了实现这一推荐过程，需要存储大量客户的访问信息，对于用户量巨大的电子商务网站，这些信息的数据量是非常庞大的。推荐系统是大数据非常典型的应用，只有基于对大量数据的分析，推荐系统才能准确地获得用户的兴趣点。一些推荐系统甚至会结合用户社会网络来实现推荐，这就需要对更大的数据集进行分析，从而挖掘出数据之间广泛的关联性。推荐系统使大量看似无用的用户访问信息产生了巨大的商业价值，这就是大数据的魅力。

### 三、搜索引擎系统

搜索引擎是大家最为熟悉的大数据系统，成立于1998年的谷歌和成立于2000年的百度在简洁的用户界面下面隐藏着世界上最大规模的大数据系统。搜索引擎是简单与复杂的完美结合，目前最为常用的开源系统Hadoop就是按照谷歌的系统架构设计的。图1-9所示为百度搜索页面。

为了有效地完成互联网上数量巨大的信息的收集、分类和处理工作，搜索引擎系统大多是基于集群架构的。中国出现较早的搜索引擎还有北大天网搜索，天网搜索在早期是由几百台PC搭建的机群构建的，这一思路也被谷歌所采用，谷歌由于早期搜索利润的微薄只能利用廉价服务器来实现。每一次搜索请求可能都会有大量的服务响应，搜索引擎是一个典型而成熟的大数据系统，它的发展历程为大数据研究积累了宝贵的经验。2003年在北京大学召开了第一届全国搜索引擎和网上信息挖掘学术研讨会，大大推动了搜索引擎在国内的技术发展。搜索引擎与数据挖掘技术的结合预示着大数据时代的逐步到来，从某种意义上可以将这次会议作为中国在大数据领域的第一次重要学术会议，如图1-10所示。当时百度还未上市，但派出不少工程师参加了此次会议。

图1-9 百度搜索引擎

图1-10 首届全国搜索引擎和网上信息挖掘学术研讨会合影

### 四、百度迁徙

"百度迁徙"项目是 2014 年百度利用其位置服务(Location Based Service,LBS)所获得的数据,将人们在春节期间位置移动情况用可视化的方法显示在屏幕上。这些位置信息来自于百度地图的 LBS 开放平台,通过安装在大量移动终端上的应用程序获取用户位置信息,这些数以亿计的信息通过大数据处理系统的处理可以反映全国总体的迁移情况,通过数据可视化,为春运时人们了解春运情况和决策管理机构进行管理决策提供了第一手的信息支持。这一大数据系统所提供的服务为今后政府部门的科学决策和社会科学的研究提供了新的技术手段,也是大数据进入人们生活的一个案例。

## 任务 6　大数据的发展及面临的挑战

### 【任务概述】

"大数据时代"悄然崛起,掀起了"第三次信息化浪潮",大数据技术的研究和产业发展已快速上升为国家战略,人们必须做好时刻迎接大数据的准备和接受挑战。本任务主要介绍了大数据的发展历程,大数据发展现状,大数据与云计算、物联网三者之间的关系,以及在应用大数据过程中所必然会遇到的难题。

### 【支撑知识】

#### 一、大数据的发展历程

(1)以年代来划分

以年代或技术里程碑来划分,可以认为大数据的发展历程经历了 3 个重要阶段:萌芽期、成熟期和大规模应用期,如表 1-5 所示。

表 1-5　以年代来划分:大数据发展经历的 3 个阶段

| 阶　段 | 时　间 | 内　容 |
| --- | --- | --- |
| 第一阶段:萌芽期 | 20 世纪 90 年代到 21 世纪初 | 随着数据挖掘理论和数据库技术的逐步成熟,一批商业智能工具和知识管理技术开始被应用,如数据仓库、专家系统、知识管理系统等 |
| 第二阶段:成熟期 | 21 世纪前十年 | Web 2.0 应用的快速发展,产生了大量半结构化和非结构化数据,传统处理方法已难应付,带动了大数据技术的快速突破,大数据解决方案逐渐走向成熟,形成了并行计算与分布式系统两大核心技术,谷歌的 GFS 和 MapReduce 等大数据技术受到追捧,Hadoop 平台开始大行其道 |
| 第三阶段:大规模应用期 | 2010 年以后 | 大数据应用渗透各行各业,数据驱动决策,信息社会智能化程度大幅提高 |

(2)以数据量的大小来划分

由于大数据的发展历程是和有效存储管理日益增大的数据集的能力紧密联系在一起的,因此,每一次处理能力的提高都伴随着新数据库技术的发展,表 1-6 是以数据大小来划分的。

表 1-6  以数据大小来划分：大数据发展经历的 4 个阶段

| 阶　　段 | 时　　间 | 内　　容 |
|---|---|---|
| 第一阶段：MB~GB | 20 世纪 70 年代到 80 年代 | 当商业数据从 MB 达到 GB 量级时是最早点燃挑战"大数据"的信号，迫切需求存储数据并运行关系型数据查询以完成商业数据的分析和报告，产生了数据库计算机和可以运行在通用计算机上的数据库软件系统 |
| 第二阶段：GB~TB | 20 世纪 80 年代末期 | 单个计算机系统的存储和处理能力受限，提出了数据并行化技术思想，可实现内存共享数据库、磁盘共享数据库和无共享数据库，这些技术及系统的出现成为后来使用分治法并行化数据存储的先驱 |
| 第三阶段：TB~PB | 20 世纪 90 年代末期至今 | 进入互联网时代，PB 级的半结构化和非结构化的网页数据迅速增长，虽然并行数据库能够较好地处理结构化数据，但是对于处理半结构或非结构化数据几乎没有提供任何支持且处理能力也仅几 TB。为了应对 Web 规模的数据管理和分析挑战，谷歌提出了 GFS 和 MapReduce 编程模型，运行 GFS 和 MapReduce 的系统能够向上和向外扩展，能处理无限的数据。在此阶段，出现了著名的"第四范式"、Hadoop、Spark、NoSQL 等新兴技术 |
| 第四阶段：PB~EB | 不久的将来 | 大公司存储和分析的数据毫无疑问将在不久后从 PB 达到 EB 量级，然而，现有的技术只能处理 PB 量级的数据，目前几乎所有重要的产业界公司，如 EMC、Oracle、Microsoft、Google、Amazon 和 Facebook 等都开始启动各自的大数据项目。但迄今为止仍没有出现革命性的新技术能够处理更大的数据集 |

## 二、大数据的发展现状

随着大数据的快速发展，大数据成为信息时代的一大新兴产业，并引起了国内外政府、学术界和产业界的高度关注。

早在 2009 年，联合国就启动了"全球脉动计划"，拟通过大数据推动落后地区的发展，2012 年 1 月的世界经济论坛年会也把"大数据、大影响"作为重要议题之一。在美国，从 2009 年至今，美国政府数据库（Data.gov）全面开放了大量政府原始数据集，大数据已成为美国国家创新战略、国家安全战略及国家信息网络安全战略的交叉领域和核心领域。2012 年 3 月，美国政府提出"大数据研究和发展倡议"，发起全球开放政府数据运动，并投资 2 亿美元促进大数据核心技术研究和应用，涉及 NSF、DARPA 等 6 个政府部门和机构，把大数据放在重要的战略位置。英国政府也将大数据作为重点发展的科技领域，在发展 8 类高新技术的 6 亿英镑投资中，大数据的注资占三成。2014 年 7 月，欧盟委员会也呼吁各成员国积极发展大数据，迎接"大数据"时代，并将采取具体措施发展大数据业务。例如，建立大数据领域的公私合作关系；依托"地平线 2020"科研规划，创建开放式数据孵化器；成立多个超级计算中心；在成员国创建数据处理设施网络。

在中国，政府、学术界和产业界对大数据的研究和应用也相当重视，纷纷启动了相应的研究计划。在 2012 年，科技部"十二五"规划除了部署关于物联网、云计算的相关专项外，还专门发布了《"十二五"国家科技计划信息技术领域 2013 年度备选项目征集指南》，其中的"先进计算"板块明确提出"面向大数据的先进存储结构及关键技术"，并制订了面向大数据的研究计划和专项基金，如国家"973 计划""863 计划"及国家自然科学基金等。

欧美等发达国家对大数据的探索和发展已走在世界前列，我国也不甘示弱，除了政府组织外，国内还有不少知名企业或组织也成立了大数据产品团队和实验室，力争在大数据产业竞争中占据领先地位。目前，全球各国政府都纷纷将大数据发展提升至战略高度，大力促进大数据产业健康平稳快速发展。

### 三、大数据与云计算、物联网的关系

云计算、大数据和物联网代表了 IT 领域最新的技术发展趋势，三者相辅相成，既有联系又有区别。云计算最初主要包含两类含义：一类是以谷歌的 GFS 和 MapReduce 为代表的大规模分布式并行计算技术；另一类是以亚马逊的虚拟机和对象存储为代表的"按需租用"的商业模式。但是，随着大数据概念的提出，云计算中的分布式计算技术开始更多地被列入大数据技术，而人们提到云计算时，更多指的是底层基础 IT 资源的整合优化，以及以服务的方式提供 IT 资源的商业模式，如 IaaS、PaaS、SaaS。从云计算和大数据概念的诞生到现在，二者之间的关系非常微妙，既密不可分，又千差万别。因此，不能把云计算和大数据割裂开来作为截然不同的两类技术来看待。此外，物联网也是和云计算、大数据相伴相生的技术，图 1-11 描述了三者之间的联系与区别。

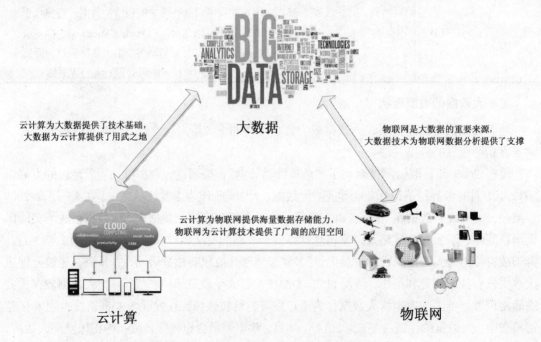

图 1-11　云计算、大数据和物联网三者之间的关系

（1）大数据、云计算和物联网的联系

从整体上看，大数据、云计算和物联网这三者是相辅相成的。大数据根植于云计算，大数据分析的很多技术都来源于云计算，云计算的分布式数据存储和管理系统（包括分布式文

件系统和分布式数据库系统）提供了海量数据的存储和管理能力，分布式并行处理框架 MapReduce 提供了海量数据分析能力，没有这些云计算技术的支撑，大数据分析就无从谈起。反之，大数据为云计算提供了"用武之地"，没有大数据这个"练兵场"，云计算技术就算再先进，也不能很好地发挥出它的应用价值。物联网的传感器源源不断产生的大量数据，构成了大数据的重要数据来源，没有物联网的飞速发展，就不会带来数据产生方式的变革，即由数据人工生产阶段转向数据自动化产生阶段；物联网还需要借助于云计算和大数据技术，实现物联网大数据的存储、分析和处理。三者的有机结合，标志着"大数据时代"的到来。

（2）大数据、云计算和物联网的区别

大数据侧重于对海量数据的存储、处理与分析，从海量数据中发现价值，服务于生产和生活；云计算本质上旨在整合和优化各种 IT 资源并通过网络以服务的方式，廉价地提供给用户；物联网的发展目标是实现物物相连，应用创新是物联网发展的核心。

云计算、大数据和物联网三者已经彼此渗透、相互融合，在很多应用场合都可以同时看到三者的身影。在未来，三者仍会继续相互促进、相互影响，更好地服务于社会生产和生活的各个领域。

### 四、大数据面临的挑战

尽管大数据是社会各界都高度关注的话题，但时下大数据从底层的处理系统到高层的分析手段都存在许多问题，也面临一系列挑战。例如，信息系统正由"数据围着处理器转"向"处理能力围着数据转"转变，系统结构设计的出发点要从重视单任务的完成时间转变为提高系统吞吐率和并发处理能力，即以数据为中心的计算系统基本思路，实现数据搬运由"大象搬木头"（少量强核处理复杂任务）转变为"蚂蚁搬大米"（大量弱核处理简单任务）的过程。表 1-7 描述了大数据处理流程中所面临的问题及挑战。

表 1-7　大数据处理流程所面临的挑战

| 研究的主题 | 面临的挑战 |
| --- | --- |
| 大数据预处理及集成 | 广泛的异构性、时空特性、数据质量 |
| 大数据分析 | 先有数据后有模式、动态增长、先验知识的缺乏、实时性 |
| 大数据硬件处理平台 | 硬件异构性、新硬件 |
| 性能测试基准 | 系统复杂性高、案例多样性、数据规模庞大、系统的快速演变 |
| 隐私保护 | 隐性数据的暴露、数据公开与保护、数据动态性 |
| 大数据管理的易用性 | 可视化、人机交互、数据起源技术、海量元数据的高效管理 |
| 大数据的能耗 | 低功耗、新能源 |

所面临的这些问题及挑战除了有由大数据自身的特征导致的，也有由当前大数据分析模型与方法引起的，也有大数据处理系统所隐含的。

（1）数据复杂性带来的挑战

大数据的涌现使人们处理计算问题时获得了前所未有的大规模样本，但同时也不得不面对更加复杂的数据对象，典型的特性是类型和模式多样、关联关系繁杂、质量良莠不齐。大数据内在的复杂性（包括类型复杂性、结构复杂性和模式复杂性）使得数据的感知、表达、理解和计算等多个环节面临着巨大的挑战，导致了传统全量数据计算模式下时空维度上计算复杂度的激增,传统的数据分析与挖掘任务如检索、主题发现、语义和情感分析等变得异常困难。然而,目前人们对大数据复杂性的内在机理及其背后的物理意义缺乏理解,对大数据的分布与协作关联等规律认识不足,对大数据的复杂性和计算复杂性的内在联系缺乏深刻理解,加上缺少面向领域的大数据处理知识,极大地制约了人们对大数据高效计算模型和方法的设计能力。

(2)计算复杂性带来的挑战

大数据多源异构、规模巨大、快速多变等特性使得传统的机器学习、信息检索、数据挖掘等计算方法不能有效地支持大数据的处理、分析和计算。特别地,大数据计算不能像小样本数据集那样依赖于对全局数据的统计分析和迭代计算,需要突破传统计算对数据的独立同分布和采样充分性的假设。在求解大数据的问题时,需要重新审视和研究它的可计算性、计算复杂性和求解算法。因此,研究面向大数据的新型高效计算范式,改变人们对数据计算的本质看法,提供处理和分析大数据的基本方法,支持价值驱动的特定领域应用,是大数据计算的核心问题。而大数据样本量充分,内在关联关系密切而复杂,价值密度分布极不均衡,这些特征对研究大数据的可计算性及建立新型计算范式提供了机遇,同时也提出了挑战。

(3)系统复杂性带来的挑战

针对不同数据类型与应用的大数据处理系统是支持大数据科学研究的基础平台。对于规模巨大、结构复杂、价值稀疏的大数据,其处理也面临计算复杂度高、任务周期长、实时性要求强等难题。大数据及其处理的这些难点不仅对大数据处理系统的系统架构、计算框架、处理方法提出了新的挑战,更对大数据处理系统的运行效率及单位能耗提出了苛刻的要求,要求大数据处理系统必须具有高效能的特点。对于以高效能为目标的大数据处理系统的系统架构设计、计算框架设计、处理方法设计和测试基准设计研究,其基础是大数据处理系统的效能评价与优化问题研究。这些问题的解决可奠定大数据处理系统设计、实现、测试与优化的基本准则,是构建能效优化的分布式存储和处理的硬件及软件系统架构的重要依据和基础,因此,是大数据分析处理必须解决的关键问题。

# 【同步训练】

## 一、简答题

(1)何谓大数据?大数据的基本特征有哪些?
(2)简述科学研究的 4 种范式。
(3)简述数据产生方式经历了哪几个阶段。
(4)试述大数据对社会发展的重要影响主要体现在哪些方面。
(5)简述大数据计算模式。
(6)简述大数据处理的基本流程。

## 二、论述题

(1)举一个发生在身边的大数据具体应用的例子。
(2)论述大数据、云计算和物联网三者之间存在何种关联。
(3)论述目前大数据所面临的问题及挑战有哪些。

# PART 2 项目二
# Ubuntu 及服务安装配置

## 【项目介绍】

安装 Ubuntu Server 和 Ubuntu Desktop 是本项目的重点,本项目还需要进行 APT 软件源的设置、FTP 服务和 MySQL 服务的安装与配置等。

本项目分以下 4 个任务:
- 任务 1　安装 Ubuntu Server
- 任务 2　搭建 FTP 系统
- 任务 3　搭建 MySQL 数据库系统
- 任务 4　安装 Ubuntu Desktop

## 【学习目标】

### 一、知识目标

- 掌握 Ubuntu Server 和 Ubuntu Desktop 的安装。
- 掌握 APT 软件源的设置。
- 掌握 VSFTPD 服务器的安装与配置。
- 掌握 MySQL 服务器安装。
- 掌握 MySQL 基本命令的使用。

### 二、能力目标

- 能够安装 Ubuntu Server 和 Ubuntu Desktop。
- 能够进行 Ubuntu Server 的安全设置。
- 能够设置 APT 软件源。
- 能够安装与配置 VSFTPD 服务器。
- 能够安装与使用 MySQL。

# 任务 1  安装 Ubuntu Server

## 【任务概述】

Ubuntu Server 是非常流行的 Linux 操作系统，是大数据平台不可或缺的系统。本任务主要完成 Ubuntu Server 的安装、网络 IP 地址配置、主机名配置、Hosts 配置、APT-Get 配置、OpenSSH 的安装。

## 【支撑知识】

### 一、Ubuntu 简介

Ubuntu 是基于 Debian GNU/Linux，支持 X86、AMD64（即 X64）和 PowerPC 架构，由全球化的专业开发团队（Canonical Ltd）打造的开源 GNU/Linux 操作系统。Ubuntu 的名称来自非洲南部祖鲁语或豪萨语的 "Ubuntu" 一词，意思是 "人性""我的存在是因为大家的存在"。Ubuntu 基于 Debian 发行版和 GNOME 桌面环境，与 Debian 的不同在于，它每 6 个月会发布一个新版本。Ubuntu Server 16.10 支持 3 种主要的体系架构：Intel X86、AMD64 和 PowerPC。Ubuntu 版本下载的国内镜像网站如下：

http://mirrors.163.com/ubuntu-releases/

http://mirrors.aliyun.com/ubuntu-releases/

http://mirrors.yun-idc.com/ubuntu-releases/

### 二、Ubuntu 目录

Linux 和 UNIX 文件系统被组织成一个有层次的树形结构。文件系统的最上层 / 为根目录。在 UNIX 和 Linux 的设计理念中，一切皆为文件——包括硬盘、分区和可插拔介质。这就意味着所有其他文件和目录（包括其他硬盘和分区）都位于根目录中。

位于根（/）目录下的常见目录如下。

/bin：重要的二进制（Binary）应用程序。

/boot：启动（Boot）配置文件。

/dev：设备（Device）文件。

/etc：配置文件、启动脚本等（Etc）。

/home：本地用户主（Home）目录。

/lib：系统库（Libraries）文件。

/lost+found：在根（/）目录下提供一个遗失+查找（lost+found）系统。

/media：挂载可移动介质（Media），诸如 CD、数码相机等。

/mnt：挂载（Mounted）文件系统。

/opt：提供一个可选的（Optional）应用程序安装目录。

/proc：特殊的动态目录，用以维护系统信息和状态，包括当前运行中进程（Processes）信息。

/root：用户（Root）主文件夹。

/sbin：重要的系统二进制（System Binaries）文件。

/sys：系统（System）文件。

/tmp：临时（Temporary）文件。

/usr：包含绝大部分所有用户（Users）都能访问的应用程序和文件。

/var：经常变化的（Variable）文件，诸如日志或数据库等。

## 三、软件包管理命令

（1）APT 命令

命令行软件包管理器 APT 提供软件包搜索、管理和信息查询等功能。它提供的功能与其他 APT 工具相同（如 APT-Get 和 APT-Cache），但是默认情况下被设置得更适合交互。

APT 用法如下。

```
apt [选项] 命令
```

常用命令如下。

list：根据名称列出软件包。

search：搜索软件包描述。

show：显示软件包细节。

install：安装软件包。

remove：移除软件包。

autoremove：卸载所有自动安装且不再使用的软件包。

update：更新可用软件包列表。

upgrade：通过安装/升级软件来更新系统。

full-upgrade：通过卸载/安装/升级来更新系统。

edit-sources：编辑软件源信息文件。

（2）APT-Get 命令

APT-Get 可以从认证软件源下载软件包及相关信息，以便安装和升级软件包，或者用于移除软件包。在这些过程中，软件包依赖会被妥善处理。APT-Get 命令是一个强大的命令行工具，用于同 Ubuntu 的 Advanced Packaging Tool（APT）一起执行诸如安装新软件包、升级已有软件包、更新包列表索引，甚至是升级整个 Ubuntu 系统等功能。

APT-Get 用法如下。

```
apt-get [选项] 命令
apt-get [选项] install|remove 软件包1 [软件包2…]
apt-get [选项] source 软件包1 [软件包2…]
```

常用命令如下。

update：取回更新的软件包列表信息。

upgrade：进行一次升级。

install：安装新的软件包（注：软件包名称是 libc6 而非 libc6.deb）。

remove：卸载软件包。

purge：卸载并清除软件包的配置。

autoremove：卸载所有自动安装且不再使用的软件包。

dist-upgrade：发布版升级。

dselect-upgrade：根据 dselect 的选择来进行升级。
build-dep：为源码包配置所需的编译依赖关系。
clean：删除所有已下载的包文件。
autoclean：删除已下载的旧包文件。
check：核对以确认系统的依赖关系的完整性。
source：下载源码包文件。
download：下载指定的二进制包到当前目录。
changelog：下载指定软件包，并显示其 changelog。

（3）APT-Cache 命令

APT-Cache 可以查询和显示已安装和可安装软件包的可用信息。它专门工作在本地的数据缓存上，而这些缓存可以通过如 APT-Get 的 update 命令来更新。如果距离上一次更新的时间太久，那么它显示的信息可能就会过时。不过作为交换，APT-Cache 不依赖当前软件源的可用性（如离线状态）。

APT-Cache 用法如下。

```
apt-cache [选项] 命令
apt-cache [选项] show 软件包1 [软件包2…]
```

常用命令如下。

showsrc：显示源文件的各项记录。
search：根据正则表达式搜索软件包列表。
depends：显示该软件包的依赖关系信息。
rdepends：显示所有依赖于该软件包的软件包名字。
show：以便于阅读的格式介绍该软件包。
pkgnames：列出所有软件包的名字。
policy：显示软件包的安装设置状态。

（4）APTITUDE 命令

APTITUDE 与 APT-Get 一样，是 Debian 及其衍生系统中功能极其强大的包管理工具。与 APT-Get 不同的是，APTITUDE 在处理依赖问题上更佳。举例来说，APTITUDE 在删除一个包时，会同时删除本身所依赖的包。

APTITUDE 用法如下。

```
aptitude [-S fname] [-u|-i]
aptitude [选项] <命令> …
```

常用操作如下。

aptitude update：更新可用的包列表。
aptitude upgrade：升级可用的包。
aptitude dist-upgrade：将系统升级到新的发行版。
aptitude install pkgname：安装包。
aptitude remove pkgname：删除包。
aptitude purge pkgname：删除包及其配置文件。

aptitude search string：搜索包。

aptitude show pkgname：显示包的详细信息。

aptitude clean：删除下载的包文件。

aptitude autoclean：仅删除过期的包文件。

### 四、OpenSSH

OpenSSH 是 Secure Shell（SSH）协议工具集中的一个自由可用的版本，用以远程控制一台计算机或在计算机之间传输文件。OpenSSH 提供一个服务器守护程序和客户端工具来保障安全、加密的远程控制和文件传输操作，以有效地取代传统的工具。OpenSSH 服务器组组件 SSHD 持续监听来自任何客户端工具的连接请求。当一个连接请求发生时，SSHD 根据客户端连接的类型来设置当前连接。使用 SSH 协议进行 FTP 传输的协议称为 SFTP（安全文件传输）。

## 【任务实施】

### 一、安装系统

① 把 Ubuntu Server 系统光盘插入服务器光驱，选择光驱引导，系统启动后提示语言选择，选择"English"和安装项目，如图 2-1 和图 2-2 所示。

图 2-1 选择引导安装语言

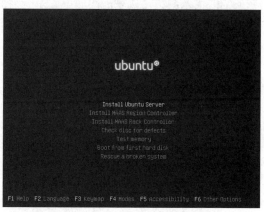

图 2-2 选择 Ubuntu 安装项目

② 选择语言，如图 2-3 所示。选择语言后出现"所选语言的安装程序编译并没有完成"提示，单击"继续"按钮安装。

图 2-3 选择语言

③ 配置键盘不需要检测键盘布局，"键盘布局所属国家"选择"Chinese"，"键盘布局"也选择"Chinese"。

④ 网络配置先不进行设置，将主机名设置为"Ubuntu"，如图 2-4 和图 2-5 所示。

图 2-4　配置网络

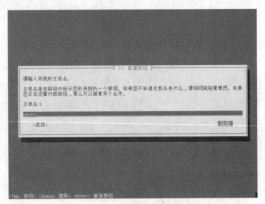

图 2-5　设置主机名

⑤ Ubuntu Server 需要创建一个普通用户，并设置密码，操作如图 2-6 和图 2-7 所示。密码需要输入两次，如果密码简单，系统会提示是否使用弱口令，不加密主目录。

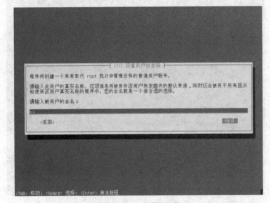

图 2-6　设置用户

图 2-7　设置密码

⑥ 磁盘分区方法选择"向导 — 使用整个磁盘并配置 LVM"，也可以使用手动，如图 2-8 所示，按 Enter 键时系统提示"所选的磁盘上的全部数据都将会被删除"。单击"是"按钮，将修改写入磁盘并配置 LVM，如图 2-9 所示。

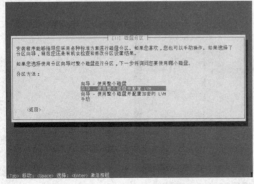

图 2-8　磁盘分区向导

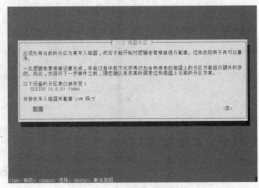

图 2-9　分区方案写入磁盘

⑦ 逻辑卷管理工具提示可用于分区向导的卷组的数量，默认就可以，如图 2-10 所示。LVM 最后生成分区信息，这些信息写入磁盘，如图 2-11 所示。磁盘分区完后开始安装系统。

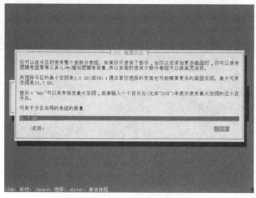
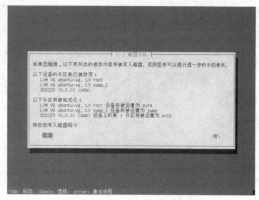

图 2-10　卷组的数量　　　　　　　　　图 2-11　LVM 分区信息

⑧ 系统提供自动更新，如图 2-12 所示。在选择软件对话框中选择"OpenSSh Server"，方便远程操作，如图 2-13 所示，软件选择后开始安装。

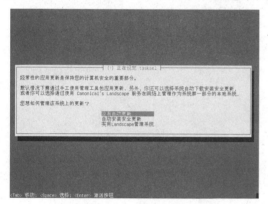
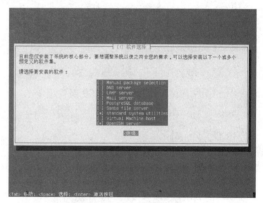

图 2-12　设定 tasksel　　　　　　　　　图 2-13　软件选择

⑨ 软件安装完成后，需要将 GRUB 安装至硬盘，如图 2-14 所示，系统安装完成后需要取出光盘，如图 2-15 所示。

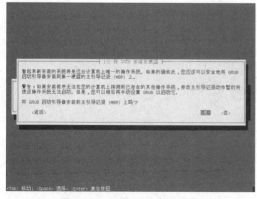
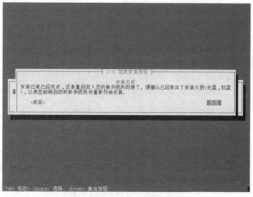

图 2-14　GRUB 安装至磁盘　　　　　　　图 2-15　安装完成提示

⑩ 系统安装完毕重启，显示 GRUB 引导提示，如图 2-16 所示。系统启动成功后，使用

普通用户 sw 登录系统，如图 2-17 所示，这时用户 root 无法登录，需要设置密码后才可以使用，使用"sudo passwd root"命令修改用户 root 密码。

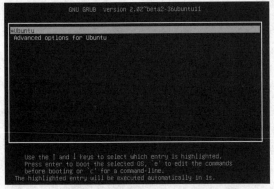

图 2-16　GRUB 引导信息　　　　　　　　　　图 2-17　用户登录

### 二、设置环境变量

① 以用户 sw 登录系统，并切换到用户 root，操作如下：

```
sw@ubuntu:~$ su - root
Password:
root@ubuntu:~#
```

注意：提示符由"$"号变成"#"号。

② 修改 /etc/profile 文件，操作如下：

```
root@ubuntu:~# vi /etc/profile
```

添加内容如下：

```
export JAVA_HOME=/opt/jdk1.8.0_121
export PATH=$JAVA_HOME/bin:$PATH
```

③ 使环境变量生效，操作如下：

```
root@ubuntu:~# source /etc/profile
```

### 三、配置网络

① 查看网络配置信息，操作如下：

```
root@ubuntu:~# ifconfig -a
ens33: flags=4098<UP,BROADCAST,RUNNING,MULTICAST>  mtu 1500
        ether 00:0c:29:df:e7:2b  txqueuelen 1000  (Ethernet)
        RX packets 0  bytes 0 (0.0 B)
        RX errors 0  dropped 0  overruns 0  frame 0
        TX packets 0  bytes 0 (0.0 B)
        TX errors 0  dropped 0  overruns 0  carrier 0  collisions 0

lo: flags=73<UP,LOOPBACK,RUNNING>  mtu 65536
        inet 127.0.0.1  netmask 255.0.0.0
        inet6 ::1  prefixlen 128  scopeid 0x10<host>
        loop  txqueuelen 1  (Local Loopback)
```

```
        RX packets 384  bytes 28640 (28.6 KB)
        RX errors 0  dropped 0  overruns 0  frame 0
        TX packets 384  bytes 28640 (28.6 KB)
        TX errors 0  dropped 0  overruns 0  carrier 0  collisions 0
```

② 配置 IP 地址，操作如下：

```
root@ubuntu:~# vi /etc/network/interfaces
```

添加内容如下：

```
auto ens33
iface ens33 inet static
address 172.25.0.10
netmask 255.255.255.0
gateway 172.25.0.254
dns-nameserver 202.96.128.86
dns-nameserver 202.96.134.133
```

提示：iface ens33 inet static　表示设置为静态 IP 地址。

iface ens33 inet dhcp　表示自动获取 IP 地址。

③ 修改主机名，操作如下：

```
root@ubuntu:~# vi /etc/hostname
master
```

④ 修改域名服务器，操作如下：

```
root@ubuntu:~# vi /etc/resolv.conf
```

修改内容如下：

```
nameserver 202.96.128.86
nameserver 202.96.134.133
```

提示：202.96.128.86 和 202.96.134.133 是 DNS 服务器提供商提供的 IP 地址，重启无效。

⑤ 永久有效需要如下操作：

```
root@ubuntu:~# vi /etc/resolvconf/resolv.conf.d/base
```

修改内容如下：

```
nameserver 202.96.128.86
nameserver 202.96.134.133
```

⑥ 修改 hosts 文件，操作如下：

```
root@ubuntu:~# vi /etc/hosts
```

修改内容如下：

```
127.0.0.1      localhost
172.25.0.10    master
172.25.0.11    slave1
172.25.0.12    slave2
172.25.0.20    sw-desktop
```

⑦ 配置网络后需要重启网络，操作如下：

```
root@ubuntu:~# systemctl restart networking
```

如果单独启动网卡，则操作如下：
```
root@ubuntu:~# ifup ens33
```
⑧ 测试网络是否连通，操作如下：
```
root@ubuntu:~# ping www.baidu.com
PING www.a.shifen.com (14.215.177.38) 56(84) bytes of data.
64 bytes from 14.215.177.38 (14.215.177.38): icmp_seq=1 ttl=128 time=9.38 ms
64 bytes from 14.215.177.38 (14.215.177.38): icmp_seq=2 ttl=128 time=9.11 ms
^C
--- www.a.shifen.com ping statistics ---
2 packets transmitted, 2 received, 0% packet loss, time 1001ms
rtt min/avg/max/mdev = 9.117/9.252/9.387/0.135 ms
```
提示：上面表示已经连通，按 Ctrl+C 组合键中断测试。

## 四、APT 软件源设置

① 挂载 Ubuntu Server 光盘，操作如下：
```
root@ubuntu:~# mkdir /media/ubuntu
root@ubuntu:~# mount /dev/cdrom /media/ubuntu
```
或
```
root@ubuntu:~# mount -o loop /root/ubuntu-16.10-server-amd64.iso /media/ubuntu
```
② 配置 APT 软件源，操作如下：
```
root@ubuntu:~# cp /etc/apt/sources.list{,.bak}
root@ubuntu:~# vi /etc/apt/sources.list
```
内容如下：
```
deb file:///media/ubuntu yakkety main restricted
```
或者
```
deb cdrom:[Ubuntu-Server 16.10 _Yakkety Yak_ - Release amd64 (20161012.1)]/ yakkety main restricted
```
注意：yakkety 为 Ubuntu Server 16.10 的版本名称，main 为系统安装文件中 pool 目录下的文件夹名字。

③ 更新 APT 软件源，操作如下：
```
root@ubuntu:~# apt-get update
```
④ 查看软件包，操作如下：
```
root@ubuntu:~# apt list openssh*
```

## 五、安装 OpenSSH

① 安装软件，操作如下：
```
root@ubuntu:~# apt-get install openssh-server openssh-client
```
② 启动 SSH 服务，操作如下：
```
root@ubuntu:~# systemctl start ssh
root@ubuntu:~# systemctl enable ssh
```
③ 关闭防火墙，操作如下：

```
root@ubuntu:~# ufw disable
```

④ 测试 SSH 服务，操作如下：

```
root@ubuntu:~# ssh sw@master
sw@master's password:
```

注意：使用 SSH 登录，用户 root 默认是禁止的。如果允许用户 root 操作，则修改配置如下：

```
root@ubuntu:~# vi /etc/ssh/sshd_config
```

修改内容如下：

```
PermitRootLogin prohibit-password
```

改为：

```
PermitRootLogin yes
```

重启 SSH：

```
root@ubuntu:~# systemctl restart ssh
```

⑤ 查看 SSH 服务，操作如下：

```
root@ubuntu:~# systemctl status ssh
```

# 任务 2　搭建 FTP 系统

## 【任务概述】

为了方便数据共享，计划搭建 FTP 系统，只允许匿名账户下载信息，默认用户目录为 /opt/ftp。允许本地用户 sw 读写数据，并将用户 sw 锁定在自己的宿主目录中。

## 【支撑知识】

### 一、FTP 简介

FTP 的全称是 File Transfer Protocol（文件传输协议），就是专门用来传输文件的协议。FTP 服务器可以向用户提供上传和下载文件服务。FTP 协议在运作时要使用两个 TCP 连接。在 FTP 会话中，会存在两个独立的 TCP 连接，一个被称为控制连接（Control Connection），另一个是数据连接（Data Connection）。

### 二、FTP 两种工作模式

（1）Standard 模式（即 PORT 或 Active 模式，主动方式）

客户端发送 PORT 命令到 FTP 服务器。客户端首先与 FTP 服务器的 TCP 21 端口建立连接，通过该通道发送命令。客户端需要接收数据时，在该通道发送 PORT 命令（包含客户端用什么端口接收数据）。传送数据时，服务器端通过 TCP 20 端口连接到客户端指定的端口发送数据，如图 2-18 所示。

（2）Passive 模式（即 PASV，被动方式）

客户端发送 PASV 命令到 FTP 服务器。建立控制通道时，与 Standard 模式类似。建立连接后，发送 Pasv 命令。FTP 服务器收到 Pasv 命令后，随机打开一个高端端口（端口号大于 1024），并通知客户端在该端口上传送数据；客户端连接 FTP 服务器该端口，FTP 服务器通过该端口进行数据传送，如图 2-19 所示。

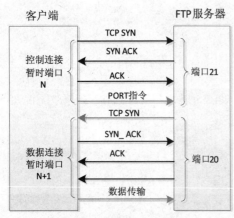
图 2-18　FTP 主动工作模式

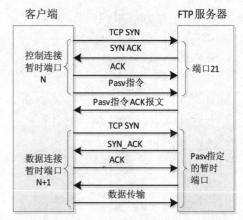
图 2-19　FTP 被动工作模式

### 三、VSFTPD

VSFTPD 是一个安全、稳定、高性能的开源 FTP 服务器软件，适用于多种 UNIX 和 Linux 系统。VSFTPD 的全称是 Very Secure FTP Daemon，中文翻译就是"非常安全的 FTP"。

（1）VSFTPD 提供的配置文件

```
/etc/vsftpd.conf          #主配置文件
/etc/ftpusers             #文件内的用户被禁止
/etc/chroot_list          #改变宿主目录例外用户
/etc/user_list            #文件内的用户被禁止或允许，由 userlist_deny 参数决定
```

（2）VSFTPD 主要配置参数

```
anonymous_enable=YES      #允许匿名用户登录
local_enable=YES          #允许本地用户登录
write_enable=YES          #允许写，全局性设置
anon_upload_enable=YES    #允许匿名用户上传文件，必须使 write_enable=YES
anon_mkdir_write_enable=YES  #允许匿名用户创建目录
chown_uploads=YES         #改变匿名用户上传文档的属主
chown_username=whoever    #设置匿名用户上传的文档的属主，建议不使用 root
anon_root=/opt/ftp        #设置匿名用户的登录根目录
chroot_local_user=YES     #本地用户只能访问宿主目录，不能切换到宿主目录之外
chroot_list_enable=YES    #表示本地用户有些例外，可以切换到宿主目录之外
                          #例外在 chroot_list_file 指定的文件中
chroot_list_file=/etc/chroot_list  #指定例外本地用户
userlist_enable=YES       #设置用户列表有效
userlist_deny=YES         #列表用户不能登录 FTP 服务器，
                          #列表用户在 userlist_file 指定文件中
userlist_file=/etc/user_list  #文件内的用户被禁止或允许
```

## 【任务实施】

### 一、安装系统

① 安装 VSFTPD，操作如下：

```
root@ubuntu:~# apt-get install vsftpd
```

② 安装 FTP 客户端，操作如下：

```
root@ubuntu:~# apt-get install ftp
```

## 二、VSFTPD 服务

① 设置参数，操作如下：

```
root@ubuntu:~# vi /etc/vsftpd.conf
```

修改参数如下：

```
anonymous_enable=NO
```
改变 YES

添加参数如下：

```
anon_root=/opt/ftp
no_anon_password=YES
chroot_local_user=YES
allow_writeable_chroot=YES
userlist_enable=YES
userlist_deny=NO
userlist_file=/etc/vsftpd.user_list
write_enable=YES
```

② 添加允许的账户，操作如下：

```
root@ubuntu:~# vi /etc/vsftpd.user_list
```

添加内容如下：

```
sw
ftp
anonymous
```

③ 开机启动 VSFTPD，操作如下：

```
root@ubuntu:~# systemctl enable vsftpd.service
```

④ 重启 VSFTPD，操作如下：

```
root@ubuntu:~# systemctl restart vsftpd.service
```

## 三、FTP 测试

① 创建目录和设置目录权限，操作如下：

```
root@ubuntu:~# mkdir /opt/ftp
root@ubuntu:~# echo "test" > /opt/ftp/test
root@ubuntu:~# chmod 755 -R /opt/ftp
```

② 查看 FTP 基本命令，操作如下：

```
root@ubuntu:~# ftp
ftp> ?
Commands may be abbreviated.  Commands are:

!           dir         mdelete     qc          site
$           disconnect  mdir        sendport    size
account     exit        mget        put         status
```

| append | form | mkdir | pwd | struct |
|---|---|---|---|---|
| ascii | get | mls | quit | system |
| bell | glob | mode | quote | sunique |
| binary | hash | modtime | recv | tenex |
| bye | help | mput | reget | tick |
| case | idle | newer | rstatus | trace |
| cd | image | nmap | rhelp | type |
| cdup | ipany | nlist | rename | user |
| chmod | ipv4 | ntrans | reset | umask |
| close | ipv6 | open | restart | verbose |
| cr | lcd | prompt | rmdir | ? |
| delete | ls | passive | runique | |
| debug | macdef | proxy | send | |

③ 以匿名账户 ftp 或 anonymous 登录 VSFTPD，操作如下：

```
root@ubuntu:~# ftp localhost
Connected to localhost.
220 (vsFTPd 3.0.3)
Name (localhost:sw): ftp
230 Login successful.
Remote system type is UNIX.
Using binary mode to transfer files.
ftp> ls
200 PORT command successful. Consider using PASV.
150 Here comes the directory listing.
-rwxr-xr-x    1 0        0               5 Jan 23 15:13 test
226 Directory send OK.
ftp> pwd
257 "/" is the current directory
ftp> mkdir abc
550 Permission denied.
ftp> get test
local: test remote: test
200 PORT command successful. Consider using PASV.
150 Opening BINARY mode data connection for test (5 bytes).
226 Transfer complete.
5 bytes received in 0.00 secs (15.1640 kB/s)
ftp>
```

④ 以用户 sw 登录 VSFTPD，操作如下：

```
root@ubuntu:~# ftp localhost
Connected to localhost.
220 (vsFTPd 3.0.3)
```

```
Name (localhost:sw): sw
331 Please specify the password.
Password:
230 Login successful.
Remote system type is UNIX.
Using binary mode to transfer files.
ftp> pwd
257 "/" is the current directory
ftp> mkdir abc
257 "/abc" created
ftp> put test
local: test remote: test
200 PORT command successful. Consider using PASV.
150 Ok to send data.
226 Transfer complete.
5 bytes sent in 0.00 secs (95.7414 kB/s)
ftp> ls
200 PORT command successful. Consider using PASV.
150 Here comes the directory listing.
drwx------    2 1000     1000         4096 Jan 23 16:33 abc
-rw-------    1 1000     1000            5 Jan 23 23:41 test
226 Directory send OK.
ftp>
```

## 任务 3　搭建 MySQL 数据库系统

### 【任务概述】

MySQL 数据库用来存放 Hive 元数据，也用来存放大数据分析结果等数据。本任务需要完成 MySQL 数据库系统的安装、MySQL 数据库初始化、用户授权、创建数据库和表、插入查询数据、删除数据库等。

### 【支撑知识】

数据库（Database）是按照数据结构来组织、存储和管理数据的仓库。以关系模型来创建的数据库称为关系型数据库（Relational Database）。相类似的实体被存入表（Table）中。表是关系型数据库的核心单元，它是存储数据的地方。在表的内部，数据被分为行和列，每一行代表一个实体，每一列代表实体的一个属性，它说明数据的名称，也限定了数据的类型。在表中用一个唯一的标识符来标识每一行，这个标识符称为主键（Primary Key），外键（Foreign Key）则用来表达表与表之间的联系。

MySQL 是一个小型关系型数据库管理系统，被广泛应用在 Internet 上的中小型网站中，具有体积小、速度快、总体拥有成本低等特点，并且开放源码。MySQL 的功能十分强大，是一个真正的多用户、多线程 SQL 数据库服务程序，它是一种客户/服务器系统，由一个服务器守护程序 mysqld，以及很多客户程序和函数库组成。MySQL 支持多种字符集，经常使用的字符集有 GBK 和 UTF8。MySQL 数据库系统的常用基本操作如表 2-1 所示。

表 2-1　MySQL 常用操作

| 常用操作 | 基本命令 |
| --- | --- |
| 设置用户密码 | mysqladmin -u 用户 password 密码 |
| 修改用户密码 | mysqladmin -u 用户 -p 旧密码 password 新密码 |
| 导出数据库 | mysqldump -u 用户 -p 密码 数据库 > 数据库.sql |
| 导入数据库 | mysql -u 用户 -p 密码 --default-character-set=字符集 数据库 < 数据库.sql |
| 创建数据库 | CREATE DATABASE 数据库 |
| 打开数据库 | USE 数据库; |
| 创建表 | CREATE TABLE table_name (column_name column_type , column_name column_type, … primary key(column_name,…)) |
| 插入记录 | INSERT INTO 表名（列1, 列2, …） VALUE(值1, 值2, …) |
| 显示数据库 | SHOW DATABASES |
| 显示数据表 | SHOW TABLES |
| 显示表结构 | DESCRIBE 表名 或 SHOW COLUMNS FROM 表名 |
| 删除表 | DROP TABLE 表名 |
| 删除数据库 | DROP DATABASE 数据库 |
| 查询数据 | SELECT * FROM 表名 |
| 导出表 | SELECT * FROM 表名 INTO OUTFILE '表名.txt' |
| 导入表 | LOAD DATA INFILE '表名.txt' INTO TABLE 表名 |
| 导入表 | mysqlimport -u 用户 -p 密码 数据库 '表名.txt' |
| 用户授权 | GRANT ALL PRIVILEGES ON 数据库.* TO '用户'@'%' IDENTIFIED BY '密码'<br>GRANT ALL PRIVILEGES ON *.* TO '用户'@'localhost' IDENTIFIED BY '密码' WITH GRANT OPTION |

## 【任务实施】

### 一、安装 MySQL

① 安装 MySQL-Client，操作如下：

```
root@ubuntu:~# apt-get install mysql-client
```

② 安装 MySQL-Server，操作如下：

```
root@ubuntu:~# apt-get install mysql-server
```

在安装过程两次提示输入用户 root 密码，按要求输入即可。

## 二、设置 MySQL 服务

① 修改 MySQL-Server 参数，操作如下：

```
root@ubuntu:~# vi /etc/mysql/mysql.conf.d/mysqld.cnf
```

修改参数如下：

```
bind-address = 127.0.0.1 改为 bind-address = 0.0.0.0
```

添加内容如下：

```
default-storage-engine = innodb
innodb_file_per_table
collation-server = utf8_general_ci
init-connect = 'SET NAMES utf8'
character-set-server = utf8
```

② 设置开机启动服务，操作如下：

```
root@ubuntu:~# systemctl enable mysql.service
```

③ 启动 MySQL 服务，操作如下：

```
root@ubuntu:~# systemctl start mysql.service
```

④ 初始化 MySQL，操作如下：

```
root@ubuntu:~# mysql_secure_installation
Enter password for user root：输入密码
VALIDATE PASSWORD PLUGIN can be used to test passwords
and improve security. It checks the strength of password
and allows the users to set only those passwords which are
secure enough. Would you like to setup VALIDATE PASSWORD plugin?
Press y|Y for Yes, any other key for No：直接按 Enter 键
Using existing password for root.
Change the password for root ? ((Press y|Y for Yes, any other key for No) : Y
New password：输入新密码
Re-enter new password：重新输入密码
Remove anonymous users? (Press y|Y for Yes, any other key for No) : Y
Disallow root login remotely? (Press y|Y for Yes, any other key for No) : Y
Remove test database and access to it? (Press y|Y for Yes, any other key for No) : Y
Reload privilege tables now? (Press y|Y for Yes, any other key for No) : Y
All done!
```

## 三、MySQL 基本操作

① 给 MySQL 的用户 hive 授权，允许访问所有数据库和远程访问，操作如下：

```
root@ubuntu:~# mysql -uroot -p123456
mysql> GRANT ALL PRIVILEGES ON *.* TO 'hive'@'%' IDENTIFIED BY '123456';
Query OK, 0 rows affected, 1 warning (0.01 sec)
mysql> GRANT ALL PRIVILEGES ON *.* TO 'hive'@'localhost' IDENTIFIED BY '123456';
```

```
Query OK, 0 rows affected, 1 warning (0.00 sec)
mysql> exit
Bye
```

② 以用户 hive 登录 MySQL，创建 hive 数据库，操作如下：

```
root@ubuntu:~# mysql -h172.25.0.10 -uhive -p123456
mysql> create database hive;
Query OK, 1 row affected (0.00 sec)
mysql> show databases;
+--------------------+
| Database           |
+--------------------+
| information_schema |
| hive               |
| mysql              |
| performance_schema |
| sys                |
+--------------------+
5 rows in set (0.00 sec)
mysql> use hive;
Database changed
```

③ 创建表 testtab，插入两条记录并查询该记录，操作如下：

```
mysql> create table testtab(id int not null auto_increment,name char(20),
price double,address varchar(50),primary key(id));
Query OK, 0 rows affected (0.07 sec)
mysql> show tables;
+----------------+
| Tables_in_hive |
+----------------+
| testtab        |
+----------------+
1 row in set (0.01 sec)
mysql> describe testtab;
+---------+-------------+------+-----+---------+----------------+
| Field   | Type        | Null | Key | Default | Extra          |
+---------+-------------+------+-----+---------+----------------+
| id      | int(11)     | NO   | PRI | NULL    | auto_increment |
| name    | char(20)    | YES  |     | NULL    |                |
| price   | double      | YES  |     | NULL    |                |
| address | varchar(50) | YES  |     | NULL    |                |
+---------+-------------+------+-----+---------+----------------+
4 rows in set (0.00 sec)
```

```
mysql> insert into testtab(id,name,price,address)
       value(100,"九阳豆浆机",378,"浙江省杭州市");
Query OK, 1 row affected (0.10 sec)
mysql> insert into testtab(id,name,price,address)
       value(101,"华为手机",2688,"广东省深圳市");
Query OK, 1 row affected (0.01 sec)
mysql> select * from testtab;
+-----+-------------+-------+--------------------+
| id  | name        | price | address            |
+-----+-------------+-------+--------------------+
| 100 | 九阳豆浆机   |  378  | 浙江省杭州市        |
| 101 | 华为手机     | 2688  | 广东省深圳市        |
+-----+-------------+-------+--------------------+
2 rows in set (0.00 sec)
```

④ 删除表及删除数据库，操作如下：

```
mysql> drop table testtab;
Query OK, 0 rows affected (0.01 sec)
mysql> drop database hive;
Query OK, 0 rows affected (0.00 sec)
```

## 任务 4　安装 Ubuntu Desktop

### 【任务概述】

大数据应用开发是大数据技术应用不可或缺的环境，开发环境主要使用 Ubuntu Desktop 客户端。本任务主要完成客户机 Ubuntu Desktop 的安装、网络配置、OpenSSH 安装。

### 【支撑知识】

#### 一、Ubuntu Desktop

Ubuntu Desktop 是 Ubuntu 的桌面版本，GNOME 是 Ubuntu 的默认桌面环境。GNOME（GNU Network Object Model Environment，GNU 网络对象模型环境）是一个国际性的项目，为开发完整的、由自由软件组成的桌面环境而努力。桌面环境，即图形用户界面，是计算机系统中最外层的软件。GNOME 项目的目标包括创建软件开发框架，选择桌面应用程序，编写负责引导应用软件的程序、文件句柄、窗口和任务管理器等。来自世界各地的社区成员将 GNOME 翻译为各种语言，让使用不同语言的人们都能享用 GNOME。

Ubuntu Desktop 主要用于日常的办公和开发。重视人机交互体验，集成了比较强大的 Desktop 桌面及桌面应用程序。系统提供大量的应用软件，非常适合学术研究或互联网开发相关的工作。

#### 二、DPKG 工具

DPKG 是 Debian Package 的简写，是为 Debian 专门开发的套件管理系统，方便软件的安装、更新及移除。所有源自 Debian 的 Linux 发行版都使用 DPKG，如 Ubuntu、Knoppix 等。DPKG 本身是一个底层的工具。上层的工具，如 APT，被用于从远程获取软件包及处理复杂的软件包关系。DPKG 常用命令如表 2-2 所示。

表 2-2 DPKG 常用命令

| 功　　能 | 命　　令 |
| --- | --- |
| 安装一个 Debian 软件包 | dpkg –i <package.deb> |
| 列出<package.deb>的内容 | dpkg –c <package.deb> |
| 从<package.deb>中提取包信息 | dpkg –I <package.deb> |
| 删除软件包（保留其配置信息） | dpkg –r <package> |
| 删除一个包（包括配置信息） | dpkg –P <package> |
| 列出<package>安装的所有文件清单 | dpkg –L <package> |
| 查找只有部分安装的软件包信息 | dpkg –C <package> |
| 显示已安装包的信息 | dpkg –s <package> |
| 安装一个目录下面所有的软件包 | dpkg –R |
| 释放软件包，但是不进行配置 | dpkg --unpack package_file |
| 重新配置和释放软件包 | dpkg –configure package_file |

## 【任务实施】

### 一、安装系统

① 把 Ubuntu Desktop 系统光盘插入光驱，选择光驱引导，系统启动后提示语言选择，如图 2-20 所示，选择"中文(简体)"，如图 2-21 所示。

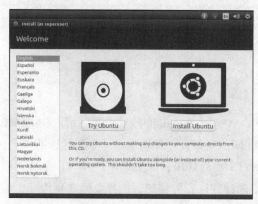

图 2-20 安装语言提示

图 2-21 选择"中文(简体)"

② 第三方软件暂时不安装，单击"继续"按钮，如图 2-22 所示。清除整个磁盘并安装 Ubuntu，如图 2-23 所示，单击"现在安装"按钮后提示"将改动写入磁盘吗？"，单击"继续"按钮即可。

③ 区域直接输入"Shanghai"或在地图上选择。"键盘布局"选择"Chinese"。

④ Ubuntu Desktop 需要创建一个普通用户，根据要求输入用户 sw 和密码等信息，如图 2-24 所示。单击"继续"按钮，开始安装软件，如图 2-25 所示。

图 2-22 第三方软件安装提示

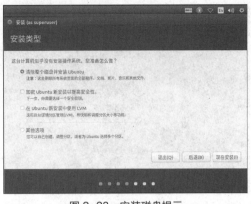

图 2-23 安装磁盘提示

图 2-24 创建用户

图 2-25 安装系统

⑤ 安装完成后，单击"现在重启"按钮，系统重启后输入用户 sw 的密码进入系统，如图 2-26 所示。进入系统后 Ubuntu 桌面如图 2-27 所示。

图 2-26 用户登录界面

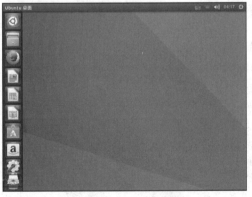

图 2-27 Ubuntu 桌面

⑥ 按 Ctrl+Alt+T 组合键打开终端窗口，如图 2-28 所示。单击右上角的图标，选择"关机…"可以关闭所有程序并关机，如图 2-29 所示。

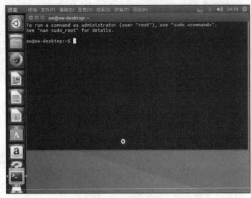
图 2-28 终端窗口

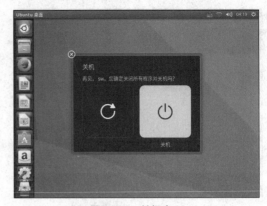
图 2-29 关机窗口

## 二、设置用户 root

① 设置用户 root 密码，并切换到 root 用户，操作如下：

sw@sw-desktop:~$ sudo passwd root

[sudo] sw 的密码：

输入新的 UNIX 密码：

重新输入新的 UNIX 密码：

passwd: 已经成功更新密码

sw@sw-desktop:~$ su - root

密码：

root@sw-desktop:~#

② 设置用户 root 直接登录，操作如下：

root@sw-desktop:~# vi /usr/share/lightdm/lightdm.conf.d/50-ubuntu.conf

增加一行

greeter-show-manual-login=true

注意：重新启动才生效。

## 三、配置网络

① 查看网络配置信息，操作如下：

root@sw-desktop:~# ifconfig -a

ens33: flags=4163<UP,BROADCAST,RUNNING,MULTICAST>  mtu 1500

    ether 00:0c:29:f9:1d:e8  txqueuelen 1000  (以太网)

    RX packets 213  bytes 13203 (13.2 KB)

    RX errors 0  dropped 0  overruns 0  frame 0

    TX packets 137  bytes 21935 (21.9 KB)

    TX errors 0  dropped 0  overruns 0  carrier 0  collisions 0

lo: flags=73<UP,LOOPBACK,RUNNING>  mtu 65536

    inet 127.0.0.1  netmask 255.0.0.0

    inet6 ::1  prefixlen 128  scopeid 0x10<host>

    loop  txqueuelen 1  (本地环回)

```
        RX packets 3108  bytes 191488 (191.4 KB)
        RX errors 0  dropped 0  overruns 0  frame 0
        TX packets 3108  bytes 28640 (191.4 KB)
        TX errors 0  dropped 0  overruns 0  carrier 0  collisions 0
```

② 配置 IP 地址，操作如下：

```
root@sw-desktop:~# vi /etc/network/interfaces
```

添加内容如下：

```
auto ens33
iface ens33 inet static
address 172.25.0.20
netmask 255.255.255.0
gateway 172.25.0.254
dns-nameserver 202.96.128.86
dns-nameserver 202.96.134.133
```

③ 修改主机名，操作如下：

```
root@sw-desktop:~# vi /etc/hostname
sw-desktop
```

④ 修改域名服务器，操作如下：

```
root@sw-desktop:~# vi /etc/resolv.conf
```

修改内容如下：

```
nameserver 202.96.128.86
nameserver 202.96.134.133
```

永久有效需要如下操作：

```
root@sw-desktop:~# vi /etc/resolvconf/resolv.conf.d/base
```

修改内容如下：

```
nameserver 202.96.128.86
nameserver 202.96.134.133
```

⑤ 修改 hosts 文件，操作如下：

```
root@sw-desktop:~# vi /etc/hosts
```

修改内容如下：

```
127.0.0.1      localhost
172.25.0.10    master
172.25.0.11    slave1
172.25.0.12    slave2
172.25.0.20    sw-desktop
```

⑥ 配置网络后需要重启网络，操作如下：

```
root@sw-desktop:~# systemctl restart networking
```

⑦ 测试网络是否连通，操作如下：

```
root@sw-desktop:~# ping www.baidu.com
PING www.a.shifen.com (14.215.177.37) 56(84) bytes of data.
```

```
64 bytes from 14.215.177.37 (14.215.177.37): icmp_seq=1 ttl=128 time=19.0 ms
64 bytes from 14.215.177.37 (14.215.177.37): icmp_seq=2 ttl=128 time=18.6 ms
^C
--- www.a.shifen.com ping statistics ---
2 packets transmitted, 2 received, 0% packet loss, time 1002ms
rtt min/avg/max/mdev = 18.662/18.833/19.004/0.171 ms
```

提示：上述表示已经连通，按 Ctrl+C 组合键中断测试。

### 四、安装 OpenSSH

① 从 Master 节点复制 OpenSSH 包，操作如下：

```
root@sw-desktop:~# scp -r 172.25.0.10:/media/ubuntu/pool/main/o/openssh /root
```

② 安装 OpenSSH 包，操作如下：

```
root@sw-desktop:~# cd /root/openssh
root@sw-desktop:~/openssh# dpkg -i openssh-sftp-server_7.3p1-1_amd64.deb
root@sw-desktop:~/openssh# dpkg -i openssh-server_7.3p1-1_amd64.deb
```

③ 启动 SSH 服务，操作如下：

```
root@sw-desktop:~/openssh# systemctl start ssh
root@sw-desktop:~/openssh# systemctl enable ssh
```

④ 测试 SSH，操作如下：

```
root@sw-desktop:~# ssh sw@sw-desktop
The authenticity of host 'sw-desktop (127.0.1.1)' can't be established.
ECDSA key fingerprint is SHA256:0DQxizwbFFTx4KgZUi+bfKqSaHLuZORWcwtFj6OyTlg.
Are you sure you want to continue connecting (yes/no)? yes
Warning: Permanently added 'sw-desktop' (ECDSA) to the list of known hosts.
sw@sw-desktop's password:
Welcome to Ubuntu 16.10 (GNU/Linux 4.8.0-22-generic x86_64)

 * Documentation:  https://help.ubuntu.com
 * Management:     https://landscape.canonical.com
 * Support:        https://ubuntu.com/advantage

0 个可升级软件包。
0 个安全更新。

Last login: Thu Jan 26 13:02:12 2017 from 172.25.0.1
sw@sw-desktop:~$ exit
注销
Connection to sw-desktop closed.
root@sw-desktop:~#
```

# 【同步训练】

## 一、简答题

（1）简述 Ubuntu Server 和 Ubuntu Desktop 的区别。

（2）简述软件包管理命令 APT、APT-Get、APTITUDE 有什么异同点。

（3）OpenSSH 是什么？

## 二、操作题

（1）下载 Ubuntu Server 系统包，并在虚拟机上安装。

（2）安装 VSFTPD 并进行配置，禁止用户 root 和 sw 登录。

（3）安装 APTITUDE 软件，并使用 APTITUDE 安装 OpenSSH。

# 项目三 Hadoop 集群部署

## 【项目介绍】

Hadoop 集群部署需要规划和设计集群节点,需要完成 Hadoop 系统的配置,集群节点需要时间同步、配置 SSH 证书登录,Hadoop 集群部署成功后使用 Hadoop Shell 对 HDFS 进行文件操作。

本项目分以下 3 个任务:
- 任务 1  构建集群系统
- 任务 2  SSH 证书登录
- 任务 3  Hadoop 部署与使用

## 【学习目标】

### 一、知识目标

- 熟悉集群系统的安装配置。
- 掌握集群系统的时间同步。
- 掌握 SSH 证书登录设置。
- 掌握 Hadoop 系统的安装。
- 掌握 Hadoop 的配置。
- 掌握 Hadoop 的启动和检查。
- 掌握 Hadoop Shell 命令的使用。

### 二、能力目标

- 能够安装配置集群系统。
- 能够进行集群系统的时间同步。
- 能够设置 SSH 证书登录。
- 能够安装配置 Hadoop 集群系统。
- 能够进行 Hadoop 系统运维。
- 能够使用 Hadoop Shell 命令。

## 任务1 构建集群系统

**【任务概述】**

集群 Hadoop 系统需要硬件和操作系统的支撑,本任务要求设计一个集群系统,要进行集群节点规划、操作系统安装、网络配置,集群节点之间还要进行时间同步。

**【支撑知识】**

### 一、集群技术

集群技术是一种较新的技术,通过集群技术,可以在付出较低成本的情况下获得在性能、可靠性、灵活性等方面的相对较高的收益,其任务调度则是集群系统中的核心技术。集群是一组相互独立的、通过高速网络互联的计算机,它们构成了一个组,并以单一系统的模式加以管理。一个客户与集群相互作用时,集群像一个独立的服务器。集群配置用于提高可用性和可缩放性。集群的优点如下:

(1)提高性能

一些计算密集型应用,如天气预报、核试验模拟等,需要计算机有很强的运算处理能力,现有的技术,即使普通的大型机器计算也很难胜任。这时,一般都使用计算机集群技术,集中几十台甚至上百台计算机的运算能力来满足要求。提高处理性能一直是集群技术研究的重要目标之一。

(2)降低成本

通常一套较好的集群配置,其软硬件开销很昂贵。但与价值上百万美元的专用超级计算机相比,已属相当便宜。在达到同样性能的条件下,采用计算机集群比采用同等运算能力的大型计算机具有更高的性价比。

(3)提高可扩展性

用户若想扩展系统能力,不得不购买更高性能的服务器,才能获得额外所需的 CPU 和存储器。如果采用集群技术,则只需要将新的服务器加入集群中即可,对于客户来说,服务无论从连续性还是性能上都几乎没有变化,好像系统在不知不觉中完成了升级。

(4)增强可靠性

集群技术使系统在故障发生时仍可以继续工作,将系统停运时间减到最小。集群系统在提高系统可靠性的同时,也大大减小了故障损失。

### 二、NTP

NTP 协议的全称为网络时间协议(Network Time Protocol)。它的目的是在国际互联网上传递统一、标准的时间。具体的实现方案是在网络上指定若干时钟源网站,为用户提供授时服务,并且这些网站间应该能够相互比对,提高准确度。局域网内所有的 PC、服务器和其他设备通过网络与时间服务器保持同步,NTP 协议自动判断网络延时,并给得到的数据进行时间补偿。

NTP 在 Linux 下有两种时钟同步方式,分别为直接同步和平滑同步。

(1)直接同步

使用 ntpdate 命令进行同步,直接进行时间变更。假设服务器上存在一个 12 点运行的任务，当前服务器时间是 13 点，但标准时间时 11 点，使用此命令可能会造成任务重复执行。因此，使用 ntpdate 同步可能会引发风险，该命令也多用于配置时钟同步服务第一次同步时间时使用。

（2）平滑同步

使用 ntpd 命令进行时钟同步，可以保证一个时间不经历两次，它每次同步时间的偏移量不会太陡，是慢慢来的，正因为这样，ntpd 平滑同步可能耗费的时间比较长。

## 【任务实施】

### 一、集群系统的规划

集群节点一般选择普通硬件厂商生产的标准化、广泛有效的硬件。由于 Hadoop 的主体用 Java 写成，能在安装了 JVM 的平台上运行。但仍有部分代码需要在 UNIX 环境下执行，因而不适合在非 UNIX 平台上运行。本任务使用 3 台服务器作为集群节点，一台 PC 作为客户机，集群系统规划如表 3-1 所示，拓扑图如图 3-1 所示。

表 3-1 集群系统规划

| 主机名 | IP 地址 | 网 关 | 角 色 | 操作系统 |
| --- | --- | --- | --- | --- |
| master | 172.25.0.10/24 | 172.25.0.254 | Master | Ubuntu Server |
| slave1 | 172.25.0.11/24 | 172.25.0.254 | Slave | Ubuntu Server |
| slave2 | 172.25.0.12/24 | 172.25.0.254 | Slave | Ubuntu Server |
| sw-desktop | 172.25.0.20/24 | 172.25.0.254 | Desktop | Ubuntu Desktop |

图 3-1 集群系统拓扑图

### 二、集群系统的安装与配置

各集群节点安装 Ubuntu Server 系统，PC 上安装 Ubuntu Desktop 系统。各节点配置基本一样，下面以 master 节点为例，以系统安装用户 sw 登录 master 节点。

① 配置网络，操作如下：

```
sw@ubuntu:~$ sudo vi /etc/network/interfaces
[sudo] password for sw:
```

添加如下内容：
```
auto ens33
iface ens33 inet static
address 172.25.0.10
netmask 255.255.255.0
gateway 172.25.0.254
```

```
dns-nameserver 202.96.128.86
dns-nameserver 202.96.134.133
```

② 设置主机名，操作如下：
```
sw@ubuntu:~$ sudo vi /etc/hostname
```

内容如下：
```
master
```

③ 设置 hosts，操作如下：
```
sw@ubuntu:~$ sudo vi /etc/hosts
```

内容如下：
```
127.0.0.1       localhost
172.25.0.10     master
172.25.0.11     slave1
172.25.0.12     slave2
172.25.0.20     sw-desktop
```

④ 关闭防火墙，操作如下：
```
sw@ubuntu:~$ sudo ufw disable
```

⑤ 重启网络，操作如下：
```
sw@ubuntu:~$ sudo systemctl restart networking
```

### 三、设置 FTP 服务器与 APT 软件源

① 在 Master 节点上搭建 FTP 服务器，操作如下：
```
sw@master:~$ sudo mkdir /media/ubuntu
sw@master:~$ sudo mount /dev/cdrom /media/ubuntu
sw@master:~$ sudo dpkg -i /media/ubuntu/pool/main/libe/libeatmydata/libeatmydata1_105-3_amd64.deb
sw@master:~$ sudo dpkg -i /media/ubuntu/pool/main/v/vsftpd/vsftpd_3.0.3-7_amd64.deb
```

② 配置 FTP 服务器，操作如下：
```
sw@master:~$ sudo vi /etc/vsftpd.conf
```

修改内容如下：
`anonymous_enable=NO` 改为 `anonymous_enable=YES`

并添加内容如下：
```
anon_root=/media/ubuntu/
```

③ 重启 FTP 服务器，操作如下：
```
sw@master:~$ sudo systemctl restart vsftpd
sw@master:~$ sudo systemctl enable vsftpd
```

④ 各集群节点设置 APT 源，操作如下：
```
sw@master:~$ sudo vi /etc/apt/sources.list
```

内容如下：
```
deb ftp://master/ yakkety main restricted
```

⑤ 各集群节点进行更新，操作如下：

```
sw@master:~$ sudo apt-get update
```

## 四、时间同步

① 各集群节点安装 NTP 软件，操作如下：
```
sw@master:~$ sudo apt-get install ntp
```

② Master 节点设置 NTP，操作如下：
```
sw@master:~$ sudo vi /etc/ntp.conf
```

删除默认配置 restrict 两个选项 nopeer 和 noquery，删除后如下：
```
restrict -4 default kod notrap nomodify limited
restrict -6 default kod notrap nomodify limited
```

添加内容如下：
```
server 127.127.1.0
fudge 127.127.1.0 stratum 10
```

③ Slave 节点设置 NTP，操作如下：
```
sw@master:~$ sudo vi /etc/ntp.conf
```

添加内容如下：
```
server master iburst
```

④ 各集群节点重启 NTP，操作如下：
```
sw@master:~$ sudo systemctl restart ntp
sw@master:~$ sudo systemctl enable ntp
```

⑤ 验证 Master 节点的 NTP，操作如下：
```
sw@master:~$ ntpq -c peers
     remote           refid      st t when poll reach   delay   offset  jitter
==============================================================================
 0.ubuntu.pool.n .POOL.          16 p    -   64    0   0.000    0.000   0.000
 1.ubuntu.pool.n .POOL.          16 p    -   64    0   0.000    0.000   0.000
 2.ubuntu.pool.n .POOL.          16 p    -   64    0   0.000    0.000   0.000
 3.ubuntu.pool.n .POOL.          16 p    -   64    0   0.000    0.000   0.000
 ntp.ubuntu.com  .POOL.          16 p    -   64    0   0.000    0.000   0.000
 LOCAL(0)        .LOCL.          10 l    8   64    1   0.000    0.000   0.000

sw@master:~$ ntpq -c assoc

ind assid status  conf reach auth condition  last_event cnt
===========================================================
  1  7269  8811   yes  none  none reject     mobilize   1
  2  7270  8811   yes  none  none reject     mobilize   1
  3  7271  8811   yes  none  none reject     mobilize   1
  4  7272  8811   yes  none  none reject     mobilize   1
  5  7273  8811   yes  none  none reject     mobilize   1
  6  7274  9024   yes  yes   none reject     reachable  2
```

等待几分钟，操作如下：

```
sw@master:~$ ntpq -c peers
     remote           refid      st t when poll reach   delay   offset  jitter
==============================================================================
 0.ubuntu.pool.n .POOL.          16 p    -   64    0    0.000    0.000   0.000
 1.ubuntu.pool.n .POOL.          16 p    -   64    0    0.000    0.000   0.000
 2.ubuntu.pool.n .POOL.          16 p    -   64    0    0.000    0.000   0.000
 3.ubuntu.pool.n .POOL.          16 p    -   64    0    0.000    0.000   0.000
 ntp.ubuntu.com  .POOL.          16 p    -   64    0    0.000    0.000   0.000
*LOCAL(0)        .LOCL.          10 l   14   64  377    0.000    0.000   0.000

sw@master:~$ ntpq -c assoc

ind assid status  conf reach auth condition  last_event cnt
===========================================================
  1  7269  8811   yes  none  none   reject    mobilize  1
  2  7270  8811   yes  none  none   reject    mobilize  1
  3  7271  8811   yes  none  none   reject    mobilize  1
  4  7272  8811   yes  none  none   reject    mobilize  1
  5  7273  8811   yes  none  none   reject    mobilize  1
  6  7274  963a   yes  yes   none  sys.peer   sys_peer  3
```

⑥ 验证 Slave 节点的 NTP，操作如下：

```
sw@slave1:~$ ntpq -c peers
     remote           refid      st t when poll reach   delay   offset  jitter
==============================================================================
 0.ubuntu.pool.n .POOL.          16 p    -   64    0    0.000    0.000   0.000
 1.ubuntu.pool.n .POOL.          16 p    -   64    0    0.000    0.000   0.000
 2.ubuntu.pool.n .POOL.          16 p    -   64    0    0.000    0.000   0.000
 3.ubuntu.pool.n .POOL.          16 p    -   64    0    0.000    0.000   0.000
 ntp.ubuntu.com  .POOL.          16 p    -   64    0    0.000    0.000   0.000
 master          .INIT.          16 u    -   64    0    0.000    0.000   0.000

sw@slave1:~$ ntpq -c assoc

ind assid status  conf reach auth condition  last_event cnt
===========================================================
  1 46424  8811   yes  none  none   reject    mobilize  1
  2 46425  8811   yes  none  none   reject    mobilize  1
  3 46426  8811   yes  none  none   reject    mobilize  1
  4 46427  8811   yes  none  none   reject    mobilize  1
  5 46428  8811   yes  none  none   reject    mobilize  1
```

```
6  46429  8011    yes    no   none    reject     mobilize  1
```

等待几分钟,操作如下:

```
sw@slave1:~$ ntpq -c peers
     remote           refid      st t when poll reach   delay   offset  jitter
==============================================================================
 0.ubuntu.pool.n .POOL.           16 p    -   64    0   0.000    0.000   0.000
 1.ubuntu.pool.n .POOL.           16 p    -   64    0   0.000    0.000   0.000
 2.ubuntu.pool.n .POOL.           16 p    -   64    0   0.000    0.000   0.000
 3.ubuntu.pool.n .POOL.           16 p    -   64    0   0.000    0.000   0.000
 ntp.ubuntu.com  .POOL.           16 p    -   64    0   0.000    0.000   0.000
*master          LOCAL(0)         11 u   67   64  177   0.237   -0.009   4.527

sw@slave1:~$ ntpq -c assoc

ind assid status  conf reach auth condition  last_event cnt
===========================================================
 1 46424  8811    yes  none  none   reject     mobilize  1
 2 46425  8811    yes  none  none   reject     mobilize  1
 3 46426  8811    yes  none  none   reject     mobilize  1
 4 46427  8811    yes  none  none   reject     mobilize  1
 5 46428  8811    yes  none  none   reject     mobilize  1
 6 46429  963a    yes  yes   none  sys.peer    sys_peer  3
```

这时使用 date 命令查看时间,时间已经同步。

## 任务 2  SSH 证书登录

### 【任务概述】

集群 Hadoop 系统中 Hadoop 需要对 Linux 系统进行脚本控制,需要使用 SSH 免密码登录。为了区分 Hadoop 和本机上的其他服务,最好单独创建用户 hadoop。本任务主要完成用户 hadoop 的创建和 SSH 证书登录设置。

### 【支撑知识】

#### 一、SSH 简介

SSH 为 Secure Shell 的缩写,由 IETF 的网络小组所制定。SSH 为建立在应用层基础上的安全协议。SSH 是目前较可靠,专为远程登录会话和其他网络服务提供安全性的协议。利用 SSH 协议可以有效防止远程管理过程中的信息泄露问题。SSH 最初是 UNIX 系统上的一个程序,后来又迅速扩展到其他操作平台。SSH 在正确使用时可弥补网络中的漏洞。SSH 客户端适用于多种平台。几乎所有 UNIX 平台——HP-UX、Linux、AIX、Solaris、Digital UNIX、Irix,以及其他平台,都可运行 SSH。

SSH 主要由如下 3 部分组成。

（1）传输层协议（SSH-TRANS）

传输层协议提供了服务器认证、保密性及完整性。此外，它有时还提供压缩功能。SSH-TRANS 通常运行在 TCP/IP 连接上，也可用于其他可靠数据流上。SSH-TRANS 提供了强力的加密技术、密码主机认证及完整性保护。该协议中的认证基于主机，并且该协议不执行用户认证。更高层的用户认证协议可以设计为在此协议之上。

（2）用户认证协议（SSH-USERAUTH）

用户认证协议用于向服务器提供客户端用户鉴别功能。它运行在传输层协议 SSH-TRANS 上面。当 SSH-USERAUTH 开始后，它从低层协议那里接收会话标识符（从第一次密钥交换中的交换哈希）。会话标识符唯一标识此会话并且适用于标记以证明私钥的所有权。SSH-USERAUTH 也需要知道低层协议是否提供保密性保护。

（3）连接协议（SSH-CONNECT）

连接协议将多个加密隧道分成逻辑通道。它运行在用户认证协议上。它提供了交互式登录话路、远程命令执行、转发 TCP/IP 连接和转发 X11 连接。

SSH 分为有密码登录和证书登录。考虑到安全性因素，一般都是采用证书登录，即每次登录无须输入密码。如果是密码登录，很容易遭受外来的攻击。

## 二、证书登录

证书是已有的 SSH 公钥认证系统的扩展，可应用于任何已有的公钥和私钥对，也可以用于任何当前 SSH 支持的认证方法。

证书登录过程如下：

（1）客户端生成证书的私钥和公钥对

私钥放在客户端，公钥上传到服务端（远程登录端）。一般为了安全，防止黑客复制客户端的私钥，客户端在生成私钥时，会设置一个密码，以后每次登录 SSH 服务器时，客户端都要输入密码解开私钥。

（2）服务器添加信用公钥

把客户端生成的公钥上传到 SSH 服务器，添加到指定的文件中，这样就完成 SSH 证书登录的配置了。假设客户端想通过私钥登录其他 SSH 服务器，可以把公钥上传到其他 SSH 服务器。

## 【任务实施】

### 一、创建用户

① 各集群节点创建用户 hadoop，UID、GID 设为一样，操作如下：

```
sw@master:~$ sudo groupadd -g 730 hadoop
sw@master:~$ sudo useradd -u 730 -g 730 -m -s /bin/bash hadoop
```

② 添加用户 hadoop 到用户组 sudo，操作如下：

```
sw@master:~$ sudo gpasswd -a hadoop sudo
```

③ 设置用户 hadoop 的密码，操作如下：

```
sw@master:~$ sudo passwd hadoop
Enter new UNIX password:
Retype new UNIX password:
passwd: password updated successfully
```

### 二、证书操作

① 生成证书的私钥和公钥对,有 RSA 和 DSA 两种算法,一般使用 RSA 即可。以用户 hadoop 登录 Master 节点,生成证书密钥对,操作如下:

```
hadoop@master:~$ ssh-keygen -t rsa
Generating public/private rsa key pair.
Enter file in which to save the key (/home/hadoop/.ssh/id_rsa):
Created directory '/home/hadoop/.ssh'.
Enter passphrase (empty for no passphrase):
Enter same passphrase again:
Your identification has been saved in /home/hadoop/.ssh/id_rsa.
Your public key has been saved in /home/hadoop/.ssh/id_rsa.pub.
The key fingerprint is:
SHA256:hd8RyRLTaoU9gH1Wap6XtKtJnVSAPvfPsDAv5dEAFsw hadoop@master
The key's randomart image is:
+--,-[RSA 2048]----+
|        o+Xo=o   |
|        ..+.Eo . |
|         . .O+o. .|
|          oo+o+o+ |
|        S.. +o=+ |
|           o++oo|
|           .*+=.|
|           ..o+ o|
|            o.   |
+----[SHA256]-----+
```

② 复制公钥到各集群节点,操作如下:

```
hadoop@master:~$ ssh-copy-id -i .ssh/id_rsa.pub master
hadoop@master:~$ ssh-copy-id -i .ssh/id_rsa.pub slave1
hadoop@master:~$ ssh-copy-id -i .ssh/id_rsa.pub slave2
```

③ SSH 证书登录测试,操作如下:

```
hadoop@master:~$ ssh master
hadoop@master:~$ exit
hadoop@master:~$ ssh slave1
hadoop@slave1:~$ exit
hadoop@master:~$ ssh slave2
hadoop@slave2:~$ exit
```

## 任务 3  Hadoop 部署与使用

### 【任务概述】

在集群系统上构建 Hadoop 系统,尽管单机上可以运行 HDFS 和 MapReduce,但要运行大数据处理，必须在集群系统上运行。本任务需要在集群系统上完成 Hadoop 的安装与配置，Hadoop 运行管理，使用 Hadoop Shell 对 HDFS 进行操作（包括目录创建、文件上传、列目录、查看文件内容等）。

## 【支撑知识】

### 一、Hadoop 简介

Hadoop 是 Apache 软件基金会旗下的一个开源分布式计算平台。以 Hadoop 分布式文件系统（Hadoop Distributed File System，HDFS）和 MapReduce（Google MapReduce 的开源实现）为核心的 Hadoop 为用户提供了系统底层细节透明的分布式基础架构。

对于 Hadoop 的集群来讲，可以分成两大类角色：Master 和 Slave。一个 HDFS 集群是由一个 NameNode 和若干个 DataNode 组成的。其中，NameNode 作为主服务器，管理文件系统的命名空间和客户端对文件系统的访问操作；集群中的 DataNode 管理存储的数据。MapReduce 框架是由一个单独运行在主节点上的 JobTracker 和运行在每个集群从节点的 TaskTracker 共同组成的。主节点负责调度构成一个作业的所有任务，这些任务分布在不同的从节点上。主节点监控它们的执行情况，并且重新执行之前的失败任务；从节点仅负责由主节点指派的任务。当一个 Job 被提交时，JobTracker 接收到提交作业和配置信息之后，就会将配置信息等分发给从节点，同时调度任务并监控 TaskTracker 的执行。

HDFS 和 MapReduce 共同组成了 Hadoop 分布式系统体系结构的核心。HDFS 在集群上实现分布式文件系统，MapReduce 在集群上实现了分布式计算和任务处理。HDFS 在 MapReduce 任务处理过程中提供了文件操作和存储等支持，MapReduce 在 HDFS 的基础上实现了任务的分发、跟踪、执行等工作，并收集结果，二者相互作用，完成了 Hadoop 分布式集群的主要任务。

（1）Hadoop 的优点

Hadoop 是一个能够对大量数据进行分布式处理的软件框架。Hadoop 以一种可靠、高效、可伸缩的方式进行数据处理。

① Hadoop 是可靠的，因为它假设计算元素和存储会失败，因此，它维护多个工作数据副本，确保能够针对失败的节点重新分布处理。

② Hadoop 是高效的，因为它以并行的方式工作，通过并行处理加快处理速度。

③ Hadoop 是可伸缩的，能够处理 PB 级数据。

④ Hadoop 是低成本的，它依赖于社区服务，任何人都可以使用。

Hadoop 带有用 Java 语言编写的框架，因此，运行在 Linux 生产平台上是非常理想的。Hadoop 上的应用程序也可以使用其他语言编写，如 C++。

（2）Hadoop 的构成

Hadoop 由许多元素构成,如图 3-2 所示。其底部是 Hadoop Distributed File System（HDFS），它存储 Hadoop 集群中所有存储节点上的文件。HDFS 的上一层是 MapReduce 引擎，该引擎由 JobTrackers 和 TaskTrackers 组成。Hadoop 分布式计算平台最核心的是分布式文件系统（HDFS）、MapReduce 处理过程，以及数据仓库工具 Hive 和分布式数据库 HBase。

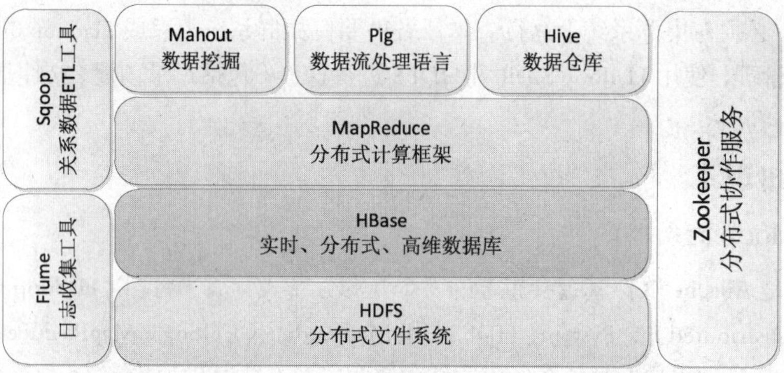

图 3-2 Hadoop 的构成

## 二、Hadoop 分布式文件系统

HDFS 是 Hadoop Distributed File System 的简称，即分布式文件系统。HDFS 设计理念之一就是让它能运行在普通的硬件之上，即便硬件出现故障，也可以通过容错策略来保证数据的高可用。

（1）HDFS 的主要设计理念

① 存储超大文件。这里的"超大文件"是指几百 MB、GB 甚至 TB 级别的文件。

② 最高效的访问模式是一次写入、多次读取（流式数据访问）。HDFS 存储的数据集作为 Hadoop 的分析对象，在数据集生成后，长时间在此数据集上进行各种分析，每次分析都将涉及该数据集的大部分数据甚至全部数据，因此，读取整个数据集的时间延迟比读取第一条记录的时间延迟更重要。

③ 运行在普通廉价的服务器上。HDFS 设计理念之一就是让它能运行在普通的硬件之上，即便硬件出现故障，也可以通过容错策略来保证数据的高可用。

（2）HDFS 架构

HDFS 采用 Master/Slave 架构，如图 3-3 所示。一个 HDFS 集群是由一个 NameNode 和一定数目的 DataNodes 组成的。NameNode 是一个中心服务器，负责管理文件系统的名字空间（Namespace），以及客户端对文件的访问。集群中的 DataNode 一般是一个节点，负责管理它所在节点上的存储。HDFS 暴露了文件系统的名字空间，用户能够以文件的形式在上面存储数据。从内部看，一个文件其实被分成一个或多个数据块（Block），这些块存储在一组 DataNode 上。NameNode 执行文件系统的名字空间操作；如打开、关闭、重命名文件或目录。它也负责确定 Block 到具体 DataNode 节点的映射。DataNode 负责处理文件系统客户端的读/写请求。在 NameNode 的统一调度下进行 Block 的创建、删除和复制。大文件会被分割成多个 Block 进行存储，Block 大小默认为 128MB。每一个 Block 会在多个 DataNode 上存储多份副本，默认是 3 份。

① NameSpace 和 NameNode。HDFS 支持传统的层次型文件组织结构。用户或者应用程序可以创建目录，然后将文件保存在这些目录中。文件系统名字空间的层次结构和大多数现有的文件系统类似：用户可以创建、删除、移动或重命名文件。当前，HDFS 不支持用户磁盘配额和访问权限控制，也不支持硬链接和软链接，但是 HDFS 架构并不妨碍实现这些特性。

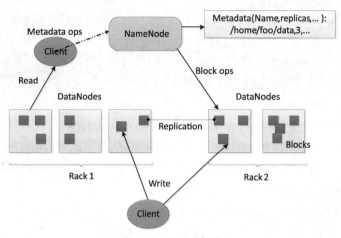

图 3-3　HDFS 架构

NameNode 负责维护文件系统的名字空间（NameSpace），任何对文件系统名字空间或属性的修改都将被 NameNode 记录下来。应用程序可以设置 HDFS 保存的文件的副本数目。文件副本的数目称为文件的副本系数，这个信息也是由 NameNode 保存的。NameNode 通常在 HDFS 实例中的单独机器上运行。对于任何对文件系统元数据产生修改的操作，NameNode 都会使用一种称为 editlog 的事务日志记录下来。例如，在 HDFS 中创建一个文件，NameNode 就会在 editlog 中插入一条记录来表示；同样，修改文件的副本系数也将往 editlog 插入一条记录。NameNode 在本地操作系统的文件系统中存储这个 editlog。整个文件系统的名字空间，包括数据块到文件的映射、文件的属性等，都存储在一个称为 fsimage 的文件中，这个文件也是放在 NameNode 所在的本地文件系统上。

NameNode 在内存中保存着整个文件系统的名字空间和文件数据块映射（BlockMap）的映像。这个关键的元数据结构设计得很紧凑，因而，一个有 4GB 内存的 NameNode 足够支撑大量的文件和目录。当 NameNode 启动时，它从硬盘中读取 editlog 和 fsimage，将所有 editlog 中的事务作用在内存中的 fsimage 上，并将这个新版本的 fsimage 从内存中保存到本地磁盘上，然后删除旧的 editlog，因为这个旧的 editlog 的事务都已经作用在 fsimage 上了。这个过程称为一个检查点（CheckPoint）。在当前实现中，检查点只发生在 NameNode 启动时，在不久的将来将实现支持周期性的检查点。

NameNode 中有两个很重要的文件：fsimage 和 edits。

fsimage 是元数据镜像文件（保存文件系统的目录树）。fsimage 文件是 Hadoop 文件系统元数据的一个永久性的检查点，其中包含 Hadoop 文件系统中的所有目录和文件 idnode 的序列化信息。fsimage 文件不会在写操作后马上更新，因为 fsimage 写非常慢。

edits 是元数据操作日志（记录每次保存 fsimage 之后到下次保存之间的所有 HDFS 操作）。edits 文件存放的是 Hadoop 文件系统的所有更新操作的路径,文件系统客户端执行的所有写操作首先会被记录到 edits 文件中。

内存中保存了最新的元数据信息（fsimage 和 edits）。如果 edits 过大，会导致 NameNode 重启速度慢，Secondary NameNode 负责定期合并它们。

NameNode 的目录结构如下：

```
${dfs.name.dir}/current/VERSION
                       /edits
                       /fsimage
                       /fstime
```

② Secondary NameNode。NameNode 存储文件系统的变化作为日志追加在本地的一个 edits 文件中。当一个 NameNode 启动时，它从一个映像文件 fsimage 读取 HDFS 的状态，使用来自 edits 日志文件的 edits 操作。然后它将新的 HDFS 状态写入 fsimage 中，并使用一个空的 edits 文件开始正常操作。由于 NameNode 只有在启动阶段才合并 fsimage 和 edits，因此，在一个繁忙的集群中 edits 日志文件会随着时间变得非常大，一个副作用是一个更大的 edits 文件会使 NameNode 在下次重新启动时需要更长的时间。

Secondary NameNode 定期合并 fsimage 和 edits 日志文件，并保持 edits 日志文件大小在一定限度。它通常和 NameNode 运行在不同的机器上，内存需求和 NameNode 相同。

检查点进程开始由 Secondary NameNode 两个配置参数控制。

dfs.Namenode.checkpoint.period 默认被设置为1个小时,指定连续两个检查点间的最大延迟。

dfs.Namenode.checkpoint.txns 默认被设置为 1 个小时，定义了 uncheckpointed 事务在 NameNode 中的数量，这将迫使有一个紧急检查点，即使检查点时间间隔尚未达到。

Secondary NameNode 保存最新检查点的目录与 NameNode 的目录结构相同。所以，NameNode 可以在需要的时候读取 Secondary NameNode 上的检查点镜像。

Secondary NameNode 定期合并 fsimage 和 edits 日志，将 edits 日志文件大小控制在一个限度下。Secondary NameNode 定期为 Primary NameNode 内存中的文件系统元数据创建检查点。创建检查点的步骤如下（如图 3-4 所示）：

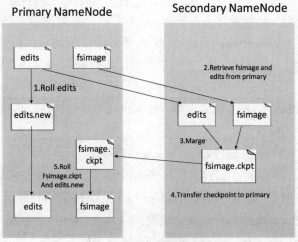

图 3-4 创建检查点的过程

- Secondary NameNode 请求 Primary NameNode 停止使用 edits 文件，暂时将新的写操作记录到一个新的 edits 文件中。
- Secondary NameNode 从 Primary NameNode 获取 fsimage 和 edits 文件（采用 HTTP GET）。

- Secondary NameNode 将 fsimage 文件加载到内存，逐一执行 edits 文件中的操作，创建新的 fsimage 文件。
- Secondary NameNode 将新的 fsimage 文件发送回 NameNode（采用 HTTP POST）。
- Primary NameNode 用从 Secondary NameNode 接受新的 fsimage 文件替换旧的 fsimage 文件，用第一个产生的 edits 文件替换旧的 edits 文件。同时更新 fstime 文件来记录检查点执行的时间。

③ DataNode。DataNode 通常在 HDFS 实例中的单独机器上运行。Hadoop 集群包含一个 NameNode 和大量 DataNode。DataNode 通常以机架的形式组织，机架通过一个交换机将所有系统连接起来。Hadoop 的一个假设是：机架内部节点之间的传输速度快于机架间节点的传输速度。

DataNode 响应来自 HDFS 客户机的读/写请求，还响应来自 NameNode 的创建、删除和复制块的命令。NameNode 依赖来自每个 DataNode 的定期心跳（Heartbeat）消息。每条消息都包含一个块报告，NameNode 可以根据这个报告验证块映射和其他文件系统元数据。如果 DataNode 不能发送心跳消息，NameNode 将采取修复措施，重新复制在该节点上丢失的块。

④ CheckPoint Node。在 Hadoop-2.x 版本之前只存在 SecondaryNameNode，没有 CheckPoint Node、Backup Node 的概念，在 2.x 版本中引入了后两者，增强了对 NameNode 的同步和备份。

NameNode 在 NameSpace 保存两个文件：fsimage 和 edits，fsimage 记录 NameSpace 最后一个检查点，edits 记录上一次检查点开始的变化日志。NameNode 启动时，合并 fsimage 和 edits 日志，用来提供一个文件系统元数据的最新视图。NameNode 之后用 HDFS 新的状态重写 fsimage 并且开始新的日志记录。

CheckPoint Node 周期性地创建 NameSpace 的检查点，它从活动的 NameNode 下载 fsimage 和 edit 日志，在本地合并，并把合并后新的 fsimage 上传到活动的 NameNode。CheckPoint Node 以与 NameNode 相同的目录结构存储最新的 CheckPoint。新的检查点时刻准备好在 NameNode 需要时对其进行读取。CheckPoint 通常和 NameNode 运行在不同的主机上，因为运行时所需要的内存要保证和 NameNode 同样优先。Checkpoint Node 用 bin/hdfs namenode –checkpoint 在节点指定配置文件启动。

本地 CheckPoint（或 Backup）节点和它的 Web 接口可以通过两个参数配置：dfs.namenode.checkpoint.period 和 dfs.Namenode.checkpoint.txns，CheckPoint 节点存储最新检查点，目录与 NameNode 的目录结构相同。所以，NameNode 可以在需要的时候读取检查点镜像。

⑤ Backup Node。Backup Node 不但提供了同 CheckPoint Node 一样的 CheckPoint 功能，而且还维持一个内存，及时复制文件系统的 NameSpace，同步到当前工作中的 NameNode 的状态。Backup Node 从 NameNode 接收文件系统 edits 并持久化到磁盘，同时还应用那些 edits 到自己的 NameSpace 内存副本，这就建立了 NameSpace 的备份。

Backup Node 不需要像 CheckPoint Node 或 Secondary NameNode 一样，为了创建检查点，需要从活动的 NameNode 上下载 fsimage 和 edits 文件，因为在它的内存中已经有了命名空间的最新状态。Backup Node 的 CheckPoint 处理效率很高，因为它只需要保存 NameSpace 到本地 fsimage 并重设 edits 文件。

Backup Node 在内存中维护一个命名空间的副本，它的 RAM 要求和 NameNode 一致。

NameNode 一次只支持一个 Backup Node，如果 Backup 节点在使用，则 CheckPoint Node 不允许注册。不久的将来进行支持使用多个 Backup Node。Backup Node 和 CheckPoint Node 在同一个管理器，用 bin/hdfs namenode –backup 启动。

Backup Node 和它的 Web 接口可以通过两个参数配置：dfs.namenode.backup.address 和 dfs.namenode.backup.http-address。

Backup Node 提供了选项支持运行 NameNode 不持久化到硬盘，代理所有的命名空间持久化状态的请求到 Backup Node。要实现这一点，启动 NameNodes 时使用 –importcheckpoint 选项，并在配置文件中将参数 dfs.namenode.edits.dir 设置为空。

（3）数据复制

HDFS 被设计成能够在一个大集群中跨机器可靠地存储超大文件。它以块序列的形式存储文件，文件中除了最后一个块，其他块都有相同的大小。为了容错，文件的所有数据块都会有副本。每个文件的数据块大小和副本系数都是可配置的。应用程序可以指定某个文件的副本数目。副本系数可以在文件创建的时候指定，也可以在之后改变，如图 3-5 所示。HDFS 中的文件是一次写的，并且任何时候都只有一个写操作。

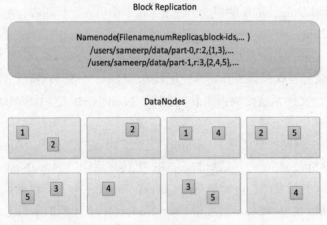

图 3-5　数据块复制

NameNode 全权管理数据块的复制，它周期性地从集群中的每个 DataNode 接收心跳信号和块状态报告（Blockreport）。接收到心跳信号意味着该 DataNode 节点工作正常。块状态报告包含了一个该 DataNode 上所有数据块的列表。

（4）副本存放

副本存放是 HDFS 可靠性和性能的关键。优化的副本存放策略是 HDFS 区分于其他大部分分布式文件系统的重要特性。这种特性需要做大量的调优，并需要经验的积累。HDFS 采用一种称为机架感知（Rack Awareness）的策略来改进数据的可靠性、可用性和网络带宽的利用率。目前实现的副本存放策略只是在这个方向上的第一步。实现这个策略的短期目标是验证它在生产环境下的有效性，观察它的行为，为实现更先进的策略打下测试和研究的基础。

大型 HDFS 实例一般运行在跨越多个机架的计算机组成的集群上，不同机架上的两台机器之间的通信需要经过交换机。在大多数情况下，同一个机架内的两台机器间的带宽会比不同机架的两台机器间的带宽大。

通过一个机架感知的过程，NameNode 可以确定每个 DataNode 所属的机架 ID。一个简单但没有优化的策略就是将副本存放在不同的机架上。这样可以有效防止当整个机架失效时数据的丢失，并且允许读数据的时候充分利用多个机架的带宽。这种策略设置可以将副本均匀分布在集群中，有利于组件失效情况下的负载均衡。但是，因为这种策略的一个写操作需要传输数据块到多个机架，所以，这增加了写的代价。

在大多数情况下，副本系数是 3，HDFS 的存放策略是将一个副本存放在本地机架的节点上，一个副本放在同一机架的另一个节点上，最后一个副本放在不同机架的节点上。这种策略减少了机架间的数据传输，这就提高了写操作的效率。机架故障的概率远小于节点故障，所以，这个策略不会影响到数据的可靠性和可用性。与此同时，因为数据块只放在两个（不是三个）不同的机架上，所以，此策略减少了读取数据时需要的网络传输总带宽。在这种策略下，副本并不是均匀分布在不同的机架上。1/3 的副本在一个节点上，2/3 的副本在一个机架上，其他副本均匀分布在剩下的机架中，这一策略在不损害数据可靠性和读取性能的情况下改进了写的性能。

（5）副本选择

为了降低整体的带宽消耗和读取延时，HDFS 会尽量让读取程序读取离它最近的副本。如果在读取程序的同一个机架上有一个副本，那么就读取该副本。如果一个 HDFS 集群跨越多个数据中心，那么客户端也将首先读本地数据中心的副本。

（6）HDFS 文件的存储

HDFS 在对一个文件进行存储时有两个重要的策略：一个是副本策略，另一个是分块策略。副本策略保证了文件存储的高可靠性，分块策略保证数据并发读/写的效率并且是 MapReduce 实现并行数据处理的基础，如图 3-6 所示。

HDFS 的分块策略：通常 HDFS 在存储一个文件时会将文件切为 64MB 大小的块来进行存储，数据块会被分别存储在不同的 DataNode 节点上，这一过程其实就是一种数据任务的切分过程，在后面对数据进行 MapReduce 操作

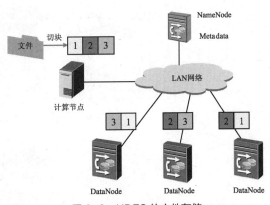

图 3-6　HDFS 的文件存储

时十分重要，同时数据被分块存储后在数据读/写时能实现对数据的并发读/写，提高数据读/写效率。随着新一代磁盘驱动器传输速率的提升，块的大小被设置得更大（如 128MB）。

HDFS 采用这样较大的文件分块策略有以下 3 个优点：

① 降低客户端与主服务器的交互代价。
② 降低网络负载。
③ 减少主服务器中元数据的数量。

HDFS 的副本策略：HDFS 对数据块典型的副本策略为 3 个副本，第一个副本存放在本地节点，第二个副本存放在同一个机架的另一个节点，第三个本副本存放在不同机架上的另一个节点。这样的副本策略保证了在 HDFS 文件系统中存储的文件具有很高的可靠性。

一个文件写入 HDFS 的基本过程可以描述如下：写入操作首先由 NameNode 为该文件创

建一个新的记录，该记录为文件分配存储节点包括文件的分块存储信息，在写入时系统会对文件进行分块，文件写入的客户端获得存储位置的信息后直接与指定的 DataNode 进行数据通信，将文件块按 NameNode 分配的位置写入指定的 DataNode，数据块在写入时不再通过 NameNode，因此，NameNode 不会成为数据通信的瓶颈。

（7）HDFS 文件的读取

HDFS 读取文件的过程如下（如图 3-7 所示）：

① 使用 HDFS 提供的客户端开发库 Client，向远程的 NameNode 发起 RPC 请求。

② NameNode 会视情况返回文件的部分或者全部 Block 列表，对于每个 Block，NameNode 都会返回有该 Block 副本的 DataNode 地址。

③ 客户端开发库 Client 会选取离客户端最接近的 DataNode 来读取 Block；如果客户端本身就是 DataNode，那么将从本地直接获取数据。

④ 读取完当前 Block 的数据后，关闭与当前的 DataNode 连接，并为读取下一个 Block 寻找最佳的 DataNode。

⑤ 当读完列表的 Block 后，且文件读取还没有结束，客户端会继续向 NameNode 获取下一批的 Block 列表。

⑥ 读取完一个 Block 都会进行 CheckSum 验证，如果读取 DataNode 时出现错误，客户端会通知 NameNode，然后从下一个拥有该 Block 副本的 DataNode 继续读。

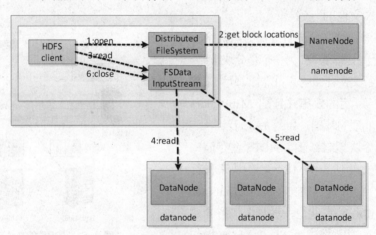

图 3-7　客户端读取 HDFS 中的数据

（8）HDFS 文件的写入

HDFS 写入文件的过程比读取较为复杂，写文件如下（如图 3-8 所示）：

① 使用 HDFS 提供的客户端开发库 Client，向远程的 NameNode 发起 RPC 请求。

② NameNode 会检查要创建的文件是否已经存在，创建者是否有权限进行操作，成功则会为文件创建一个记录，否则会让客户端抛出异常。

③ 当客户端开始写入文件时，开发库会将文件切分成多个 Packets，并在内部以数据队列（Data Queue）的形式管理这些 Packets，并向 NameNode 申请新的 Blocks，获取用来存储 Replicas 的合适的 DataNodes 列表，列表的大小根据在 NameNode 中对 Replication 的设置而定。

④ 开始以 Pipeline（管道）的形式将 Packet 写入所有的 Replicas 中。开发库把 Packet 以

流的方式写入第一个 DataNode，该 DataNode 把该 Packet 存储之后，再将其传递给在此 Pipeline 中的下一个 DataNode，直到最后一个 DataNode，这种写数据的方式呈流水线的形式。

⑤ 最后一个 DataNode 成功存储之后会返回一个 ack packet，在 Pipeline 中传递至客户端，在客户端的开发库内部维护着"ack queue"，成功收到 DataNode 返回的 ack packet 后会从"ack queue"移除相应的 Packet。

⑥ 如果传输过程中有某个 DataNode 出现了故障，那么当前的 Pipeline 会被关闭，出现故障的 DataNode 会从当前的 Pipeline 中移除，剩余的 Block 会从剩下的 DataNode 中继续以 Pipeline 的形式传输，同时 NameNode 会分配一个新的 DataNode，保持 Replicas 设定的数量。

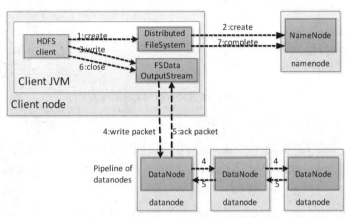

图 3-8 客户端将数据写入 HDFS

### 三、Apache Hadoop YARN

从 2012 年 8 月开始，Apache Hadoop YARN 成了 Apache Hadoop 的一项子工程。自此 Apache Hadoop 由下面 4 个子工程组成。

- Hadoop Comon：核心库。
- Hadoop HDFS：分布式存储系统。
- Hadoop MapReduce：MapReduce 模型的开源实现。
- Hadoop YARN：新一代 Hadoop 数据处理框架。

概括来说，Hadoop YARN 的目的是使得 Hadoop 数据处理能力超越 MapReduce。众所周知，Hadoop HDFS 是 Hadoop 的数据存储层，Hadoop MapReduce 是数据处理层。然而，MapReduce 已经不能满足今天广泛的数据处理需求，如实时/准实时计算、图计算等。而 Hadoop YARN 提供了一个更加通用的资源管理和分布式应用框架。在这个框架上，用户可以根据自己的需求，实现定制化的数据处理应用。而 Hadoop MapReduce 也是 YARN 上的一个应用。我们将会看到 MPI、图处理、在线服务等（如 Spark、Storm、HBase）都会和 Hadoop MapReduce 一样成为 YARN 上的应用。

（1）Apache Hadoop MapReduce 架构

传统的 Apache Hadoop MapReduce 系统由 JobTracker 和 TaskTracker 组成。其中 JobTracker 是 Master，只有一个；TaskTracker 是 Slaves，每个节点部署一个，如图 3-9 所示。

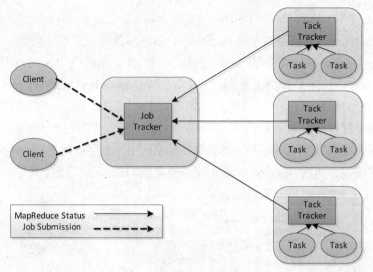

图 3-9 Apache Hadoop MapReduce 架构

JobTracker 负责资源管理（通过管理 TaskTracker 节点），追踪资源消费/释放，以及 Job 的生命周期管理（调度 Job 的每个 Task，追踪 Task 进度，为 Task 提供容错等）。而 TaskTracker 的职责很简单，依次启动和停止由 JobTracker 分配的 Task，并且周期性地向 JobTracker 汇报 Task 进度及状态信息。

传统的 Apache Hadoop MapReduce 系统具有如下局限性。

① 单节点故障问题。
- JobTracker 是 Map/Reduce 的集中处理点，存在单点故障。
- 系统故障会丢失所有正在排队和运行的作业。

② 扩展性较差。
- 最大集群服务器数：4000（NameNode 内存限制）。
- 最大并发任务数：40000。
- 任务调度粒度太大：JobTracker。

③ 资源调度粒度太大。以 Map/Reduce 任务的数目作为资源的表示过于简单，没有考虑到 CPU/内存的占用情况。

④ 固定分配 Map 和 Reduce 的资源问题。
- 无法进行资源优化。
- 基于内存 Slot，如每个节点 10 个 Slot，没考虑任务大小。

⑤ 缺乏支持其他计算框架和服务的能力。多个计算框架集群间的资源不能共享，资源利用率低，将所有的计算框架运行在一个集群中，共享一个集群的资源，按需分配。

（2）Apache Hadoop YARN 架构

Apache Hadoop YARN 是一种新的 Hadoop 资源管理器，它是一个通用资源管理系统，可为上层应用提供统一的资源管理和调度，它的引入为集群在利用率、资源统一管理和数据共享等方面带来了巨大的好处。

YARN 的最基本思想是将 JobTracker 的两个主要职责：资源管理和 Job 调度管理分别交

给两个角色负责。一个是全局的 ResourceManager，另一个是每个应用一个的 ApplicationMaster。ResourceManager 及每个节点一个的 NodeManager 构成了新的通用系统，实现以分布式方式管理应用。

ResourceManager 是系统中仲裁应用之间资源分配的最高权威。每个应用一个的 ApplicationMaster 负责向 ResourceManager 协商资源，并与 NodeManager 协同工作来执行和管理 task。ResourceManager 有一个可插入的调度器，负责向各个应用分配资源以满足容量、组等限制。这个调度器是一个纯粹的调度器，意思是它不负责管理或追踪应用的状态，也不负责由于硬件错误或应用问题导致的 task 失败重启工作。调度器只依据应用的资源需求来执行调度工作，调度内容是一个抽象概念 Resource Container，其中包含了资源元素，如内存、CPU、网络、磁盘等。

NodeManager 是每个节点一个的 Slave，其负责启动应用的 Container，管理它们的资源使用（内存、CPU、网络、磁盘），并向 ResourceManager 汇报整体的资源使用情况。

每个应用一个的 ApplicationMaster 负责向 ResourceManager 的调度器协商合理的 Resource Container 并追踪它们的状态，管理进度。从系统角度看，ApplicationMaster 本身也是以一个普通 Container 的形式执行的，如图 3-10 所示。

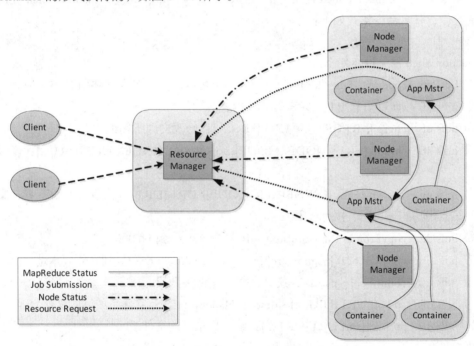

图 3-10　Apache Hadoop YARN 架构

由于 MapReduce 在计算模型方面的局限性，Hadoop 实现了更加通用的资源管理系统 YARN，并将 MapReduce 作为其一个应用。在 YARN 上可以实现多种多样计算模型的应用以满足业务需要。另外，由于 YARN 系统将 JobTracker 的主要工作进行切分，使得 Master 的压力大大减小（ResourceManager 承担的工作量远小于 JobTracker），这样 YARN 系统就可以支持更大的集群规模。

Apache Hadoop YARN 的优势如下：

- 老的框架中，JobTracker 一个很大的负担就是监控 Job 下的 tasks 的运行状况，现在这部分就交给 ApplicationMaster 做了，而 ApplicationMaster 不是全集群一个，而是每个应用一个，并且是分布在各节点上，所以，监测每一个 Job 子任务（tasks）状态的程序分布在每个节点，因而更安全、更优美。
- 大大减小了 JobTracker（也就是现在的 ResourceManager）的资源消耗，所以，可以大大增加集群的可扩展性。
- ApplicationMaster 是一个可变更的部分，用户可以对不同的编程模型写自己的 AppMst，让更多类型的编程模型能够在 Hadoop 集群中运行。
- 应用可以申请任意资源量（内存、CPU），比之前基于内存 Slot 的资源申请更合理。不会出现之前的 Map Slot/Reduce Slot 分开造成集群资源闲置的尴尬情况。
- ResourceManager 中有一个模块为 ApplicationsMasters，它监测 ApplicationMaster 的运行状况，如果出问题，会将其在其他机器上重启，从而提高了系统的可用性。

## 四、Hadoop Shell

HDFS 是存取数据的分布式文件系统，对 HDFS 的操作就是文件系统的基本操作，对 HDFS 的操作命令类似于 Linux 的 Shell 对文件的操作，如 ls、mkdir、rm 等。Hadoop Shell 提供一个基本的操作，基本命令格式如下：

```
hdfs dfs -cmd <args>
```

Hadoop Shell 常见操作如下。

① hdfs dfs -appendToFile <localsrc> ... <dst>：可同时上传多个文件到 HDFS 中。

② hdfs dfs -cat URI [URI ...]：查看文件内容。

③ hdfs dfs -chgrp [-R] GROUP URI [URI ...]：修改文件所属组。

④ hdfs dfs -chmod [-R] <MODE[,MODE]... | OCTALMODE> URI [URI ...]：修改文件权限。

⑤ hdfs dfs -chown [-R] [OWNER][:[GROUP]] URI [URI ]：修改文件所有者，文件所属组，其他用户的读、写、执行权限。

⑥ hdfs dfs -copyFromLocal <localsrc> URI：复制文件到 HDFS。

⑦ hdfs dfs -copyToLocal [-ignorecrc] [-crc] URI <localdst>：复制文件到本地。

⑧ hdfs dfs -count [-q] <paths>：统计文件及文件夹数目。

⑨ hdfs dfs -cp [-f] URI [URI ...] <dest>：Hadoop HDFS 文件系统间的文件复制。

⑩ hdfs dfs -du [-s] [-h] URI [URI ...]：统计目录下的文件及大小。

⑪ hdfs dfs -dus <args>：汇总目录下的文件总大小。

⑫ hdfs dfs -expunge：清空回收站，文件被删除时，它首先会移到临时目录.Trash/中，当超过延迟时间之后，文件才会被永久删除。

⑬ hdfs dfs -get [-ignorecrc] [-crc] <src> <localdst>：下载文件到本地。

⑭ hdfs dfs -getfacl [-R] <path>：查看 ACL （访问权限组拥有者）。

⑮ hdfs dfs -getmerge <src> <localdst> [addnl]：合并下载文件到本地。

⑯ hdfs dfs -ls <args>：查看目录。

⑰ hdfs dfs -lsr <args>：循环列出目录、子目录及文件信息。

⑱ hdfs dfs −mkdir [−p] <paths>：创建空白文件夹。
⑲ hdfs dfs −moveFromLocal <localsrc> <dst>：剪切文件到 HDFS。
⑳ hdfs dfs −moveToLocal [−crc] <src> <dst>：剪切文件到本地。
㉑ hdfs dfs −mv URI [URI ...] <dest>：剪切 HDFS 文件。
㉒ hdfs dfs −put <localsrc> ... <dst>：上传文件。
㉓ hdfs dfs −rm [−skipTrash] URI [URI ...]：删除文件/空白文件夹。
㉔ hdfs dfs −rmr [−skipTrash] URI [URI ...]：递归删除，删除文件及文件夹下的所有文件。

## 【任务实施】

### 一、Master 节点安装软件

① 下载软件包到/home/hadoop 目录下，下载网址如下：

```
http://www.oracle.com/technetwork/java/javase/downloads/index.html
http://archive.apache.org/dist/hadoop/common/stable/hadoop-2.7.3.tar.gz
```

② 以用户 hadoop 登录 Master 节点，安装 JDK 软件，操作如下：

```
hadoop@master:~$ cd /opt
hadoop@master:/opt$ sudo tar xvzf /home/hadoop/jdk-8u121-linux-x64.tar.gz
```

③ 安装 Hadoop 系统，操作如下：

```
hadoop@master:/opt$ sudo tar xvzf /home/hadoop/hadoop-2.7.3.tar.gz
```

④ 修改文件属性，操作如下：

```
hadoop@master:/opt$ sudo chown -R hadoop:hadoop jdk1.8.0_121 hadoop-2.7.3
```

### 二、Master 节点设置 Hadoop 参数

Hadoop 配置文件在/opt/hadoop-2.7.3/etc/hadoop/目录下，最重要的几个文件如表 3-2 所示。

表 3-2 Hadoop 配置文件

| 文件名称 | 格 式 | 描 述 |
| --- | --- | --- |
| hadoop−env.sh | Bash 脚本 | 记录脚本要用的环境变量，以运行 Hadoop |
| core−site.xml | Hadoop 配置 XML | Hadoop Core 的配置项，如 HDFS 和 MapReduce 常用的 I/O 设置等 |
| hdfs−site.xml | Hadoop 配置 XML | Hadoop 守护进程的配置项，包括 NameNode、辅助 NameNode 和 DataNode 等 |
| mapred−site.xml | Hadoop 配置 XML | MapReduce 守护进程的配置项 |
| yarn−site.xml | YARN 配置 XML | |
| slaves | 纯文本 | 运行 DataNode 的机器列表（每行一个） |
| hadoop−metrics.properties | Java 属性 | 控制 metrics 在 Hadoop 上如何发布的属性 |
| log4j.properties | Java 属性 | 系统日志文件、NameNode 审计日志等 |

① 修改 hadoop−env.sh 文件，操作如下：

```
hadoop@master:/opt$ cd /opt/hadoop-2.7.3/etc/hadoop
hadoop@master:/opt/hadoop-2.7.3/etc/hadoop$ vi hadoop-env.sh
```

把如下内容：
```
export JAVA_HOME=${JAVA_HOME}
```
改为
```
export JAVA_HOME=/opt/jdk1.8.0_121
```

② 修改 core-site.xml 文件，操作如下：
```
hadoop@master:/opt/hadoop-2.7.3/etc/hadoop$ vi core-site.xml
```

在<configuration>　</configuration>之间添加内容如下：
```xml
  <property>
    <name>fs.defaultFS</name>
    <value>hdfs://master:9000</value>
  </property>
  <property>
    <name>hadoop.tmp.dir</name>
    <value>/opt/hadoop-2.7.3/tmp</value>
  </property>
```

③ 修改 hdfs-site.xml 文件，操作如下：
```
hadoop@master:/opt/hadoop-2.7.3/etc/hadoop$ vi hdfs-site.xml
```

在<configuration>　</configuration>之间添加内容如下：
```xml
  <property>
    <name>dfs.namenode.secondary.http-address</name>
    <value>master:50090</value>
  </property>
  <property>
    <name>dfs.replication</name>
    <value>3</value>
  </property>
  <property>
    <name>dfs.namenode.name.dir</name>
    <value>/opt/hadoop-2.7.3/dfs/name</value>
  </property>
  <property>
    <name>dfs.datanode.data.dir</name>
    <value>/opt/hadoop-2.7.3/dfs/data</value>
  </property>
  <property>
    <name>fs.checkpoint.dir</name>
    <value>/opt/hadoop-2.7.3/dfs/namesecondary</value>
  </property>
  <property>
    <name>dfs.block.size</name>
    <value>134217728</value>
```

```
  </property>
  <property>
    <name>dfs.namenode.handler.count</name>
    <value>20</value>
  </property>
  <property>
    <name>dfs.permissions</name>
    <value>false</value>
  </property>
```

④ 新建 mapred-site.xml 文件，操作如下：

```
hadoop@master:/opt/hadoop-2.7.3/etc/hadoop$ vi mapred-site.xml
```

添加内容如下：

```xml
<?xml version="1.0"?>
<?xml-stylesheet type="text/xsl" href="configuration.xsl"?>
<configuration>
  <property>
    <name>mapreduce.framework.name</name>
    <value>yarn</value>
  </property>
  <property>
    <name>mapreduce.jobhistory.address</name>
    <value>master:10020</value>
  </property>
  <property>
    <name>mapreduce.jobhistory.webapp.address</name>
    <value>master:19888</value>
  </property>
</configuration>
```

⑤ 修改 yarn-site.xml 文件，操作如下：

```
hadoop@master:/opt/hadoop-2.7.3/etc/hadoop$ vi yarn-site.xml
```

在<configuration>    </configuration>之间添加内容如下：

```xml
<property>
  <name>yarn.resourcemanager.hostname</name>
  <value>master</value>
</property>
<property>
  <name>yarn.nodemanager.aux-services</name>
  <value>mapreduce_shuffle</value>
</property>
```

⑥ 修改 Slaves 文件，操作如下：

```
hadoop@master:/opt/hadoop-2.7.3/etc/hadoop$ vi slaves
```

内容如下：
```
slave1
slave2
```

## 三、Slave 节点安装软件

① 以用户 hadoop 登录 Slave1 节点，操作如下：
```
hadoop@slave1:~$ sudo scp -r hadoop@master:/opt/* /opt
[sudo] password for hadoop:
hadoop@master's password:
hadoop@slave1:~$ sudo chown -R hadoop:hadoop /opt/*
```

② 以用户 hadoop 登录 Slave2 节点，操作如下：
```
hadoop@slave2:~$ sudo scp -r hadoop@master:/opt/* /opt
[sudo] password for hadoop:
hadoop@master's password:
hadoop@slave2:~$ sudo chown -R hadoop:hadoop /opt/*
```

③ 各集群节点设置环境变量（包括主节点），操作如下：
```
hadoop@master:/opt$ vi /home/hadoop/.profile
```

添加内容如下：
```
export JAVA_HOME=/opt/jdk1.8.0_121
export HADOOP_HOME=/opt/hadoop-2.7.3
export HBASE_HOME=/opt/hbase-1.2.4
export ZOOKEEPER_HOME=/opt/zookeeper-3.4.9
export PATH=$JAVA_HOME/bin:$HADOOP_HOME/bin:$HADOOP_HOME/sbin:$HBASE_HOME/bin:$ZOOKEEPER_HOME/bin:$PATH
hadoop@master:/opt$ source /home/hadoop/.profile
```

## 四、Master 节点启动 Hadoop 服务

① NameNode 格式化，操作如下：
```
hadoop@master:~$ hdfs namenode -format
```

注意：首次启动需要先在 Master 节点上执行 NameNode 的格式化，如果成功，则显示 name has been successfully formatted。

② 启动 DFS，操作如下：
```
hadoop@master:~$ start-dfs.sh
Starting Namenodes on [master]
master: starting Namenode, logging to /opt/hadoop-2.7.3/logs/hadoop-hadoop-Namenode-master.out
slave2: starting Datanode, logging to /opt/hadoop-2.7.3/logs/hadoop-hadoop-Datanode-slave2.out
slave1: starting Datanode, logging to /opt/hadoop-2.7.3/logs/hadoop-hadoop-Datanode-slave1.out
Starting secondary Namenodes [master]
```

```
master: starting secondaryNamenode, logging to /opt/hadoop-2.7.3/logs/
hadoop-hadoop-secondaryNamenode-master.out
```

③ 启动 YARN，操作如下：

```
hadoop@master:~$ start-yarn.sh
starting yarn daemons
starting resourcemanager, logging to /opt/hadoop-2.7.3/logs/yarn-hadoop-
resourcemanager-master.out
slave2: starting nodemanager, logging to /opt/hadoop-2.7.3/logs/yarn-
hadoop-nodemanager-slave2.out
slave1: starting nodemanager, logging to /opt/hadoop-2.7.3/logs/yarn-
hadoop-nodemanager-slave1.out
```

④ 启动 HistoryServer，操作如下：

```
hadoop@master:~$ mr-jobhistory-daemon.sh start historyserver
starting historyserver, logging to /opt/hadoop-2.7.3/logs/mapred-hadoop-
historyserver-master.out
```

## 五、验证服务

① 查看进程，通过命令 jps 查看。

Master 节点执行的结果如下：

```
hadoop@master:~$ jps
3843 ResourceManager
4133 JobHistoryServer
4229 Jps
3662 SecondaryNamenode
3407 Namenode
```

Slave1 节点执行的结果如下：

```
hadoop@slave1:~$ jps
2883 Datanode
3160 Jps
3032 NodeManager
```

Slave2 节点执行的结果如下：

```
hadoop@slave2:~$ jps
2337 Datanode
2610 Jps
2486 NodeManager
```

注意：在 Master 节点上有 NameNode、ResourceManager、SecondaryNamenode、JobHistoryServer 4 个进程，在 Slave 节点有 DataNode 和 NodeManager 两个进程，缺少任一进程表示有错误。

② 查看 DFS 报告，操作如下：

```
hadoop@slave2:~$ hdfs dfsadmin -report
Configured Capacity: 57900105728 (53.92 GB)
Present Capacity: 50957520896 (47.46 GB)
```

```
DFS Remaining: 50957471744 (47.46 GB)
DFS Used: 49152 (48 KB)
DFS Used%: 0.00%
Under replicated blocks: 0
Blocks with corrupt replicas: 0
Missing blocks: 0
Missing blocks (with replication factor 1): 0

-------------------------------------------------
Live Datanodes (2):

Name: 172.25.0.12:50010 (slave2)
……

Name: 172.25.0.11:50010 (slave1)
……
```

注意：如果 Live DataNode 不为 0，则说明集群启动成功。

③ 文件块检查，操作如下：

```
hadoop@master:~$ hdfs fsck / -files -blocks
Connecting to Namenode via http://master:50070/fsck?ugi=hadoop&files=1&blocks=1&path=%2F
FSCK started by hadoop (auth:SIMPLE) from /172.25.0.10 for path / at Sat Feb 04 13:40:27 CST 2017
/ <dir>
Status: HEALTHY
 Total size:0 B
 Total dirs:1
 Total files:    0
 Total symlinks:     0
 Total blocks (validated):   0
 Minimally replicated blocks:   0
 Over-replicated blocks:    0
 Under-replicated blocks:   0
 Mis-replicated blocks:    0
 Default replication factor:    3
 Average block replication: 0.0
 Corrupt blocks:    0
 Missing replicas:      0
 Number of data-nodes:     2
 Number of racks:      1
FSCK ended at Sat Feb 04 13:40:27 CST 2017 in 0 milliseconds

The filesystem under path '/' is HEALTHY
```

④ 打开浏览器，输入网址"http://master: 50070"，查看 NameNode 信息和 DataNode 信息，如图 3-11 和图 3-12 所示。

⑤ 打开浏览器，输入网址"http://master: 50090"，查看 SecondaryNameNode 信息，如图 3-13 所示。

⑥ 打开浏览器，输入网址"http://master: 8088"和"http://master:19888"，分别查看集群节点信息和工作历史情况，如图 3-14 和图 3-15 所示。

图 3-11 NameNode 信息

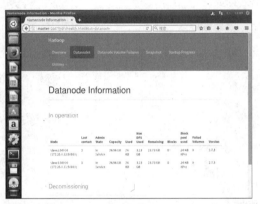

图 3-12 DataNode 信息

图 3-13 SecondaryNameNode 信息

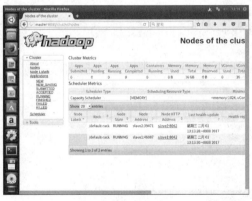

图 3-14 集群节点信息

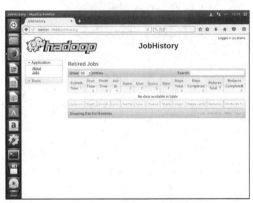

图 3-15 JobHistory

## 六、Hadoop Shell 命令

① 在 HDFS 中创建目录，操作如下：

```
hadoop@master:~$ hdfs dfs -mkdir /test
```

② 在 HDFS 中列目录，操作如下：

```
hadoop@master:~$ hdfs dfs -ls /
```

③ 上传文件到 HDFS，操作如下：

```
hadoop@master:~$ vi abc.txt
```

添加内容：
汕尾职业技术学院欢迎您！
```
hadoop@master:~$ hdfs dfs -put abc.txt /test
```
④ 查看 HDFS 文件内容，操作如下：
```
hadoop@master:~$ hdfs dfs -cat /test/abc.txt
```
汕尾职业技术学院欢迎您！

## 【同步训练】

### 一、简答题

（1）什么是集群技术？集群技术的优点是什么？

（2）NTP 是什么？有什么作用？

（3）SSH 是什么？由哪些部分组成？

（4）简述证书登录过程。

（5）HDFS 的设计理念是什么？

（6）简述 YARN 的架构。

### 二、操作题

（1）设置 SSH 证书登录。

（2）安装 NTP 软件，并设置集群节点时间同步。

（3）下载 Hadoop 软件包，并安装 Hadoop 集群系统。

（4）查看 HDFS 使用情况。

（5）自己创建一个文本文件，并上传到 HDFS。

# PART 4 项目四 MapReduce 编程

## 【项目介绍】

MapReduce 编程与传统编程不同，MapReduce 程序是并行化程序，需要理解并行化运行机制和 MapReduce 工作流程，MapReduce 编程可以使用多种开发语言，需要 MapReduce 开发环境。本项目完成单词计数的分析与编写，编写查找天气数据中每年最高温度程序，使用 Eclipse 编写和调试 MapReduce 程序，使用 Hadoop-Streaming 运行 Ruby 和 Python 编写的 MapReduce 程序。

本项目分以下 3 个任务：
- 任务 1　搭建 MapReduce 开发平台
- 任务 2　编写单词计数程序
- 任务 3　编写气象数据分析程序

## 【学习目标】

### 一、知识目标

- 熟悉 MapReduce 开发平台搭建。
- 掌握 MapReduce 工作原理。
- 了解 MapReduce 工作流程。
- 掌握单词计数程序的编写。
- 掌握气象数据分析程序的编写。
- 掌握脚本的编写。

### 二、能力目标

- 能够搭建 MapReduce 开发平台。
- 能够使用 Shell 脚本编写 MapReduce 程序。
- 能够使用 Ruby 脚本编写 MapReduce 程序。
- 能够使用 Python 脚本编写 MapReduce 程序。
- 能够编写单词计数 MapReduce 程序。
- 能够编写气象数据分析 MapReduce 程序。

## 任务1 搭建 MapReduce 开发平台

### 【任务概述】

Eclipse 工具开发 MapReduce 程序非常方便,需要图形窗口运行。本任务在 Ubuntu Desktop 客户端完成 Eclipse 及插件安装、JDK 软件安装,搭建 MapReduce 开发平台。

### 【支撑知识】

#### 一、MapReduce 简介

MapReduce 是一种用于大规模数据集离线并行运算的编程框架。MapReduce 来源于 Google,其设计思想是"移动计算,而不移动数据",以及高可靠的分布式计算,简单来说,就是"分而治之"。在 Hadoop 中,使用 HDFS 作为 MapReduce 的分布式系统。

MapReduce 框架的核心步骤主要分两部分:Map(映射)和 Reduce(归约)。当向 MapReduce 框架提交一个计算作业时,它会首先把计算作业拆分成若干个 Map 任务,然后分配到不同的节点上去执行,每一个 Map 任务处理输入数据中的一部分,当 Map 任务完成后,它会生成一些中间文件,这些中间文件将会作为 Reduce 任务的输入数据。Reduce 任务的主要目标就是把前面若干个 Map 的输出汇总到一起并输出。

相对于传统的数据库模式,MapReduce 计算框架更适合对大规模、超大规模的数据进行分析处理,且更高效、安全。MapReduce 的运用相当广泛,包括"分布 Grep、分布排序、Web 连接图反转、每台机器的词矢量、Web 访问日志分析、反向索引构建、文档聚类、用户信息、爬虫分析"等。Hadoop 就是 MapReduce 最具代表性的实现之一。

#### 二、Eclipse 开发工具

Eclipse 是著名的跨平台的自由集成开发环境(IDE)。最初主要用 Java 语言开发,通过安装不同的插件,Eclipse 可以支持不同的计算机语言,如 C++和 Python 等。Eclipse 本身只是一个框架平台,但是众多插件的支持使得 Eclipse 拥有其他功能相对固定的 IDE 软件很难具有的灵活性。许多软件开发商以 Eclipse 为框架开发自己的 IDE。

Eclipse 最初由 OTI 和 IBM 两家公司的 IDE 产品开发组创建,起始于 1999 年 4 月。IBM 提供了最初的 Eclipse 代码基础,包括 Platform、JDT 和 PDE。Eclipse 项目由 IBM 发起,围绕着 Eclipse 项目已经发展成为了一个庞大的 Eclipse 联盟,有 150 多家软件公司参与到 Eclipse 项目中,其中包括 Borland、Rational Software、Red Hat 及 Sybase 等。Eclipse 是一个开放源码项目,它其实是 Visual Age for Java 的替代品,其界面与先前的 Visual Age for Java 差不多,但由于其开放源码,任何人都可以免费得到,并可以在此基础上开发各自的插件,因此,越来越受人们关注。随后还有包括 Oracle 在内的许多大公司也纷纷加入了该项目,Eclipse 的目标是成为可进行任何语言开发的 IDE 集成者,使用者只需下载各种语言的插件即可。

Eclipse 常用快捷键如下:

Ctrl+1　　　　　　　　快速修复

Ctrl+D　　　　　　　　删除当前行

Ctrl+Alt+↓　　　　　　复制当前行到下一行

| Ctrl+Alt+↑ | 复制当前行到上一行 |
| --- | --- |
| Alt+↓ | 当前行和下面一行交互位置 |
| Alt+↑ | 当前行和上面一行交互位置 |
| Alt+← | 前一个编辑的页面 |
| Alt+→ | 下一个编辑的页面 |
| Shift+Enter | 在当前行的下一行插入空行 |
| Shift+Ctrl+Enter | 在当前行插入空行 |
| Ctrl+Q | 定位到最后编辑的地方 |
| Ctrl+L | 定位在某行 |
| Ctrl+/ | 注释当前行,再按则取消注释 |
| Ctrl+/(小键盘) | 折叠当前类中的所有代码 |
| Ctrl+×(小键盘) | 展开当前类中的所有代码 |
| Ctrl+Shift+X | 把当前选中的文本全部变为大写 |
| Ctrl+Shift+Y | 把当前选中的文本全部变为小写 |
| Ctrl+Shift+F | 格式化当前代码 |
| Ctrl+Shift+P | 定位到对应的匹配符（如{}）|

### 三、sudo 命令

sudo 是 Linux 系统管理指令，是允许系统管理员让普通用户执行一些或者全部 root 命令的工具，如 halt、reboot、su 等。这样不仅减少了 root 用户的登录和管理时间，同样也提高了安全性。sudo 不是对 shell 的一个代替，它是面向每个命令的。加入 sudo 用户组的普通用户具备执行 sudo 的权限，Ubuntu Desktop 安装时设置的普通用户已经加入用户组 sudo。

## 【任务实施】

### 一、开发平台搭建

① 在 PC 上安装 Ubuntu Desktop 系统，创建普通用户 sw。

② 以用户 sw 登录进入图形窗口，按 Ctrl+Alt+T 组合键打开终端窗口,如图 4-1 所示。

③ 设置用户 root 密码，并切换到 root 用户，操作如下：

```
sw@sw-desktop:~$ sudo passwd root
sw@sw-desktop:~$ su - root
root@sw-desktop:~#
```

图 4-1　终端窗口

④ 设置网络，操作如下：

```
root@sw-desktop:~# vi /etc/network/interfaces
```

添加内容如下：

```
auto ens33
iface ens33 inet static
```

```
address 172.25.0.20
netmask 255.255.255.0
gateway 172.25.0.254
dns-nameserver 202.96.128.86
dns-nameserver 202.96.134.133
```

⑤ 设置主机名,操作如下:

```
root@sw-desktop:~# vi /etc/hostname
```

内容如下:
```
sw-desktop
```

⑥ 设置 hosts,操作如下:
```
root@sw-desktop:~# vi /etc/hosts
```

修改内容如下:
```
127.0.0.1       localhost
172.25.0.10     master
172.25.0.11     slave1
172.25.0.12     slave2
172.25.0.20     sw-desktop
```

⑦ 重启网络,操作如下:
```
root@sw-desktop:~# systemctl restart networking
```

## 二、创建用户

① 创建用户 hadoop 并设置密码,操作如下:
```
root@sw-desktop:~# groupadd -g 730 hadoop
root@sw-desktop:~# useradd -u 730 -g 730 -m -s /bin/bash hadoop
root@sw-desktop:~# passwd hadoop
```
输入新的 UNIX 密码:
重新输入新的 UNIX 密码:
passwd: 已成功更新密码

② 将用户 hadoop 加到 sudo 组中,操作如下:
```
root@sw-desktop:~# gpasswd -a hadoop sudo
```

③ 设置环境变量,操作如下:
```
root@sw-desktop:~# vi /home/hadoop/.profile
```

添加内容如下:
```
export JAVA_HOME=/opt/jdk1.8.0_121
export HADOOP_HOME=/opt/hadoop-2.7.3
export HBASE_HOME=/opt/hbase-1.2.4
export ZOOKEEPER_HOME=/opt/zookeeper-3.4.9
export PATH=$JAVA_HOME/bin:$HADOOP_HOME/bin:$HADOOP_HOME/sbin:$HBASE_HOME/bin:$ZOOKEEPER_HOME/bin:$PATH
```

## 三、安装开发软件

① 下载 Eclipse 及插件,下载地址如下:

```
http://mirrors.yun-idc.com/eclipse/technology/epp/downloads/release/
https://github.com/winghc/hadoop2x-eclipse-plugin
https://codeload.github.com/winghc/hadoop2x-eclipse-plugin/zip/master
```

② 下载 Eclipse 开发包到/root 目录下，安装 Eclipse 开发包，操作如下：

```
root@sw-desktop:~ # cd /opt
root@sw-desktop:/opt# tar xvzf /root/eclipse-java-luna-SR2-linux-gtk-x86_64.tar.gz
```

③ 下载并解压 hadoop-eclipse-plugin-2.6.0.jar 到/opt/eclipse/plugins/目录下。

④ 从 Master 节点复制 JDK 和 Hadoop 系统，操作如下：

```
root@sw-desktop:~# scp -r hadoop@master:/opt/* /opt
The authenticity of host 'master (172.25.0.10)' can't be established.
ECDSA key fingerprint is SHA256:H8Av0KuJwlAZvw+bcp2NPH44iyJpcVTvpD5E3a6xINA.
Are you sure you want to continue connecting (yes/no)? yes
Warning: Permanently added 'master' (ECDSA) to the list of known hosts.
hadoop@master's password:
```

⑤ 修改文件属性，操作如下：

```
root@sw-desktop:~# chown -R hadoop:hadoop /opt/eclipse /opt/hadoop* /opt/jdk*
```

### 四、运行 Eclipse

① 以用户 hadoop 登录图形桌面，打开终端，运行 Eclipse，操作如下：

```
hadoop@sw-desktop:~$ /opt/eclipse/eclipse &
```

② 输入 Eclipse 工作目录，如图 4-2 所示。

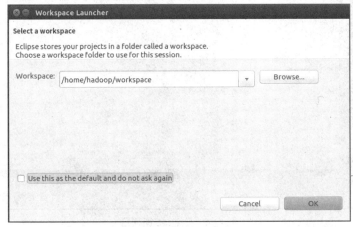

图 4-2　Eclipse 工作空间

③ 在 Eclipse 窗口中打开"Windows"→"Preferences"，窗口左侧有"Hodoop Map/Reduce"选项，表示插件安装成功，如图 4-3 所示。

④ 在 Eclipse 运行界面打开"File"→"New"→"Project"，有"Map/Reduce Project"选项，表示安装成功，如图 4-4 所示，至此，MapReduce 开发平台搭建完成。

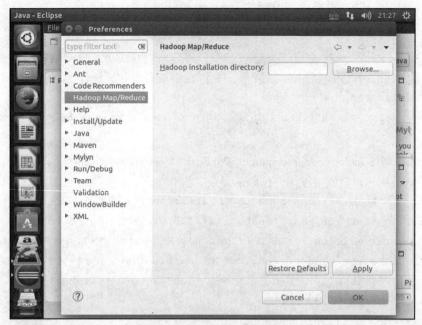

图 4-3　Eclipse 参数选择

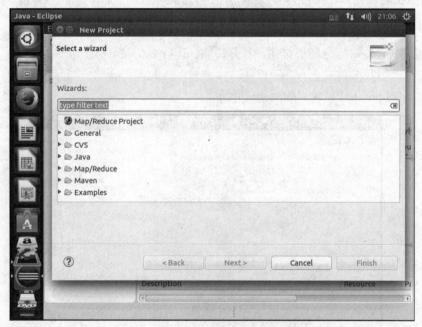

图 4-4　Eclipse 新项目

# 任务 2　编写单词计数程序

## 【任务概述】

单词计数是 MapReduce 最典型的案例，是 MapReduce 编程入门。本任务使用示例程序 hadoop-mapreduce-examples-2.7.3.jar 完成单词计数，在了解单词计数的基础上编写 WordCount 程序，并进行调试运行，实现单词计数。

## 【任务分析】

单词计数是一个典型的 MapReduce 的应用，如图 4-5 所示。

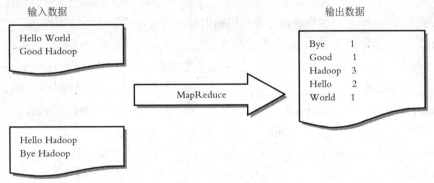

图 4-5　单词计数结果

从数据输入到结果输出，需要经过 Split、Map、Reduce，Split 由系统自动完成，Map 和 Reduce 需要编写的实现功能，Map 接收由 Split 输出的<k1,v1>计算输出<k2,v2>，Reduce 接收<k2,v2>计算输出<k3,v3>，如图 4-6 所示。

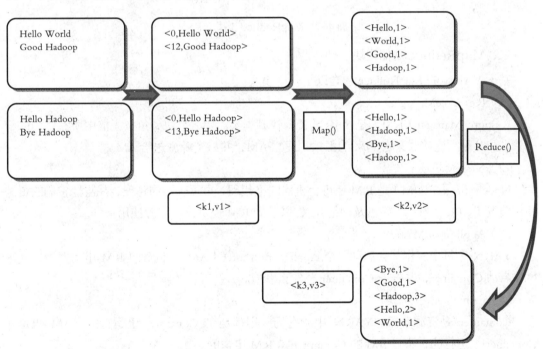

图 4-6　单词计数流程

## 【支撑知识】

### 一、MapReduce 工作原理

MapReduce 是一种编程模型，用于大规模数据集（大于 1TB）的并行运算。概念 Map 和 Reduce，以及它们的主要思想都是从函数式编程语言里借来的，还有从矢量编程语言里借来的特性。MapReduce 极大地方便了编程人员在不会分布式并行编程的情况下，将自己的程序

运行在分布式系统上。

一个 MapReduce 的作业（Job）通常会把输入的数据集切分为若干独立的数据块（Block），由 Map 任务（Task）以完全并行的方式处理它们。框架会对 Map 的输出先进行排序，然后把结果输入给 Reduce 任务。通常 Job 的输入和输出都会被存储在文件系统中。

在 Hadoop 中，每个 MapReduce 任务都被初始化为一个 Job，每个 Job 又可以分为四个主要阶段：Split、Map、Shuffle 及 Reduce，其中最重要的两个阶段是 Map 阶段和 Reduce 阶段。这两个阶段分别用两个函数表示，即 map 函数和 reduce 函数。map 函数接收一个<key,value>形式的输入，然后产生一个<key,value>形式的中间输出；reduce 函数接收一个<key,value(List of value)>形式的输入，然后对这个 value 集合进行处理并输出结果。

适合用 MapReduce 来处理的数据集（或任务）有一个基本要求：待处理的数据集可以分解成许多小的数据集，而且每一个小数据集都可以完全并行地进行处理。MapReduce 工作流程如图 4-7 所示。

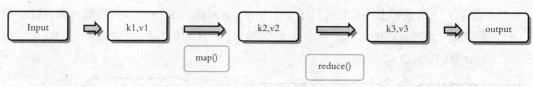

图 4-7 MapReduce 工作流程

## 二、MapReduce 功能模块

基于 YARN 的 MapReduce 包含以下几个模块。

（1）ResourceManager

ResourceManager（RM）是 YARN 资源控制框架的中心模块，负责集群中所有资源的统一管理和分配。它接收来自 NM 的汇报，建立 AM，并将资源派送给 AM。

（2）NodeManager

NodeManager（NM）是 RM 在每台机器的上代理，负责容器的管理，并监控它们的资源使用情况（CPU、内存、磁盘及网络等），以及向 RM 提供这些资源使用报告。

（3）ApplicationMaster

YARN 中每个应用都会启动一个 ApplicationMaster（AM），负责向 RM 申请资源，请求 NM 启动 Container，并告诉 Container 做什么事情。

（4）Container

Container 是资源容器。YARN 中所有的应用都是在 Container 上运行的。AM 也是在 Container 上运行的，不过 AM 的 Container 是 RM 申请的。

- Container 是 YARN 中资源的抽象，它封装了某个节点上一定量的资源（CPU 和内存两类资源）。
- Container 是由 AM 向 RM 申请的，由 RM 中的资源调度器异步分配给 AM。
- Container 的运行是由 AM 向资源所在的 NM 发起的，Container 运行时需提供内部执行的任务命令（可以是任何命令，如 Java、Python、C++进程启动命令均可），以及该命令执行所需的环境变量和外部资源（如词典文件、可执行文件、jar 包等）。

另外，一个应用程序所需的 Container 分为如下两大类。

- 运行 AM 的 Container：这是由 RM（向内部的资源调度器）申请和启动的，用户提交应用程序时，可指定唯一的 AM 所需的资源。
- 运行各类任务的 Container：这是由 AM 向 RM 申请的，并由 AM 与 NM 通信以启动之。

以上两类 Container 可能在任意节点上，它们的位置通常是随机的，即 AM 可能与它管理的任务运行在一个节点上。

### 三、MapReduce 执行流程

基于 YARN 的 MapReduce 执行流程如图 4-8 所示。

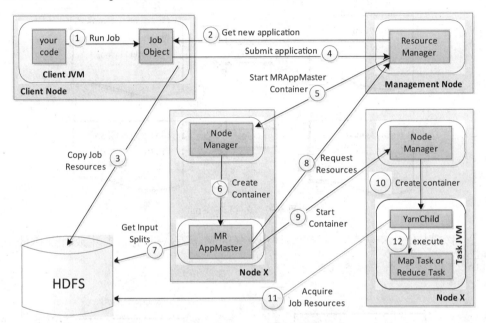

图 4-8　MapReduce 执行流程

① JobClient 通过 Job 对象提交 MapReduce 作业。

② JobClient 与 Resource Manager 交互获取应用的 Job ID。

③ JobClient 把与任务有关的资源都上传到 HDFS，从而使其他模块可以访问和使用。

④ JobClient 把应用提交给 Resource Manager。

⑤ Resource Manager 选择一个有足够资源的 Node Manager 并请求为 MRAppMaster 创建一个容器。

⑥ Node Manager 为 MRAppMaster 分配容器；MRAppMaster 将运行和协调 MapReduce 作业。

⑦ MRAppMaster 从 HDFS 获取所需要的资源，如 Input Splits，这些资源是在第③步的时候上传到 HDFS 的。

⑧ MRAppMaster 与 Resource Manager 协商获取资源；Resource Manager 将选择一个资源最富裕的 Node Manager。

⑨ MRAppMaster 命令 NodeManager 启动 Map 和 Reduce 任务。

⑩ NodeManager 创建 YarnChild 容器来协调和运行任务。

⑪ YarnChild 从 HDFS 获取有关作业资源用来执行 Map 和 Reduce 任务。

⑫ YarnChild 运行 Map 和 Reduce 任务。

### 四、单词计数 MapReduce 执行过程

MapReduce 的执行过程大致分为 Map（映射）→Shuffle（排序）→Combine（组合）→Reduce（归约），单词计数过程如下。

① 将文件拆分成 Splits（片），并将每个 Split 分割成<Key,Value>对，如图 4-9 所示。该过程由 MapReduce 自动完成。

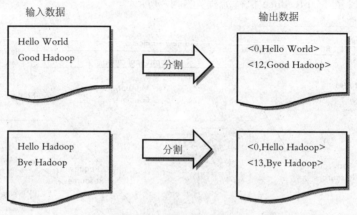

图 4-9　文件拆分 Split

② 将分割好的<Key,Value>对交给用户定义的 Map 方法进行处理，生成新的<Key,Value>对，如图 4-10 所示。

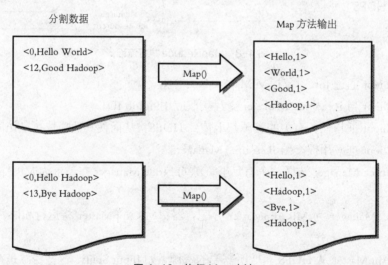

图 4-10　执行 Map 方法

③ Map 输出<Key,Value>对后，Mapper 会按照 Key 值进行 Shuffle（混洗），并执行 Combine 过程，将 Key 相同的值相加，得到 Mapper 的结果，如图 4-11 所示。

④ Reducer 先对从 Mapper 接收的数据进行排序，再交由用户自定义的 Reduce 方法进行处理，得到新的<Key,Value>对作为输出结果，如图 4-12 所示。

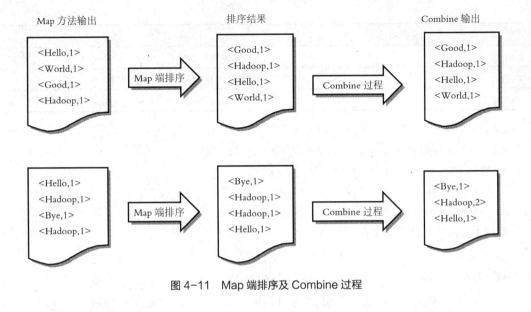

图 4-11 Map 端排序及 Combine 过程

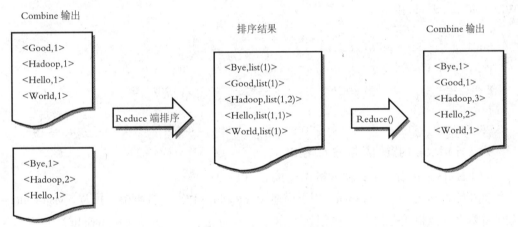

图 4-12 Reduce 端排序及输出结果

### 五、Hadoop 数据类型

（1）Hadoop 内置的数据类型

Hadoop 提供表 4-1 所示的内置的数据类型。这些数据类型都实现了 WritableComparable 接口，以便用这些类型定义的数据可以被序列化进行网络传输和文件存储，以及进行大小比较。

表 4-1 Hadoop 数据类型

| 数据类型 | 含　　义 |
| --- | --- |
| BooleanWritable | 标准布尔型数值 |
| ByteWritable | 单字节数值 |
| DoubleWritable | 双字节数 |
| FloatWritable | 浮点数 |
| IntWritable | 整型数 |

续表

| 数据类型 | 含义 |
|---|---|
| LongWritable | 长整型数 |
| Text | 使用 UTF-8 格式存储的文本 |
| NullWritable | 当<Key,Value>中的 key 或 value 为空时使用 |

```
//简单类型
IntWritable iw = new IntWritable(1);
System.out.println( iw.get() );  // 1
BooleanWritable bw = new BooleanWritable(true);
System.out.println( bw.get() );  // true
```

（2）Hadoop 用户自定义的数据类型

自定义数据类型时，需满足如下两个基本要求：

① 实现 Writable 接口，以便该数据能被序列化后完成网络传输或文件输入/输出。

② 如果该数据需要作为主键 Key 使用，或需要比较数值大小时，则需要实现 WritableComparable 接口。

```
//Hadoop - Writable
public interface Writable {
void write(DataOutput out) throws IOException;
void readFields(DataInput in) throws IOException;
}
public interface WritableComparable<T> extends Writable, Comparable<T> { }
```

### 六、Hadoop 内置的数据输入/输出

（1）Hadoop 内置的数据输入格式和 RecordReader

数据输入格式（InputFormat）用于描述 MapReduce 作业的数据输入规范。MapReduce 框架依靠数据输入格式完成输入规范检查、对数据文件进行输入分块（InputSplit），以及提供从输入分块中将数据记录逐一读出，并转换为 Map 过程的输入键值对等功能。

Hadoop 提供了丰富的内置数据输入格式，最常用的数据输入格式包括 TextInputFormat 和 KeyValueInputFormat。

TextInputFormat 是系统默认的数据输入格式，可以将文本文件分块并逐行读入以便 Map 节点进行处理。读入一行时，所产生的主键 Key 就是当前行在整个文本文件中的字节偏移位置，而 Value 就是该行的内容。

```
//TextInputFormat
public class TextInputFormat extends FileInputFormat<LongWritable, Text> {
  @Override
  public RecordReader<LongWritable, Text>
    createRecordReader(InputSplit split,
                TaskAttemptContext context) {
    String delimiter = context.getConfiguration().get(
       "textinputformat.record.delimiter");
```

```
    byte[] recordDelimiterBytes = null;
    if (null != delimiter)
      recordDelimiterBytes = delimiter.getBytes(Charsets.UTF_8);
    return new LineRecordReader(recordDelimiterBytes);
  }
  //....
}
```

KeyValueTextInputFormat 是另一个常用的数据输入格式，可将一个按照<Key,Value>格式逐行存放的文本文件逐行读出，并自动解析生成相应的 Key 和 Value。

```
//KeyValueTextInputFormat
public class KeyValueTextInputFormat extends FileInputFormat<Text, Text> {
  // ...
  public RecordReader<Text,Text> createRecordReader(InputSplit genericSplit,
    TaskAttemptContext context) throws IOException {

    context.setStatus(genericSplit.toString());
    return new KeyValueLineRecordReader(context.getConfiguration());
  }
}
```

RecordReader：对于一个数据输入格式，都需要有一个对应的 RecordReader，主要用于将一个文件中的数据记录拆分成具体的键值对。TextInputFormat 的默认 RecordReader 是 LineRecordReader，而 KeyValueTextInputFormat 的默认 RecordReader 是 KeyValueLineRecordReader。

（2）Hadoop 内置的数据输出格式与 RecordWriter

数据输出格式（OutputFormat）用于描述 MapReduce 作业的数据输出规范。MapReduce 框架依靠数据输出格式完成输出规范检查，以及提供作业结果数据输出功能。

同样，最常用的数据输出格式是 TextOutputFormat，也是系统默认的数据输出格式，可以将计算结果以 "Key + \t + Value" 的形式逐行输出到文本文件中。

与数据输入格式类似，数据输出格式也提供一个对应的 RecordWriter，以便系统明确输出结果写入到文件中的具体格式。TextInputFormat 的默认 RecordWriter 是 LineRecordWriter，其实际操作是将结果数据以 "Key + \t + Value" 的形式输出到文本文件中。

```
//TextOutputFormat
public class TextOutputFormat<K, V> extends FileOutputFormat<K, V> {
  protected static class LineRecordWriter<K, V> extends RecordWriter<K, V> {
    // ...
    public LineRecordWriter(DataOutputStream out, String keyValueSeparator) {
      //...
    }
    public LineRecordWriter(DataOutputStream out) {
      this(out, "\t");
    }
    private void writeObject(Object o) throws IOException {
```

```
      // ...
    }
    public synchronized void write(K key, V value) throws IOException {
      //...
      out.write(newline);
    }
  }
  public RecordWriter<K, V> getRecordWriter(TaskAttemptContext job
                  ) throws IOException, InterruptedException {
    // ...
  }
}
```

### 七、Hadoop 程序运行

Hadoop 程序运行，命令格式如下：

```
hadoop jar x.jar ×.MainClassName inputPath outputPath
```

其中，x.jar 表示运行的包，×.MainClassName 表示运行的类，inputPath 表示 HDFS 输入目录，outputPath 表示 HDFS 输出目录。

## 【任务实施】

### 一、运行单词计数实例

① 新建文件 sw1.txt 和 sw2.txt。

文件 sw1.txt 的内容如下：
```
This is the first hadoop test!
```

文件 sw2.txt 的内容如下：
```
Hello world
www.swvtc.cn welcome
My name is hadoop
```

② 在 HDFS 上创建目录，并上传数据，操作如下：
```
hadoop@sw-desktop:~$ hdfs dfs -mkdir /test
hadoop@sw-desktop:~$ hdfs dfs -put sw*.txt /test
hadoop@sw-desktop:~$ hdfs dfs -ls /test/
Found 2 items
-rw-r--r--   3 hadoop supergroup         31 2017-02-01 22:06 /test/sw1.txt
-rw-r--r--   3 hadoop supergroup         51 2017-02-01 22:06 /test/sw2.txt
```

③ 运行单词统计实例，操作如下：
```
hadoop@sw-desktop:~$ cd /opt/hadoop-2.7.3/share/hadoop/mapreduce/
hadoop@sw-desktop...$ hadoop jar hadoop-mapreduce-examples-2.7.3.jar wordcount /test /out1
```

④ 查看结果，操作如下：
```
hadoop@sw-desktop:~$ hdfs dfs -text /out1/part-r-00000
Hello   1
My      1
```

```
This      1
first     1
hadoop    2
is        2
name      1
test!     1
the       1
welcome   1
world     1
www.swvtc.cn    1
```

二、编写单词计数程序

① 运行 Eclipse，执行菜单栏中的"File"→"New"→"Other…"命令，选择"Map/Reduce Project"，如图 4-13 所示。

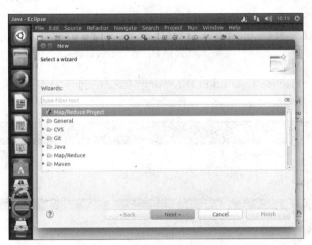

图 4-13　新建 MapReduce 项目

② 输入项目名称"WordCount"，选择"Configure Hadoop install directory…"，如图 4-14 所示。

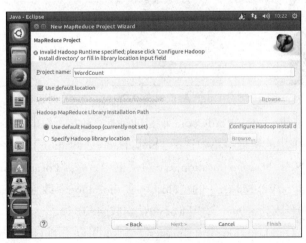

图 4-14　WordCount 项目

③ 选择 Hadoop 安装目录，直接输入"/opt/hadoop-2.7.3"或者单击"Browse…"按钮进行选择，如图 4-15 所示。

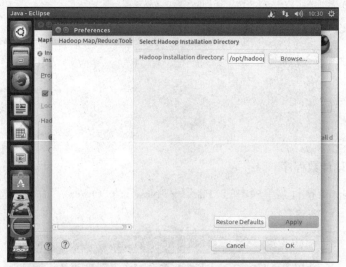

图 4-15　选择 Hadoop 安装目录

④ 单击"Select Hadoop Installation Directory"窗口中的"OK"按钮，关闭当前窗口，再单击"Finish"按钮，弹出图 4-16 所示的对话框。

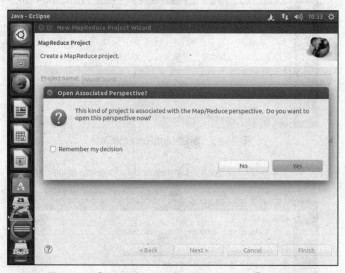

图 4-16　"Open Associated Perspective"对话框

⑤ 单击"Yes"按钮，完成 WordCount 项目的创建，在右下角的"Map/Reduce Locations"上单击鼠标右键，在弹出的快捷菜单中选择"New Hadoop location…"命令，如图 4-17 所示。

⑥ 在弹出的"New Hadoop location…"对话框中，在"Location name"文本框中输入"master"，"Map/Reduce(V2) Master"的"Host"原"localhost"改为"master"，"DFS Master"的"Port"原"50040"改为"9000"，其他不变，如图 4-18 所示。

⑦ 在左侧展开"DFS Locations"→"master"节点，可以看到 HDFS 上的目录，如图 4-19 所示。

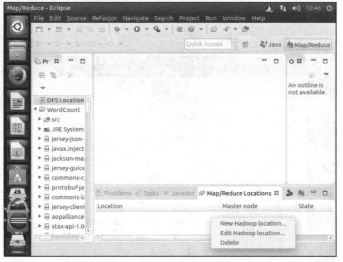

图 4-17 选择 "New Hadoop location…" 命令

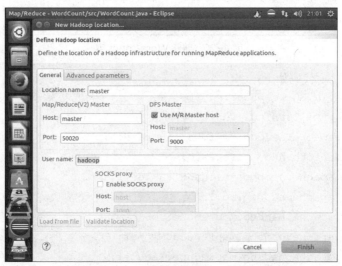

图 4-18 Define Hadoop location

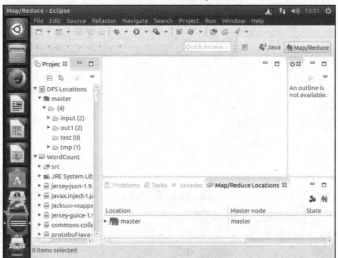

图 4-19 DFS locations

⑧ 在"WordCount"→"src"节点上单击鼠标右键,在弹出的快捷菜单中选择"New"→"Class"命令,新建"WordCount"类,如图4-20所示。

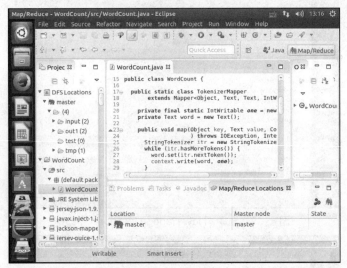

图4-20 新建WordCount类

WordCound.java代码如下:

```java
import java.io.IOException;
import java.util.StringTokenizer;
import org.apache.hadoop.conf.Configuration;
import org.apache.hadoop.fs.Path;
import org.apache.hadoop.io.IntWritable;
import org.apache.hadoop.io.Text;
import org.apache.hadoop.mapreduce.Job;
import org.apache.hadoop.mapreduce.Mapper;
import org.apache.hadoop.mapreduce.Reducer;
import org.apache.hadoop.mapreduce.lib.input.FileInputFormat;
import org.apache.hadoop.mapreduce.lib.output.FileOutputFormat;
import org.apache.hadoop.util.GenericOptionsParser;
public class WordCount {
  public static class TokenizerMapper
       extends Mapper<Object, Text, Text, IntWritable>{
    private final static IntWritable one = new IntWritable(1);
    private Text word = new Text();
    public void map(Object key, Text value, Context context
                ) throws IOException, InterruptedException {
      StringTokenizer itr = new StringTokenizer(value.toString());
      while (itr.hasMoreTokens()) {
        word.set(itr.nextToken());
        context.write(word, one);
```

```java
      }
    }
    public static class IntSumReducer
         extends Reducer<Text,IntWritable,Text,IntWritable> {
      private IntWritable result = new IntWritable();
      public void reduce(Text key, Iterable<IntWritable> values,
                      Context context
                      ) throws IOException, InterruptedException {
        int sum = 0;
        for (IntWritable val : values) {
          sum += val.get();
        }
        result.set(sum);
        context.write(key, result);
      }
    }
    public static void main(String[] args) throws Exception {
      Configuration conf = new Configuration();
      String[] otherArgs = new GenericOptionsParser(conf, args).getRemainingArgs();
      if (otherArgs.length < 2) {
        System.err.println("Usage: wordcount <in> [<in>...] <out>");
        System.exit(2);
      }
      Job job = Job.getInstance(conf, "word count");
      job.setJarByClass(WordCount.class);
      job.setMapperClass(TokenizerMapper.class);
      job.setCombinerClass(IntSumReducer.class);
      job.setReducerClass(IntSumReducer.class);
      job.setOutputKeyClass(Text.class);
      job.setOutputValueClass(IntWritable.class);
      for (int i = 0; i < otherArgs.length - 1; ++i) {
        FileInputFormat.addInputPath(job, new Path(otherArgs[i]));
      }
      FileOutputFormat.setOutputPath(job,
        new Path(otherArgs[otherArgs.length - 1]));
      System.exit(job.waitForCompletion(true) ? 0 : 1);
    }
  }
```

⑨ 执行菜单中的"Run"→"Run Configurations…"命令,弹出对话框,双击"Java Application"生成"WordCount",选择"Arguments"选项卡,在"Program arguments"文本框中输入两行参

数:"hdfs://master:9000/input"和"hdfs://master:9000/output",如图 4-21 所示。

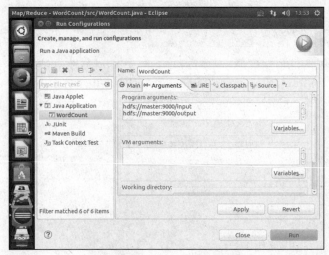

图 4-21 WordCount 运行参数

⑩ 按 Ctrl+Alt+T 组合键打开新终端窗口,上传 sw1.txt 和 sw2.txt 到 HDFS 的/input 目录下,操作如下:

```
hadoop@sw-desktop:~$ hdfs dfs -rm /input/*
hadoop@sw-desktop:~$ hdfs dfs -put sw*.txt /input
```

⑪ 单击"Run"按钮,运行"WordCount",刷新或重新连接左窗口的"master",单击文件"part-r-00000",运行结果如图 4-22 所示,单词计数完成。

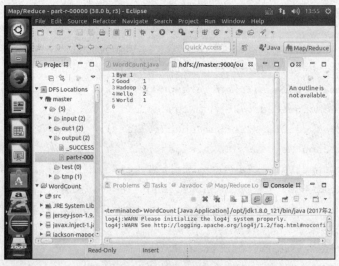

图 4-22 WordCount 运行结果

## 任务 3 编写气象数据分析程序

### 【任务概述】

气象数据分析主要是查找天气数据中每年最高温度,也是比较典型的 MapReduce 例子。

本任务主要完成天气预报数据分析，编写 Shell 脚本、Ruby 脚本、Python 脚本，使用 Linux 管道模拟 MapReduce 过程；最后使用 Eclipse 工具编写分析程序，并调试运行。

# 【任务分析】

## 一、气象数据

① 气象基本数据如下：

```
0067011990999991950051507004+68750+023550FM-12+038299999V0203301N0067122
0001CN9999999N9+00001+99999999999
    0043011990999991950051512004+68750+023550FM-12+038299999V0203201N0067122
0001CN9999999N9+00221+99999999999
    0043011990999991950051518004+68750+023550FM-12+038299999V0203201N0026122
0001CN9999999N9-00111+99999999999
    0043012650999991949032412004+62300+010750FM-12+048599999V0202701N0046122
0001CN0500001N9+01111+99999999999
    0043012650999991949032418004+62300+010750FM-12+048599999V0202701N0046122
0001CN0500001N9+00781+99999999999
```

② 数据格式。气象数据来自美国国家气候数据中心（NCDC），数据定义如下：

```
0067
011990              # 美国空军气象局标识
99999               # 空军与海军气象局标识
19500515            # 观察日期
0700                # 观察时间
4 +68750            # 纬度（度数 × 1000）
+023550             # 经度（度数 × 1000）
FM-12
+0382               # 海拔（米）
99999
V020
330                 # 风向（度）
1                   # 质量代码
N
0067
1
22000               # 云层高度（米）
1                   # 质量代码
C
N
999999              # 可见度（米）
9                   # 质量代码
N
9 +0000             # 气温（摄氏度 × 10）
```

| | | |
|---|---|---|
| 1 | # | 质量代码 |
| +9999 | # | 露点温度（摄氏度 × 10） |
| 9 | # | 质量代码 |
| 99999 | # | 气压（百帕 × 10） |
| 9 | # | 质量代码 |

其中：9 表示缺失记录。

## 二、数据分析过程

① 输入气象数据，省略号表示缺少部分。

0067011990999991950051507004...9999999N9+00001+99999999999...
0043011990999991950051512004...9999999N9+00221+99999999999...
0043011990999991950051518004...9999999N9-00111+99999999999...
0043012650999991949032412004...0500001N9+01111+99999999999...
0043012650999991949032418004...0500001N9+00781+99999999999...

② 以<键,值>对的方式作为 map 函数的输入。

(0,   0067011990999991950051507004...9999999N9+00001+99999999999...)
(106, 0043011990999991950051512004...9999999N9+00221+99999999999...)
(212, 0043011990999991950051518004...9999999N9-00111+99999999999...)
(318, 0043012650999991949032412004...0500001N9+01111+99999999999...)
(424, 0043012650999991949032418004...0500001N9+00781+99999999999...)

③ map 函数的功能仅限于提取年份和气温信息，以<键,值>对输出。

(1950, 0)
(1950, 22)
(1950, -11)
(1949, 111)
(1949, 78)

④ map 函数的输出经由 MapReduce 框架处理后，最后发送到 reduce 函数。

(1949, [111, 78])
(1950, [0, 22, -11])

⑤ reduce 函数现在要做的是遍历整个列表并从中找出最大的读数。

(1949, 111)
(1950, 22)

⑥ 整个数据流如图 4-23 所示，在图的底部是 UNIX 管线，用于模拟整个 MapReduce 的流程。

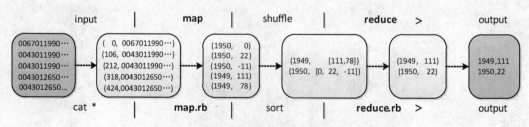

图 4-23  MapReduce 的逻辑数据流

## 【支撑知识】

### 一、Hadoop 数据流

Hadoop 将 MapReduce 的输入数据划分成等长的小数据块,称为输入分片(Input Split)或分片。Hadoop 为每个分片构建一个 map 任务,并由该任务来运行用户自定义的 map 函数,从而处理分片中的每条记录。

拥有许多分片就意味着处理每个分片的时间与处理整个输入的时间相比是比较小的。因此,如果并行处理每个分片,且分片是小块的数据,那么处理过程将有一个更好的负载平衡。另一方面,分片不能太小,如果分片太小,那么管理分片的总时间和 map 任务创建的总时间将决定作业执行的总时间。对于大多数作业,一个理想的分片大小往往是一个 HDFS 块的大小,默认是 64 MB。

Hadoop 在存储有输入数据的节点上运行 map 任务,可以获得最佳性能,这就是所谓的"数据本地化优化(data locality optimization)"。但是,有时对于一个 map 任务的输入来说,存储有某个 HDFS 数据块备份的 3 个节点可能正在运行其他 map 任务,此时作业调度需要在 3 个备份中的某个数据寻求同个机架中空闲的机器来运行该 map 任务。仅仅在非常偶然的情况下,会使用其他机架中的机器运行该 map 任务,这将导致机架与机架之间的网络传输。图 4-24 显示了这 3 种可能。

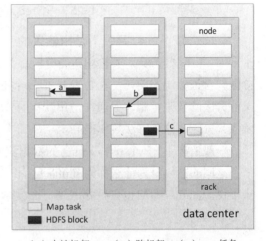

(a)本地机架　(b)跨机架　(c)map 任务

图 4-24　本地数据

现在应该清楚为什么最佳分片的大小应该与块大小相同,因为它是确保可以存储在单个节点上的最大输入块的大小。如果分片跨越两个数据块,那么对于任何一个 HDFS 节点,基本上都不可能同时存储这两个数据块,因此,分片中的部分数据需要通过网络传输到 map 任务节点。与使用本地数据运行整个 map 任务相比,这种方法显然效率更低。

map 任务将其输出写入本地硬盘,而非 HDFS。这是为什么?因为 map 的输出是中间结果,该中间结果由 reduce 任务处理后才产生最终输出结果,而且一旦作业完成,map 的输出结果就可以删除。因此,如果把它存储在 HDFS 中并实现备份,难免有些小题大做。如果该节点上运行的 map 任务在将 map 中间结果传送给 reduce 任务之前失败,Hadoop 将在另一个节点上重新运行这个 map 任务以再次构建 map 中间结果。

reduce 任务并不具备数据本地化的优势——单个 reduce 任务的输入通常来自所有 mapper 的输出。本例中,仅有一个 reduce 任务,其输入是所有 map 任务的输出。因此,排序过的 map 输出需要通过网络发送到运行 reduce 任务的节点,数据在 reduce 端合并,然后由 reduce 函数处理,reduce 的输出通常存储在 HDFS 中。

一个 reduce 任务的完整数据流如图 4-25 所示,虚线框表示节点,虚线箭头表示节点内部的数据传输,实线箭头表示节点之间的数据传输。

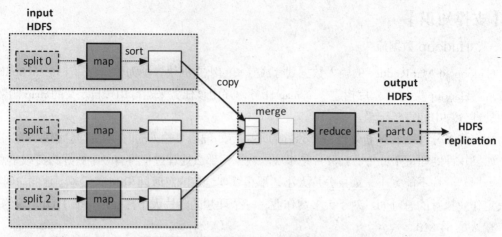

图 4-25　一个 reduce 任务的 MapReduce 数据流

reduce 任务的数目并不是由输入的大小来决定，而是单独具体指定的。如果有多个 reducer，map 任务会对其输出进行分区，为每个 reduce 任务创建一个分区（Partition）。每个分区包含许多键（及其关联的值），但每个键的记录都在同一个分区中。分区可以通过用户定义的分区函数来控制，但通常是用默认的分区器（Partitioner，也称"分区函数"）通过 Hash 函数来分区，这种方法效率很高。

一般情况下，多个 reduce 任务的数据流如图 4-26 所示，此图清楚地表明了 map 和 reduce 任务之间的数据流为什么称为 shuffle（混洗），因为每个 reduce 任务的输入都由许多 map 任务来提供。shuffle 其实比图 4-26 所显示的更复杂，并且调整它可能会对作业的执行时间产生很大的影响。

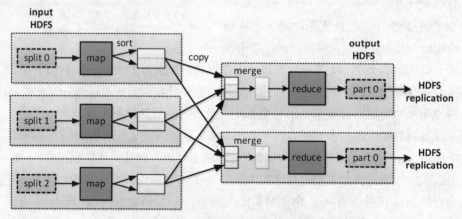

图 4-26　多个 reduce 任务的 MapReduce 数据流

有时候可能不存在 reduce 任务，就不需要 shuffle，因为处理可以并行进行。在这种情况下，唯一的非本地节点数据传输是当 map 任务写入到 HDFS 中，如图 4-27 所示。

二、Combiner

在集群上 MapReduce 作业的数量被可用带宽限制，因此，要保证 map 和 reduce 任务之间传输的代价是最小的。Hadoop 允许用户声明一个 Combiner，运行在 map 的输出上——该函数的输出作为 reduce 函数的输入。由于 Combiner 是一个优化方法，所以，Hadoop 不保证对于某个 map 的输出记录是否调用该方法或者调用该方法有多少次。换言之，不调用该方法或者调用该方法多次，reducer 的输出结果都一样。

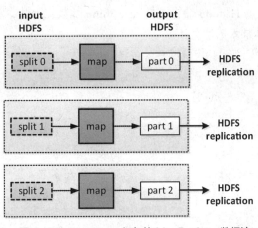

图 4-27　无 reduce 任务的 MapReduce 数据流

Combiner 的规则制约着可用的函数类型。假设计算最高气温的例子，1950 年的读数由两个 map 任务处理（因为它们在不同的分片中），第一个 map 的输出如下：

```
(1950, 0)
(1950, 20)
(1950, 10)
```

第二个 map 的输出如下：

```
(1950, 25)
(1950, 15)
```

reduce 函数被调用时，输入如下：

```
(1950, [0, 20, 10, 25, 15])
```

因为 25 为该列数据中最大的，所以，它的输出如下：

```
(1950, 25)
```

可以像使用 reduce 函数那样，使用 Combiner 找出每个 map 任务输出结果中的最高气温。如此一来，reduce 函数调用时将被传入以下数据：

```
(1950, [20, 25])
```

reduce 输出的结果和以前一样。更简单地说，通过下面的表达式来说明气温数值的函数调用：

```
max (0, 20, 10, 25, 15) = max (max (0, 20, 10), max (25, 15)) = max (20, 25) = 25
```

并非所有函数都具有该属性，有此属性的函数称为 commutative 和 associative，有时也将它们称为 distributive，但是 Combiner 不能取代 reduce 函数。例如，计算气温为 0, 20, 10, 25, 15 的平均气温就不能用平均数作为 Combiner，因为用 Combiner 计算后 mean（mean（0, 20, 10），mean（25, 15））= mean（10, 20）= 15；而用 reduce 计算，结果为 mean（0, 20, 10, 25, 15）= 14，两种计算方法结果不一致，用 reduce 计算出的结果才是正确解。

Combiner 的规则限制着可用的函数类型。从求最高气温和平均气温的例子中，就很好地说明了这点，求最高气温可以使用 Combiner，但是求平均气温时再使用就可能出错。因此，Combiner 并不能取代 reduce 函数。虽然它可以帮助减少 map 和 reduce 之间的数据传输量，但是是否在 MapReduce 作业中使用 Combiner 还是需要慎重考虑。

### 三、Hadoop Streaming

Hadoop 提供了一个 API 来运行 MapReduce，并允许用除 Java 以外的语言来编写自己的 map 和 reduce 函数。Hadoop 流使用 UNIX 标准流作为 Hadoop 和程序之间的接口，所以，可以使用任何语言，只要编写的 MapReduce 程序能够读取标准输入，并写入到标准输出。流适用于文字处理，在文本模式下使用时，它有一个面向行的数据视图。

Hadoop 命令支持 Streaming 函数，因此，需要知道 Streaming JAR 文件流与 jar 选项指定。其基本语法如下：

```
$HADOOP_PREFIX/bin/hadoop jar hadoop-streaming.jar [options]
Options:
  -input          <路径>                         #为 Map 步骤提供输入 DFS 文件
  -output         <路径>                         #为 Reduce 步骤提供输出 DFS 目录
  -mapper         <命令|Java 类名> 选项          #mapper 运行命令
  -combiner       <命令|Java 类名> 选项          #combiner 运行命令
  -reducer        <命令|Java 类名> 选项          #reducer 运行命令
  -file           <文件> 选项                    #文件/目录加载 Job jar 文件（不建议使用）
                                                 #一般使用选择 "-files" 代替
  -inputformat    <TextInputFormat（默认）|SequenceFileAsTextInputFormat|Java 类名>
                  选项                           #输入格式类
  -outputformat   <TextOutputFormat（默认）|Java 类名>
                  选项                           #输出格式类
  -partitioner    <Java 类名> 选项               #分区类
  -numReduceTasks <数字> 选项                    #reduce 任务数
  -inputreader    <spec> 选项                    #recordreader spec 输入
  -cmdenv         <n>=<v>选项                    #通过 env.var 的流命令
  -mapdebug       <命令> 选项                    #运行 map 任务失败的脚本
  -reducedebug    <命令> 选项                    #运行 reduce 任务失败的脚本
  -io             <标识> 选项                    #从 mapper/reducer 命令输出
                                                 #用于输入格式
  -lazyOutput     选项                           #松散创建输出
  -background     选项                           #提交任务但不等待任务完成
  -verbose        选项                           #打印详细输出
  -info           选项                           #打印详细语法
  -help           选项                           #打印帮助信息
```

## 【任务实施】

### 一、使用 Linux 脚本查找每年的最高气温

① 下载 1901 与 1902 两个气象数据文件到 /home/hadoop 目录下。

② 编写脚本，操作如下：

```
hadoop@sw-desktop:~$ vi max_temp.sh
for year in /home/hadoop/19*
do
  echo -n 'basename $year'"\t"
  cat $year | \
    awk '{ temp = substr($0, 88, 5) + 0;
           q = substr($0, 93, 1);
       if ( temp!=9999 && q ~ /[01459]/ && temp > max )  max = temp }
```

```
    END { print max }'
done
```

天气数据第 88~92 位为气温，9999 为缺失记录；第 93 位为质量代码，01459 有效，其他数字为可疑或错误。

③ 运行脚本，结果如下：

```
hadoop@sw-desktop:~$ sh max_temp.sh
1901    317
1902    244
```

## 二、编写 MapReduce 程序查找每年的最高气温

① 打开 Eclipse，执行菜单栏中的"File"→"Project"命令，选择"Map/Reduce Project"选项，新建"MaxTemp"项目，选择"Configure Hadoop install directory…"，如图 4-28 所示。

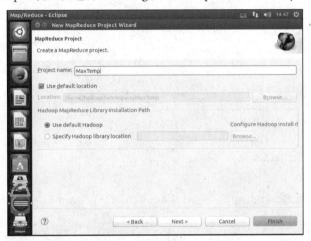

图 4-28　新建 MapReduce 项目

② 在"MaxTemp"→"src"上单击鼠标右键，新建 MaxTemp 类，如图 4-29 所示。

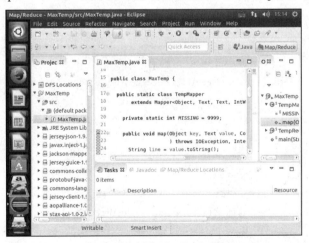

图 4-29　新建 MaxTemp 类

MaxTemp 代码如下：

```
import java.io.IOException;
```

```java
import java.util.StringTokenizer;
import org.apache.hadoop.conf.Configuration;
import org.apache.hadoop.fs.Path;
import org.apache.hadoop.io.IntWritable;
import org.apache.hadoop.io.Text;
import org.apache.hadoop.mapreduce.Job;
import org.apache.hadoop.mapreduce.Mapper;
import org.apache.hadoop.mapreduce.Reducer;
import org.apache.hadoop.mapreduce.lib.input.FileInputFormat;
import org.apache.hadoop.mapreduce.lib.output.FileOutputFormat;
import org.apache.hadoop.util.GenericOptionsParser;
public class MaxTemp {
  public static class TempMapper
       extends Mapper<Object, Text, Text, IntWritable>{
    private static int MISSING = 9999;
    public void map(Object key, Text value, Context context
                ) throws IOException, InterruptedException {
      String line = value.toString();
      String year = line.substring(15,19);        //日期在15~19位
      int airTemperature;
      if (line.charAt(87) == '+') {               //判断正负号
        airTemperature = Integer.parseInt(line.substring(88, 92));
      } else {
        airTemperature = Integer.parseInt(line.substring(87, 92));
      }
      String quality = line.substring(92, 93);    //质量代码
      if (airTemperature != MISSING && quality.matches("[01459]")) {
        context.write(new Text(year), new IntWritable(airTemperature));
      }
    }
  }
  public static class TempReducer
       extends Reducer<Text,IntWritable,Text,IntWritable> {
    private IntWritable result = new IntWritable();
    public void reduce(Text key, Iterable<IntWritable> values,
                  Context context
                  ) throws IOException, InterruptedException {
      int maxValue = Integer.MIN_VALUE;
      for (IntWritable value : values) {
        maxValue = Math.max(maxValue, value.get());
      }
      context.write(key, new IntWritable(maxValue));
    }
  }
```

```java
public static void main(String[] args) throws Exception {
  Configuration conf = new Configuration();
  String[] otherArgs = new GenericOptionsParser(conf, args).getRemainingArgs();
  if (otherArgs.length < 2) {
    System.err.println("Usage: MaxTemp <in> [<in>...] <out>");
    System.exit(2);
  }
  Job job = Job.getInstance(conf, "Max Temperature");
  job.setJarByClass(MaxTemp.class);
  job.setMapperClass(TempMapper.class);
  job.setCombinerClass(TempReducer.class);
  job.setReducerClass(TempReducer.class);
  job.setOutputKeyClass(Text.class);
  job.setOutputValueClass(IntWritable.class);
  for (int i = 0; i < otherArgs.length - 1; ++i) {
    FileInputFormat.addInputPath(job, new Path(otherArgs[i]));
  }
  FileOutputFormat.setOutputPath(job,
    new Path(otherArgs[otherArgs.length - 1]));
  System.exit(job.waitForCompletion(true) ? 0 : 1);
  }
}
```

③ 上传1901和1902两个气象数据文件到HDFS的/input中，操作如下：

```
hadoop@sw-desktop:~$ hdfs dfs -rm /input/*
hadoop@sw-desktop:~$ hdfs dfs -put 19* /input
```

④ 执行菜单栏中的"Run"→"Run Configurations…"命令，打开"Run Configurations"窗口，右击"Java Application"，在列表框中选择"New"生成"MaxTemp"，单击"Arguments"选项卡，在"Program arguments"文本框中输入两行参数："hdfs://master:9000/input"和"hdfs://master:9000/output"，如图4-30所示。

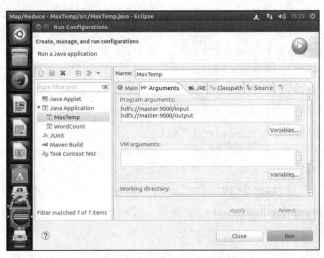

图4-30　MaxTemp 运行参数

⑤ 单击"Run"按钮，运行"MaxTemp"，查看运行结果，如图 4-31 所示。在运行前必须先删除 HDFS 中的/output 目录。

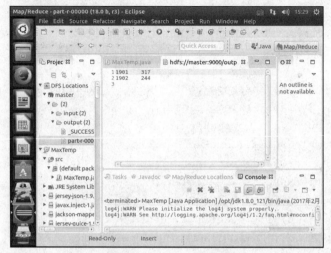

图 4-31 MaxTemp 运行结果

### 三、编写 Hadoop Streaming 程序查找每年的最高气温

准备数据：将含有 1949 年和 1950 年部分气象数据的 sample.txt 文件下载到/home/hadoop 目录下。

（1）Ruby 脚本

① 下载 Ruby 软件，下载地址如下：

```
http://mirrors.aliyun.com/ubuntu/pool/main/r/
http://mirrors.aliyun.com/ubuntu/pool/main/g/gmp/
```

② 安装 Ruby 软件，操作如下：

```
hadoop@sw-desktop:~$ sudo dpkg -i rubygems-integration_1.11_all.deb
hadoop@sw-desktop:~$ sudo dpkg -i libreadline6_6.3-8ubuntu2_amd64.deb
hadoop@sw-desktop:~$ sudo dpkg -i ruby-power-assert_0.3.0-1_all.deb
hadoop@sw-desktop:~$ sudo dpkg -i ruby-test-unit_3.1.7-2_all.deb
hadoop@sw-desktop:~$ sudo dpkg -i libruby2.2_2.2.3-1_amd64.deb
hadoop@sw-desktop:~$ sudo dpkg -i ruby2.2_2.2.3-1_amd64.deb
hadoop@sw-desktop:~$ sudo dpkg -i  libruby2.2-dbg_2.2.3-1_amd64.deb
hadoop@sw-desktop:~$ sudo dpkg -i libgmpxx4ldbl_6.1.1+dfsg-1_amd64.deb
hadoop@sw-desktop:~$ sudo dpkg -i libgmp-dev_6.1.1+dfsg-1_amd64.deb
hadoop@sw-desktop:~$ sudo dpkg -i ruby2.2-dev_2.2.3-1_amd64.deb
hadoop@sw-desktop:~$ sudo cp -a /usr/bin/ruby2.2 /usr/bin/ruby
hadoop@sw-desktop:~$ ruby -v
ruby 2.2.3p173 (2015-08-18) [x86_64-linux-gnu]
```

③ 用 Ruby 编写查找最高气温的 map 函数，操作如下：

```
hadoop@sw-desktop:~$ vi max_temp_map.rb
#!/usr/bin/env ruby
```

```
STDIN.each_line do |line|
  val = line
  year, temp, q = val[15,4], val[87,5], val[92,1]
  puts "#{year}\t#{temp}" if (temp != "+9999" && q =~ /[01459]/)
end
```

④ 使用 Linux 管道测试 map 函数，操作如下：

```
hadoop@sw-desktop:~$ chmod +x max_temp_map.rb
hadoop@sw-desktop:~$ cat sample.txt | ./max_temp_map.rb
1950    +0000
1950    +0022
1950    -0011
1949    +0111
1949    +0078
```

⑤ 用 Ruby 编写查找最高气温的 reduce 函数，操作如下：

```
hadoop@sw-desktop:~$ vi max_temp_reduce.rb
#!/usr/bin/env ruby
last_key, max_val = nil, -1000000
STDIN.each_line do |line|
  key, val = line.split("\t")
  if last_key && last_key != key
    puts "#{last_key}\t#{max_val}"
    last_key, max_val = key, val.to_i
  else
    last_key, max_val = key, [max_val, val.to_i].max
  end
end
puts "#{last_key}\t#{max_val}" if last_key
```

⑥ 使用 Linux 管道模拟整个 MapReduce 过程，操作如下：

```
hadoop@sw-desktop:~$ chmod +x max_temp_reduce.rb
hadoop@sw-desktop:~$ cat sample.txt|./max_temp_map.rb|sort|./max_temp_reduce.rb
1949    111
1950    22
```

⑦ 使用 Hadoop-Streaming 运行，操作如下：

```
hadoop@sw-desktop:~$ hdfs dfs -mkdir /ncdc
hadoop@sw-desktop:~$ hdfs dfs -put sample.txt /ncdc
hadoop@sw-desktop:~$ hadoop jar \
/opt/hadoop-2.7.3/share/hadoop/tools/lib/hadoop-streaming-2.7.3.jar \
-input /ncdc \
-output /ncdc_out \
-mapper max_temp_map.rb \
-reducer max_temp_reduce.rb \
```

```
-file max_temp_map.rb \
-file max_temp_reduce.rb
hadoop@sw-desktop:~$ hdfs dfs -cat /ncdc_out/part-00000
1949    111
1950    22
```

mapper 和 reducer 也可以使用系统软件，操作如下：

```
hadoop@sw-desktop:~$ hadoop jar \
/opt/hadoop-2.7.3/share/hadoop/tools/lib/hadoop-streaming-2.7.3.jar \
-input /ncdc \
-output /ncdc_out \
-mapper /bin/cat \
-reducer /usr/bin/wc
hadoop@sw-desktop:~$ hdfs dfs -cat /ncdc_out/part-00000
    5     5     535
```

（2）Python 脚本

① 用 Python 编写查找最高气温的 map 函数，操作如下：

```
hadoop@sw-desktop:~$ vi max_temp_map.py
#!/usr/bin/env python
import re
import sys
for line in sys.stdin:
  val = line.strip()
  (year, temp, q) = (val[15:19], val[87:92], val[92:93])
  if (temp != "+9999" and re.match("[01459]", q)):
    print "%s\t%s" % (year, temp)
```

② 使用 Linux 管道测试 map 函数，操作如下：

```
hadoop@sw-desktop:~$ chmod +x max_temp_map.py
hadoop@sw-desktop:~$ cat sample.txt | ./max_temp_map.py
1950    +0000
1950    +0022
1950    -0011
1949    +0111
1949    +0078
```

③ 用 Python 编写查找最高气温的 reduce 函数，操作如下：

```
hadoop@sw-desktop:~$ vi max_temp_reduce.py
#!/usr/bin/env python
import sys
(last_key, max_val) = (None, -sys.maxint)
for line in sys.stdin:
  (key, val) = line.strip().split("\t")
  if last_key and last_key != key:
```

```
    print "%s\t%s" % (last_key, max_val)
    (last_key, max_val) = (key, int(val))
  else:
    (last_key, max_val) = (key, max(max_val, int(val)))
if last_key:
  print "%s\t%s" % (last_key, max_val)
```

④ 使用 Linux 管道模拟整个 MapReduce 过程，操作如下：

```
hadoop@sw-desktop:~$ chmod +x max_temp_reduce.py
hadoop@sw-desktop:~$ cat sample.txt|./max_temp_map.py|sort|./max_temp_reduce.py
1949    111
1950    22
```

⑤ 使用 Hadoop-Streaming 运行，操作如下：

```
hadoop@sw-desktop:~$ hdfs dfs -rm -r -f /ncdc_out
hadoop@sw-desktop:~$ hadoop jar \
/opt/hadoop-2.7.3/share/hadoop/tools/lib/hadoop-streaming-2.7.3.jar \
-input /ncdc \
-output /ncdc_out \
-mapper max_temp_map.py \
-reducer max_temp_reduce.py \
-file max_temp_map.py \
-file max_temp_reduce.py
hadoop@sw-desktop:~$ hdfs dfs -cat /ncdc_out/part-00000
1949    111
1950    22
```

注意：在 Hadoop 集群环境下运行，各集群节点要安装 Python 软件，操作如下：

```
hadoop@master:~$ sudo apt-get install python
```

（3）Hadoop Pipes

HadoopPipes 是 Hadoop MapReduce 的 C++接口名称。不同于使用标准输入和输出来实现 map 代码和 reduce 代码之间的 Streaming，HadoopPipes 使用套接字作为 tasktracker 与 C++版本 map 函数或 reduce 函数的进程之间的通道，而未使用 JNI。

① 编写 C++程序，操作如下：

```
hadoop@sw-desktop:~$ vi max_temperature.cpp
#include <algorithm>
#include <limits.h>
#include <stdint.h>
#include <string>
#include "Pipes.hh"
#include "TemplateFactory.hh"
#include "StringUtils.hh"
class MaxTemperatureMapper : public HadoopPipes::Mapper {
public:
```

```cpp
    MaxTemperatureMapper(HadoopPipes::TaskContext& context) {
    }
    void map(HadoopPipes::MapContext& context) {
      std::string line = context.getInputValue();
      std::string year = line.substr(15, 4);
      std::string airTemperature = line.substr(87, 5);
      std::string q = line.substr(92, 1);
      if (airTemperature != "+9999" &&
          (q == "0" || q == "1" || q == "4" || q == "5" || q == "9")) {
        context.emit(year, airTemperature);
      }
    }
};
class MapTemperatureReducer : public HadoopPipes::Reducer {
public:
  MapTemperatureReducer(HadoopPipes::TaskContext& context) {
  }
  void reduce(HadoopPipes::ReduceContext& context) {
    int maxValue = INT_MIN;
    while (context.nextValue()) {
      maxValue = std::max(maxValue, HadoopUtils::toInt(context.getInputValue()));
    }
    context.emit(context.getInputKey(), HadoopUtils::toString(maxValue));
  }
};
int main(int argc, char *argv[]) {
  return HadoopPipes::runTask(HadoopPipes::TemplateFactory<MaxTemperatureMapper,
                    MapTemperatureReducer>());
}
```

② 编译代码，操作如下：

```
hadoop@sw-desktop:~$ vi makefile
HADOOP_INSTALL=/opt/hadoop-2.7.3
CC=g++
CPPFLAGS = -m64 -c -fPIC -D_GLIBCXX_USE_CXX11_ABI=0
RM = rm
SRCS = max_temperature.cpp
PROGRAM = max_temperature
INC_PATH = -I$(HADOOP_INSTALL)/include
LIB_PATH = -L$(HADOOP_INSTALL)/lib/native
LIBS = -lhadooppipes -lhadooputils -lpthread
```

```
$(PROGRAM):$(SRCS)
        $(CC) $(CPPFLAGS) $(INC_PATH) $< -Wall $(LIB_PATH) $(LIBS)  -g -O2 -o $@

.PHONY:clean
clean:
        $(RM) $(PROGRAM)
hadoop@sw-desktop:~$ make
```

③ 通过 Hadoop 运行，操作如下：

```
hadoop@sw-desktop:~$ hdfs dfs -rm -r -f /ncdc_out
hadoop@sw-desktop:~$ hdfs dfs -put max_temperature /
hadoop@sw-desktop:~$ hadoop pipes \
-D hadoop.pipes.java.recordreader=true \
-D hadoop.pipes.java.recordwriter=true \
-input /ncdc \
-output /ncdc_out \
-program /max_temperature
hadoop@sw-desktop:~$ hdfs dfs -cat /ncdc_out/part-00000
1949    111
1950    22
```

## 【同步训练】

### 一、简答题

（1）简述 MapReduce 的运行步骤。

（2）ResourceManager（RM）和 NodeManager（NM）的功能是什么？

（3）简述单词统计的 Map 过程。

（4）MapReduce 的框架核心是什么？

### 二、操作题

（1）搭建 Eclipse 开发环境。

（2）使用 Python 脚本编写单词计数 Map 程序。

（3）使用 Ruby 脚本编写每年的平均气温。

（4）使用 Eclipse 编写 MapReduce 程序，计算每年的平均气温。

# PART 5 项目五
# HBase 数据库部署与应用

## 【项目介绍】

HBase 是 Hadoop 生态圈非常重要的成员，部署 HBase 是本项目的重点；HBase Shell 命令是 HBase 交互工具，使用 HBase Shell 命令创建、修改、删除 HBase 表，完成对数据的添加、查询和删除等；编写程序实现对 HBase 的访问，编写 MapReduce 程序实现对 HBase 的操作。

本项目分以下 4 个任务：
- 任务 1　HBase 部署
- 任务 2　HBase Shell
- 任务 3　HBase 编程
- 任务 4　MapReduce 与 HBase 集成

## 【学习目标】

### 一、知识目标

- 了解 HBase 分布式数据库的特点。
- 熟悉 HBase 逻辑视图、概念视图和物理视图的关系。
- 掌握 HBase 的基本概念。
- 掌握 HBase 的系统架构。
- 了解 Zookeeper 的作用。
- 掌握 Zookeeper 的设置。
- 掌握 HBase Shell 命令。
- 掌握 HBase API 编程。
- 掌握 MapReduce 与 HBase 集成编程。

### 二、能力目标

- 能够部署 HBase 分布式数据库系统。
- 能够使用 HBase Shell 命令对表的创建、修改和删除。
- 能够使用 HBase Shell 命令对数据的插入和删除。
- 能够使用 FILTER 过滤器对数据进行过滤查询。

- 能够使用 HBase API 完成对 HBase 的编程。
- 能够编写 MapReduce 程序实现对 HBase 的操作。

# 任务 1　HBase 部署

## 【任务概述】

HBase 是运行在 Hadoop 上的分布式数据库，需要在 Hadoop 集群环境上部署，需要配置 Zookeeper，最后还需要对部署的 HBase 环境进行检验测试。

## 【支撑知识】

### 一、HBase 简介

HBase（Hadoop Database）是一个高可靠性、高性能、面向列、可伸缩的分布式存储系统，利用 HBase 技术可在普通计算机上搭建起大规模的结构化存储集群。HBase 不同于一般的关系数据库，它是一个适合于非结构化数据存储的数据库，是基于列的而不是基于行的模式。

HBase 是 Google Bigtable 的开源实现，类似于 Google Bigtable 利用 GFS 作为其文件存储系统，HBase 利用 Hadoop HDFS 作为其文件存储系统；Google 运行 MapReduce 来处理 Bigtable 中的海量数据，HBase 同样利用 Hadoop MapReduce 来处理 HBase 中的海量数据；Google Bigtable 利用 Chubby 作为协同服务，HBase 利用 Zookeeper 作为对应。

Hadoop EcoSystem 中的各层系统如图 5-1 所示。其中，HBase 位于结构化存储层，Hadoop HDFS 为 HBase 提供了高可靠性的底层存储支持，Hadoop MapReduce 为 HBase 提供了高性能的计算能力，Zookeeper 为 HBase 提供了稳定服务和 failover 机制。

此外，Pig 和 Hive 还为 HBase 提供了高层语言支持，使得在 HBase 上进行数据统计处理变得非常简单。Sqoop 则为 HBase 提供了方便的 RDBMS 数据导入功能，使得传统数据库数据向 HBase 中迁移变得非常方便。

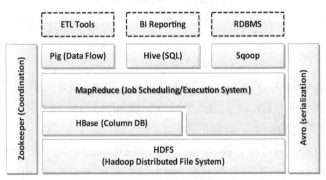

图 5-1　Hadoop 生态圈

HBase 具有以下特点。

① 面向列：面向列（族）的存储和权限控制，列（族）独立检索。

② 大表：表可以非常大，一个表可以有上亿行、上百列。

③ 稀疏：列族中的列可以动态增加，一般情况下，列比较多，一行数据只有少数的列有值，而对于空值，HBase 并不存储。因此，表可以设计得非常稀疏。

④ 非结构化：HBase 不是关系型数据库，适合存储非结构化的数据。

⑤ 数据多版本：每个单元中的数据可以有多个版本，默认情况下版本号自动分配，是单元格插入时的时间戳。

⑥ 数据类型单一：HBase 中的数据都是字符串，没有其他类型。

⑦ 数据操作简单：HBase 只有很简单的插入、查询、删除、清空等操作，表和表之间是分离的，没有复杂的表和表之间的关系。

⑧ 线性扩展：当存储空间不足时，可以通过简单地增加节点的方式进行扩展。

## 二、HBase 基本概念

（1）HBase 逻辑视图

HBase 不同于一般关系型数据库，在一般的关系数据库中，采用二维表进行数据存储，只有行和列，列属性事先定义好，而行动态扩展。HBase 一般有行键、时间戳、列族、列、行组成，列族在使用前先定义好，但是列族中的列、时间戳和行在使用时动态扩展。HBase 逻辑视图如表 5-1 所示。

表 5-1 HBase 逻辑视图

| RowKey | timestamp | name | class | course | |
|---|---|---|---|---|---|
| | | | class | Java | Python |
| 610213 | TS0 | Tom | 163Cloud | 85 | |
| | TS1 | | | | 79 |
| 610215 | TS2 | John | 173BigData | | |
| | TS3 | | | 70 | |
| | TS4 | | | 80 | 86 |

① 行键。行键（RowKey）是字节数组，任何字符串（最大长度是 64KB）都可以作为行键；表中的行根据行键进行排序，数据按照 Rowkey 的字节排序存储；所有对表的访问都要通过行键。访问 HBase Table 中的行只有 3 种方式：通过单个 RowKey 访问、通过 Rowkey 的 range、全表扫描。

② 列族。列族（Column Family，CF）必须在表定义时给出，每个 CF 可以有一个或多个列成员（Column Qualifier），列成员不需要在表定义时给出，新的列成员可以随后按需动态加入；数据按 CF 分开存储，HBase 所谓的列式存储就是根据 CF 分开存储（每个 CF 对应一个 Store），这种设计非常适合于数据分析的情形。

③ 时间戳（TimeStamp）。HBase 中通过 Row 和 Columns 确定的一个存储单元称为 Cell。每个 Cell 都保存着同一份数据的多个版本。版本通过时间戳来索引。时间戳的类型是 64 位整型。时间戳可以由 HBase 自动赋值，此时时间戳是精确到毫秒的当前系统时间。时间戳也可以由客户显式赋值。如果应用程序要避免数据版本冲突，就必须自己生成具有唯一性的时间戳。每个 Cell 中，不同版本的数据按照时间倒序排序，即最新的数据排在最前面。

为了避免数据存在过多版本造成的管理（包括存储和索引）负担，HBase 提供了两种数据版本回收方式。一是保存数据的最后 n 个版本，二是保存最近一段时间内的版本（如最近 7 天）。用户可以针对每个列族进行设置。

④ 单元格。单元格（Cell）由行键、列族、限定符、时间戳唯一决定；Cell 中的数据是没有类型的，全部以字节码形式存储。

⑤ 区域。HBase 自动把表水平（按 Row）划分成多个区域（Region），每个 Region 会保存一个表中某段连续的数据；每个表一开始只有一个 Region，随着数据不断插入表，Region 不断增大，当增大到一个阈值的时候，Region 就会等分成两个新的 Region；当表中的行不断增多，就会有越来越多的 Region。这样一张完整的表被保存在多个 Region 上；HRegion 是 HBase 中分布式存储和负载均衡的最小单元。最小单元表示不同的 HRegion 可以分布在不同的 HRegionServer 上。但一个 HRegion 不会拆分到多个 Server 上。

（2）概念视图

我们可以把 HBase 想象成一个大的映射关系，通过行键、行键和时间戳、行键和列（列族：列）就可以定位特定数据，HBase 是稀疏存储数据的，因此，某些列可以是空白的，如表 5-2 所示。

表 5-2 概念视图

| RowKey | timestamp | 列族:C1 | | 列族:C2 | |
|---|---|---|---|---|---|
| | | 列 | 值 | 列 | 值 |
| R1 | T0 | C1：1 | value1-1/1 | | |
| | T1 | C1：2 | value1-1/2 | | |
| | T2 | C1：3 | value1-1/3 | | |
| | T3 | | | C2：1 | value1-2/1 |
| | T4 | | | C2：2 | value1-2/2 |
| R2 | T5 | C1：1 | value2-1/1 | | |
| | T6 | | | C2：1 | value2-1/1 |

从表 5-2 中可以看出，该表有 R1 和 R2 两行数据，有 C2 和 C2 两个列族；在 R1 中，列族 C1 有 3 条数据，列族 C2 有 2 条数据；在 R2 中，列族 C1 有 1 条数据，列族 C2 有 1 条数据。列族中每一条数据都有对应的时间戳，时间戳越大表示数据越新。

（3）物理视图

虽然从概念视图来看，每个表格是由很多行组成的，但是在物理存储上，它是按照列来保存的，行键 R1 的物理视图如表 5-3 和表 5-4 所示。

表 5-3 物理视图 1

| RowKey | timestamp | 列族:C1 | |
|---|---|---|---|
| | | 列 | 值 |
| R1 | T0 | C1：1 | value1-1/1 |
| | T1 | C1：2 | value1-1/2 |
| | T2 | C1：3 | value1-1/3 |

需要注意的是，在概念视图上有些列是空白的，这样的列实际上并不会被存储，当请求这些空白的单元格时，会返回 NULL 值。如果在查询的时候不提供时间戳，那么会返回距离现在最近的那一个版本的数据，因为在存储的时候，数据会按照时间戳来排序。

表 5-4  物理视图 2

| RowKey | timestamp | 列族:C2 | |
| --- | --- | --- | --- |
| | | 列 | 值 |
| R1 | T3 | C2：1 | value1-2/1 |
| | T4 | C2：2 | value1-2/2 |

（4）HBase 物理存储

① Table 中所有行都按照 RowKey 的字典序排列。

② Table 在行的方向上分割为多个 Region。

③ Region 是按大小分割的，每个表开始只有一个 Region，随着数据增多，Region 不断增大，当增大到一个阈值时，Region 就会等分为两个新的 Region，之后会有越来越多的 Region。

④ Region 是 HBase 中分布式存储和负载均衡的最小单元，不同 Region 分布到不同 RegionServer 上，如图 5-2 所示。

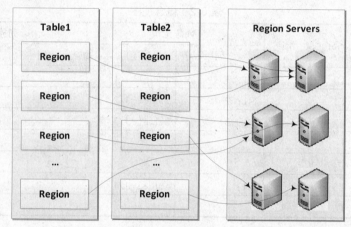

图 5-2  Region 分布

⑤ Region 虽然是分布式存储的最小单元，但并不是存储的最小单元。Region 由一个或者多个 Store 组成，每个 Store 保存一个 Columns Family；每个 Strore 又由一个 MemStore 和 0 至多个 StoreFile 组成，StoreFile 包含 HFile；MemStore 存储在内存中，StoreFile 存储在 HDFS 上，如图 5-3 所示。

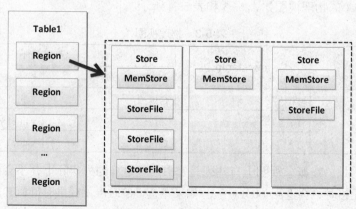

图 5-3  Region 存储结构

### 三、HBase 系统架构

HBase Client 使用 HBase 的远程过程调用协议（Remote Procedure Call Protocol，RPC）机制与 HMaster 和 HRegionServer 进行通信，对于管理类操作，Client 与 HMaster 进行 RPC；对于数据读/写类操作，Client 与 HRegionServer 进行 RPC。HBase 系统架构如图 5-4 所示。

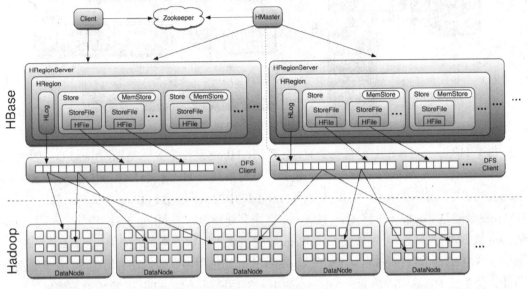

图 5-4　HBase 系统架构

（1）Zookeeper Quorum

Zookeeper Quorum 中除了存储了-ROOT-表的地址和 HMaster 的地址，HRegionServer 也会把自己以 Ephemeral 方式注册到 Zookeeper 中，使得 HMaster 可以随时感知到各个 HRegionServer 的健康状态。此外，Zookeeper 也避免了 HMaster 的单点问题。

HMaster 没有单点问题，HBase 中可以启动多个 HMaster，通过 Zookeeper 的 Master Election 机制保证总有一个 Master 运行，HMaster 在功能上主要负责 Table 和 Region 的管理工作。

① 管理用户对表的增、删、改、查操作。

② 管理 HRegionServer 的负载均衡，调整 Region 分布。

③ 在 Region Split 后，负责新 Region 的分配。

④ 在 HRegionServer 停机后，负责失效 HRegionServer 上的 Regions 迁移。

（2）HRegionServer

HRegionServer 主要负责响应用户 I/O 请求，向 HDFS 文件系统中读/写数据，是 HBase 中最核心的模块。HRegionServer 内部管理了一系列 HRegion 对象，每个 HRegion 对应了 Table 中的一个 Region，HRegion 中由多个 HStore 组成。每个 HStore 对应了 Table 中的一个 Column Family 的存储，可以看出，每个 Column Family 其实就是一个集中的存储单元，因此，最好将具备共同 I/O 特性的 Column 放在一个 Column Family 中，这样最高效。HRegionServer 的数据流如图 5-5 所示。

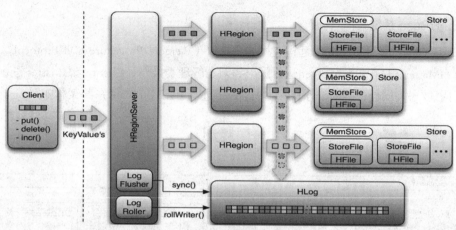

图 5-5　HRegion Server 数据流

每个 HRegionServer 中都有一个 HLog 对象，HLog 是一个实现 Write Ahead Log 的类，在每次用户操作写入 MemStore 的同时，也会写一份数据到 HLog 文件中，HLog 文件定期会滚动出新的，并删除旧的文件（已持久化到 StoreFile 中的数据）。当 HRegionServer 意外终止后，HMaster 会通过 Zookeeper 感知到，HMaster 首先会处理遗留的 HLog 文件，将其中不同 Region 的 Log 数据进行拆分，分别放到相应 Region 的目录下，然后将失效的 Region 重新分配，领取到这些 Region 的 HRegionServer 在 Load Region 的过程中，会发现有历史 HLog 需要处理，因此，会重放 HLog 中的数据到 MemStore 中，然后刷新到 StoreFiles，完成数据恢复。

（3）HStore

HStore 存储是 HBase 存储的核心，由两部分组成，一部分是 MemStore，另一部分是 StoreFiles。MemStore 是 Sorted Memory Buffer，用户写入的数据首先会放入 MemStore，当 MemStore 满了以后会刷新成一个 StoreFile（底层实现是 HFile）。当 StoreFile 文件数量增长到一定阈值时，会触发 Compact 合并操作，将多个 StoreFiles 合并成一个 StoreFile，合并过程中会进行版本合并和数据删除，因此，可以看出 HBase 其实只有增加数据，所有的更新和删除操作都是在后续的 Compact 过程中进行的，这使得用户的写操作只要进入内存中就可以立即返回，保证了 HBase I/O 的高性能。当 StoreFiles Compact 后，会逐步形成越来越大的 StoreFile，当单个 StoreFile 大小超过一定阈值后，会触发 Split 操作，同时把当前 Region Split 成两个 Region，父 Region 会下线，新 Split 出的两个子 Region 会被 HMaster 分配到相应的 HRegionServer 上，使得原先一个 Region 的压力得以分流到两个 Region 上。

HStore 在系统正常工作的前提下是没有问题的，但是在分布式系统环境中，无法避免系统出错或者宕机，因此，一旦 HRegionServer 意外退出，MemStore 中的内存数据将会丢失，这就需要引入 HLog。

（4）Catalog Table

HBase 内部保留名为-ROOT-和.META.的特殊目录表（Catalog Tble），如图 5-6 所示。它们维护着当前集群上所有区域的列表、状态和位置。-ROOT-表包含.META.表的区域列表。.META.包含所有用户空间区域（User-Space Region）的列表。表中的项使用区域名作为键。区域名由所属的表名、区域的起始行、区域的创建时间，以及对其整理进行的 MD5 哈希值（即对表名、起始行、创建时间戳进行哈希后的结果）组成。

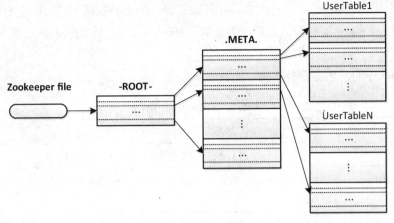

图 5-6 特殊目录表

新连接到 Zookeeper 集群上的客户端首先查找 -ROOT- 的位置,然后客户端通过 -ROOT- 获得请求行所在范围所属.META.区域的位置,客户端接着查找.META.区域获取用户空间区域所在节点及其位置,接着,客户端就可以直接和管理那个区域的 RegionServer 进行交互。

### 四、Zookeeper

(1) Zookeeper 简介

Zookeeper 是一个分布式的、开放源码的分布式应用程序协调服务,是 Google 的 Chubby 一个开源的实现,是 Hadoop 和 HBase 的重要组件。它是一个为分布式应用提供一致性服务的软件,提供的功能包括配置维护、域名服务、分布式同步、组服务等。

Zookeeper 的目标就是封装好复杂易出错的关键服务,将简单易用的接口和性能高效、功能稳定的系统提供给用户。Zookeeper 包含一个简单的原语集,分布式应用程序可以基于它实现同步服务、配置维护和命名服务等。

(2) Zookeeper 的工作原理

Zookeeper 的核心是原子广播,这个机制保证了各个 Server 之间的同步。实现这个机制的协议称为 Zab 协议。Zab 协议有两种模式,它们分别是恢复模式(选举)和广播模式(同步)。当服务启动或者在领导者崩溃后,Zab 就进入了恢复模式,当领导者被选举出来,且大多数 Server 完成了和 Leader 的状态同步以后,恢复模式就结束了。状态同步保证了 Leader 和 Server 具有相同的系统状态。

为了保证事务的顺序一致性,Zookeeper 采用了递增的事务 ID 号(zxid)来标识事务。所有的提议(Proposal)都在被提出的时候加上了 zxid。实现中 zxid 是一个 64 位的数字,它的高 32 位是 epoch 用来标识 Leader 关系是否改变,每次一个 Leader 被选出来,它都会有一个新的 epoch,标识当前属于那个 Leader 的统治时期。低 32 位用于递增计数。

每个 Server 在工作过程中有 3 种状态。

① LOOKING:当前 Server 不知道 Leader 是谁,正在搜寻。

② LEADING:当前 Server 即为选举出来的 Leader。

③ FOLLOWING:Leader 已经选举出来,当前 Server 与之同。

当 Leader 崩溃或者 Leader 失去大多数的 Follower 时,zk 进入恢复模式,恢复模式需要重新选举出一个新的 Leader，让所有的 Server 都恢复到一个正确的状态。zk 的选举算法有两种：一种是基于 basic paxos 实现的，另外一种是基于 fast paxos 算法实现的。系统默认的选举算法为 fast paxos。

Zookeeper 中的角色主要有 3 类，如表 5-5 所示。

表 5-5　Zookeeper 角色

| 角色 | | 功能 |
| --- | --- | --- |
| 领导者（Leader） | | 领导者负责进行投票的发起和决议，更新系统状态 |
| 学习者（Learner） | 跟随者（Follower） | Follower 用于接收客户请求并向客户端返回结果，在选择 Leader 中参与投票 |
| | 观察者（Observer） | Observer 可以接收端连接，将写请求转发给 Leader 节点，但 Observer 不参与投票过程，只同步 Leader 的状态。Observer 的目的是扩展系统，提高读取速度 |
| 客户端（Client） | | 请求发起方 |

（3）Zookeeper 的作用

Zookeeper 在 HBase 中的作用主要如下：

① 通过选举，保证任何时候，集群中只有一个 Master，Master 与 RegionServers 启动时会向 Zookeeper 注册。

② 存储所有 Region 的寻址入口。

③ 实时监控 RegionServer 的上线和下线信息，并实时通知给 Master。

④ 存储 HBase 的 schema 和 table 元数据。

⑤ 默认情况下，HBase 管理 Zookeeper 实例，如启动或者停止 Zookeeper。

⑥ Zookeeper 的引入使得 Master 不再是单点故障。

## 【任务实施】

### 一、Master 节点安装软件

① 下载 HBase 和 Zookeeper 软件包到 /home/hadoop 目录下，下载地址如下：

```
http://mirrors.aliyun.com/apache/hbase/1.2.4/hbase-1.2.4-bin.tar.gz
http://mirrors.aliyun.com/apache/zookeeper/zookeeper-3.4.9/zookeeper-3.4.9.tar.gz
```

② 以用户 hadoop 登录 Master 节点，安装 HBase 和 Zookeeper 软件，操作如下：

```
hadoop@master:~$ cd /opt
hadoop@master:/opt$ sudo tar xvzf /home/hadoop/hbase-1.2.4-bin.tar.gz
hadoop@master:/opt$ sudo tar xvzf /home/hadoop/zookeeper-3.4.9.tar.gz
hadoop@master:/opt$ sudo chown -R hadoop:hadoop hbase-1.2.4 zookeeper-3.4.9
```

### 二、Master 节点设置 HBase 参数

① 修改 hbase-env.sh 文件，操作如下：

```
hadoop@master:/opt$ cd /opt/hbase-1.2.4/conf
hadoop@master:/opt/hbase-1.2.4/conf$ vi hbase-env.sh
```

把如下内容：
```
# export JAVA_HOME=/usr/java/jdk1.6.0/
# export HBASE_MANAGES_ZK=true
```
改为：
```
export JAVA_HOME=/opt/jdk1.8.0_121/
export HBASE_MANAGES_ZK=false
```
其中，HBASE_MANAGES_ZK=false 时使用独立的 Zookeeper，为 true 时使用默认自带的。

② 修改 hbase-site.xml 文件，操作如下：
```
hadoop@master:/opt/hbase-1.2.4/conf$ vi hbase-site.xml
```
在<configuration>    </configuration>之间添加内容如下：
```
<property>
  <name>hbase.rootdir</name>
  <value>hdfs://master:9000/hbase</value>
</property>
<property>
  <name>hbase.cluster.distributed</name>
  <value>true</value>
</property>
<property>
  <name>hbase.zookeeper.quorum</name>
  <value>master,slave1,slave2</value>
</property>
<property>
  <name>hbase.zookeeper.property.datadir</name>
  <value>/opt/zookeeper-3.4.9/data/</value>
</property>
<property>
  <name>hbase.regionserver.handler.count</name>
  <value>20</value>
</property>
<property>
  <name>hbase.regionserver.maxlogs</name>
  <value>64</value>
</property>
<property>
  <name>hbase.hregion.max.filesize</name>
  <value>10485760</value>
</property>
```

③ 修改 regionservers，操作如下：
```
hadoop@master:/opt/hbase-1.2.4/conf$ vi regionservers
```

内容如下：
```
master
slave1
slave2
```

### 三、Master 节点设置 Zookeeper 参数

① 修改 zoo.cfg，操作如下：
```
hadoop@master:/opt/zookeeper-3.4.9/conf$ cd /opt/zookeeper-3.4.9/conf
hadoop@master:/opt/zookeeper-3.4.9/conf$ vi zoo.cfg
```

内容如下：
```
tickTime=2000
initLimit=10
syncLimit=5
dataDir=/opt/zookeeper-3.4.9/data
clientPort=2181
server.0=master:2888:3888
server.1=slave1:2888:3888
server.2=slave2:2888:3888
```

其中，server.id=host:port1:port2 表示：id 是一个数字，为 Zookeeper 节点的编号，保存在 dataDir 目录下的 myid 文件中，host 是每个 Zookeeper 节点的主机名，port1 是用于连接 Leader 的端口（2888：集群内通信端口），port2 是用于 Leader 选举的端口（3888：集群外通信端口）。

② 在 dataDir 指定的目录下创建 myid 文件，并添加相应内容，操作如下：
```
hadoop@master:~$ mkdir /opt/zookeeper-3.4.9/data
hadoop@master:~$ echo 0 > /opt/zookeeper-3.4.9/data/myid
```

③ 复制 Zookeeper 的配置文件 zoo.cfg 到 HBase，操作如下：
```
hadoop@master:~$ cp /opt/zookeeper-3.4.9/conf/zoo.cfg /opt/hbase-1.2.4/conf/
```

### 四、Slave 节点和 sw-desktop 客户端安装软件

① 以用户 hadoop 登录 slave1 节点安装软件，操作如下：
```
hadoop@slave1:~$ sudo scp -r hadoop@master:/opt/hbase-1.2.4 /opt
hadoop@slave1:~$ sudo scp -r hadoop@master:/opt/zookeeper-3.4.9 /opt
hadoop@slave1:~$ sudo chown -R hadoop:hadoop /opt/hbase-1.2.4 /opt/zookeeper-3.4.9
hadoop@slave1:~$ echo 1 > /opt/zookeeper-3.4.9/data/myid
```

② 以用户 hadoop 登录 slave2 节点安装软件，操作如下：
```
hadoop@slave2:~$ sudo scp -r hadoop@master:/opt/hbase-1.2.4 /opt
hadoop@slave2:~$ sudo scp -r hadoop@master:/opt/zookeeper-3.4.9 /opt
hadoop@slave2:~$ sudo chown -R hadoop:hadoop /opt/hbase-1.2.4 /opt/zookeeper-3.4.9
hadoop@slave2:~$ echo 2 > /opt/zookeeper-3.4.9/data/myid
```

③ 以用户 hadoop 登录 sw-desktop 客户机安装软件，操作如下：
```
hadoop@sw-desktop:~$ sudo scp -r hadoop@master:/opt/hbase-1.2.4 /opt
hadoop@sw-desktop:~$ sudo scp -r hadoop@master:/opt/zookeeper-3.4.9 /opt
```

```
hadoop@sw-desktop:~$ sudo chown -R hadoop:hadoop /opt/hbase-1.2.4 /opt/zookeeper-3.4.9
```

### 五、Zookeeper 服务

① 启动 Zookeeper 服务,操作如下。

● Master 节点:

```
hadoop@master:~$ zkServer.sh start
ZooKeeper JMX enabled by default
Using config: /opt/zookeeper-3.4.9/bin/../conf/zoo.cfg
Starting zookeeper ... STARTED
```

● Slave1 节点:

```
hadoop@slave1:~$ zkServer.sh start
ZooKeeper JMX enabled by default
Using config: /opt/zookeeper-3.4.9/bin/../conf/zoo.cfg
Starting zookeeper ... STARTED
```

● Slave2 节点:

```
hadoop@slave2:~$ zkServer.sh start
ZooKeeper JMX enabled by default
Using config: /opt/zookeeper-3.4.9/bin/../conf/zoo.cfg
Starting zookeeper ... STARTED
```

② 验证 Zookeeper 服务,操作如下。

● Master 节点:

```
hadoop@master:~$ zkServer.sh status
ZooKeeper JMX enabled by default
Using config: /opt/zookeeper-3.4.9/bin/../conf/zoo.cfg
Mode: follower
```

● Slave1 节点:

```
hadoop@slave1:~$ zkServer.sh status
ZooKeeper JMX enabled by default
Using config: /opt/zookeeper-3.4.9/bin/../conf/zoo.cfg
Mode: leader
```

● Slave2 节点:

```
hadoop@slave2:~$ zkServer.sh status
ZooKeeper JMX enabled by default
Using config: /opt/zookeeper-3.4.9/bin/../conf/zoo.cfg
Mode: follower
```

### 六、验证 HBase 服务

① Master 节点启动 HBase 服务,操作如下:

```
hadoop@master:~$ start-hbase.sh
```

② 查看进程,操作如下。

- Master 节点：

```
hadoop@master:~$ jps
3633 ResourceManager
3191 NameNode
3448 SecondaryNameNode
4568 HMaster
4761 HRegionServer
4075 QuorumPeerMain
5084 Jps
3932 JobHistoryServer
```

- Slave1 节点：

```
hadoop@slave1:~$ jps
4072 Jps
3880 HRegionServer
3592 QuorumPeerMain
3417 NodeManager
3258 DataNode
```

- Slave2 节点：

```
hadoop@slave2:~$ jps
3641 QuorumPeerMain
3466 NodeManager
3308 DataNode
3917 HRegionServer
4095 Jps
```

③ 打开浏览器，在地址栏中输入"http://master:16010"，查看 HBase Master 状态，如图 5-7 所示。

④ 打开浏览器，在地址栏中输入"http://master:16030"，查看 RegionServer 状态，如图 5-8 所示。

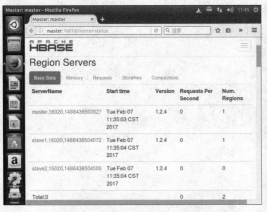

图 5-7　HBase Master 状态

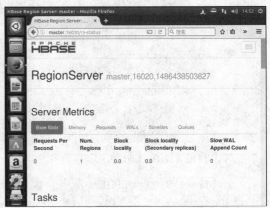

图 5-8　RegionServer 状态

# 任务 2　HBase Shell

## 【任务概述】

HBase Shell 提供操作 HBase 分布数据库的命令，本任务需要创建一个 score 表，并按照表 5-6 所示，使用 HBase Shell 命令完成表数据的添加、读取、过滤扫描和删除等操作。

## 【支撑知识】

### 一、HBase 命令

语法：hbase [<选项>] <命令> [<参数>]

选项：

```
--config DIR        #配置目录，DIR 默认为 ./conf
--hosts HOSTS       #在 regionservers 文件重写列表
```

命令：

```
shell           #运行 HBase shell
hbck            #运行 HBase 'fsck'工具 1
wal             #Write-ahead-log 分析器
hfile           #Store file 分析器
zkcli           #运行 Zookeeper shell
upgrade         #升级 HBase
master          #运行一个 HBase HMaster 节点
regionserver    #运行一个 HBase HRegionServer 节点
zookeeper       #运行一个 Zookeeper server
rest            #运行 HBase Server 复位
thrift          #运行 HBase Thrift server
thrift2         #运行 HBase Thrift2 server
clean           #运行 HBase 清理脚本
classpath       #转储类路径 CLASSPATH
mapredcp        #转储类路径 CLASSPATH（要求 MapReduce）
pe              #运行 PerformanceEvaluation
ltt             #运行 LoadTestTool 工具
version         #打印版本好
CLASSNAME       #运行一个 CLASSNAME 类
```

### 二、HBase Shell 命令

如果有 kerberos 认证，需要事先使用相应的 keytab 进行认证（使用 kinit 命令），认证成功之后再使用 HBase Shell 进入。

whoami 命令可查看当前用户，如：

```
hbase(main)> whoami
hadoop (auth:SIMPLE)
    groups: hadoop, sudo
```

HBase Shell 的基本命令如表 5-6 所示。

表 5-6  HBase Shell 基本命令

| HBase Shell 命令 | 描 述 |
|---|---|
| alter | 修改列族（Column Family）模式 |
| count | 统计表中行的数量 |
| create | 创建表 |
| describe | 显示表相关的详细信息 |
| delete | 删除指定对象的值（可以为表、行、列对应的值，另外也可以指定时间戳的值） |
| deleteall | 删除指定行的所有元素值 |
| disable | 使表无效 |
| drop | 删除表 |
| enable | 使表有效 |
| exists | 测试表是否存在 |
| exit | 退出 HBase Shell |
| get | 获取行或单元（Cell）的值 |
| incr | 增加指定表行或列的值 |
| list | 列出 HBase 中存在的所有表 |
| put | 向指向的表单元添加值 |
| tools | 列出 HBase 所支持的工具 |
| scan | 通过对表的扫描来获取对应的值 |
| status | 返回 HBase 集群的状态信息 |
| shutdown | 关闭 HBase 集群（与 exit 不同） |
| truncate | 重新创建指定表 |
| version | 返回 HBase 版本信息 |

（1）表的管理

① 查看有哪些表。

```
hbase(main)> list
```

② 创建表。

```
create <table>, {NAME => <family>, VERSIONS => <VERSIONS>}
```

例如，创建表 t1，有两个 family name：f1, f2，且版本数均为 2，代码如下：

```
hbase(main)> create 't1',{NAME => 'f1', VERSIONS => 2},{NAME => 'f2', VERSIONS => 2}
```

③ 删除表。

分两步：首先 disable，然后 drop。

例如，删除表 t1，代码如下：

```
hbase(main)> disable 't1'
hbase(main)> drop 't1'
```

④ 查看表的结构。

语法：describe <table>

例如，查看表 t1 的结构，代码如下：

```
hbase(main)> describe 't1'
```

⑤ 修改表结构。

修改表结构必须先 disable。

语法：alter 't1', {NAME => 'f1'}, {NAME => 'f2', METHOD => 'delete'}

例如，修改表 test1 的 cf 的 TTL 为 180 天，代码如下：

```
hbase(main)> disable 'test1'
hbase(main)> alter 'test1',{NAME=>'body',TTL=>'15552000'},{NAME=>'meta',TTL=>'15552000'}
hbase(main)> enable 'test1'
```

（2）权限管理

① 分配权限。

grant <user>,<permissions>,<table>,<column family>,<column qualifier> 参数后面用逗号分隔。

权限用 5 个字母表示："RWXCA"。

READ('R'), WRITE('W'), EXEC('X'), CREATE('C'), ADMIN('A')

例如，给用户 test 分配对表 t1 有读写的权限，代码如下：

```
hbase(main)> grant 'test','RW','t1'
```

② 查看权限。

语法：user_permission <table>

例如，查看表 t1 的权限列表，代码如下：

```
hbase(main)> user_permission 't1'
```

③ 收回权限。

与分配权限类似，语法：revoke <user>,<table>,<column family>,<column qualifier>

例如，收回 test 用户在表 t1 上的权限，代码如下：

```
hbase(main)> revoke 'test','t1'
```

（3）表数据的增、删、改、查

① 添加数据。

语法：put <table>,<rowkey>,<family:column>,<value>,<timestamp>

例如：给 t1 表添加一行记录：rowkey 的值为 rowkey001，family:column 的值为 f1:col1，value 的值为 value01，timestamp 的值为系统默认，语句如下：

```
hbase(main)> put 't1','rowkey001','f1:col1','value01'
```

② 查询数据。

● 查询某行记录。

语法：get <table>,<rowkey>,[<family:column>,...]

例如，查询表 t1，rowkey001 中的 f1 下的 col1 的值，代码如下：

```
hbase(main)> get 't1','rowkey001', 'f1:col1'
```

或者

```
hbase(main)> get 't1','rowkey001', {COLUMN=>'f1:col1'}
```

查询表 t1，rowke002 中的 f1 下的所有列值，代码如下：

```
hbase(main)> get 't1','rowkey001'
```

- 扫描表。

语法：scan <table>, {COLUMNS => [ <family:column>,... ], LIMIT => num}

另外，还可以添加 STARTROW、TIMERANGE 和 FITLER 等高级功能。

例如，扫描表 t1 的前 5 条数据，代码如下：

```
hbase(main)> scan 't1',{LIMIT=>5}
```

- 查询表中的数据行数。

语法：count <table>, {INTERVAL => intervalNum, CACHE => cacheNum}

INTERVAL 设置多少行显示一次及对应的 rowkey，默认 1000；CACHE 每次去取的缓存区大小，默认是 10，调整该参数可提高查询速度。

例如，查询表 t1 中的行数，每 100 条显示一次，缓存区为 500，代码如下：

```
hbase(main)> count 't1', {INTERVAL => 100, CACHE => 500}
```

③ 删除数据。

- 删除行中的某个列值。

语法：delete <table>, <rowkey>, <family:column> , <timestamp>，必须指定列名。

例如，删除表 t1，rowkey001 中的 f1:col1 的数据，代码如下：

```
hbase(main)> delete 't1','rowkey001','f1:col1'
```

注：将删除改行 f1:col1 列所有版本的数据。

- 删除行。

语法：deleteall <table>, <rowkey>, <family:column> , <timestamp>，可以不指定列名，删除整行数据。

例如，删除表 t1，rowk001 的数据，代码如下：

```
hbase(main)> deleteall 't1','rowkey001'
```

- 删除表中的所有数据。

语法： truncate <table>

其具体过程如下：disable table -> drop table -> create table

例如，删除表 t1 的所有数据，代码如下：

```
hbase(main)> truncate 't1'
```

（4）Region 管理

① 移动 region。

语法：move 'encodeRegionName', 'ServerName'

encodeRegionName 指的是 regioName 后面的编码，ServerName 指的是 master-status 的 Region Servers 列表。

示例如下：

```
hbase(main)>move '4343995a58be8e5bbc739af1e91cd72d', 'db-41.xxx.xxx.org,60020,1390274516739'
```

② 开启/关闭 region。

语法：balance_switch true|false

```
hbase(main)> balance_switch
```

③ 手动 split。

语法：split 'regionName', 'splitKey'

④ 手动触发 major compaction。

语法：

```
Compact all regions in a table:
hbase> major_compact 't1'
Compact an entire region:
hbase> major_compact 'r1'
Compact a single column family within a region:
hbase> major_compact 'r1', 'c1'
Compact a single column family within a table:
hbase> major_compact 't1', 'c1'
```

### 三、HBase 过滤器

HBase 过滤器（Filter）提供非常强大的特性来帮助用户提高其处理表中数据的效率。用户不仅可以使用 HBase 中预定义好的过滤器，而且可以实现自定义的过滤器。所有的过滤器都在服务器端生效，称为谓词下推（Predicate Push Down）。这样可以保证被过滤掉的数据不会被传送到客户端。用户可以在客户端代码实现过滤的功能，但会影响系统性能。图 5-9 描述了过滤器如何在客户端进行配置，如何在网络传输中被序列化，如何在服务端执行。

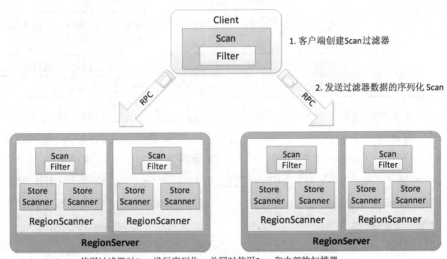

图 5-9 过滤器的工作过程

在过滤器层次结构的最底层是 Filter 接口和 FilterBase 抽象类，大部分实体过滤类一般都直接继承 FilterBase，也有一些间接继承该类，同时把定义好的过滤器实例传递给 Get 和 Scan 实例 setFilter(filter)。在实例化过滤器的时候，用户需要提供一些参数来设定过滤器的用途。其中一组特殊的过滤器继承 CompareFilter，需要用户提供至少两个特定的参数，这两个参数会被基类用于执行它的任务。因为继承 CompareFilter 的过滤器比基类 FilterBase 多了一个 compare() 方法，它需要使用传入参数定义比较操作的过程，CompareFilter 中的比较运算符如表 5-7 所示。

表 5-7 CompareFilter 中的比较运算符

| 操 作 | 描 述 |
|---|---|
| LESS | 匹配小于设定值的值 |
| LESS_OR_EQUAL | 匹配小于或等于设定值的值 |
| EQUAL | 匹配等于设定值的值 |
| NOT_EQUAL | 匹配不等于设定值的值 |
| GREATER_OR_EQUAL | 匹配大于或等于设定值的值 |
| GREATER | 匹配大于设定值的值 |
| NO_OP | 排除一切值 |

CompareFilter 所需的第二类类型是比较器（Comparator），比较器提供了多种方法来比较不同值。比较器都继承自 WritableByteArrayComparable，WritableByteArrayComparable 实现了 Writeble 和 Comparable 接口。HBase 提供的比较器如表 5-8 所示，这些比较器构造时通常只需提供一个阈值，这个值将与表中的实际值进行比较。

表 5-8 CompareFilter 中的比较器

| 比较器 | 描 述 |
|---|---|
| BinaryComparator | 使用 bytes.compareTo() 比较当前值与阈值 |
| BinaryPrefixComparator | 与上面相似，当时从前端开始前缀匹配 |
| NullComparator | 只判断当前值是不是 NULL |
| BitComparator | 通过 BitwiseOp 类提供的按位与（AND）、或（OR）、异或（XOR）操作执行按位比较 |
| RegexStringComparator | 根据一个正则表达式，在实例化这个比较器的时候去匹配表中的数据 |
| SubstringComparator | 把阈值和表中数据当做 String 实例，同时通过 contains() 操作匹配字符串 |

HBase 为筛选数据提供了一组过滤器，通过这个过滤器可以在 HBase 中的数据的多个维度（行、列、数据版本）上进行对数据的筛选操作，也就是说过滤器最终筛选的数据能够细化到具体的一个存储单元格上（由行键、列名、时间戳定位），HBase 常用的过滤器如下。

① 行过滤器（RowFilter）：筛选出匹配的所有行，对于这个过滤器的应用场景，是非常直观的，使用 BinaryComparator 可以筛选出具有某个行键的行，或者通过改变比较运算符来筛选出符合某一条件的多条数据。

② 列族过滤器（FamilyFilter）：筛选出匹配列族的数据。

③ 列名过滤器（QualifierFilter）：筛选匹配列名的数据。

④ 值过滤器（ValueFilter）：按照具体的值来筛选单元格的过滤器，这会把一行中值不能满足的单元格过滤掉。

⑤ 前缀过滤器（PrefixFilter）：筛选出具有特定前缀的行键的数据。

⑥ 列前缀过滤器（ColumnPrefixFilter）：顾名思义，它是按照列名的前缀来筛选单元格的，如果想要对返回的列的前缀加以限制，可以使用这个过滤器。

⑦ 行键过滤器（KeyOnlyFilter）：这个过滤器唯一的功能就是只返回每行的行键，值全部为空，这对于只关注于行键的应用场景来说非常合适，这样忽略掉其值就可以减少传递到客

户端的数据量，能起到一定的优化作用。

⑧ 首次行键过滤器（FirstKeyOnlyFilter）：如果想返回的结果集中只包含第一列数据，那么这个过滤器能够满足你的要求。它在找到每行的第一列之后会停止扫描，从而使扫描的性能得到一定的提升。

⑨ 单列值过滤器（SingleColumnValueFilter）：用一列的值决定这一行的数据是否被过滤。在它的具体对象上，可以调用 setFilterIfMissing(true)或者 setFilterIfMissing(false)，默认为 false，其作用是如果过滤列不存在，则这些列不存在的数据会包含在结果集中；当参数为 true 时，则不会包含。

⑩ 单列排除过滤器（SingleColumnValueExcludeFilter）：这个与 SingleColumnValueFilter 的过滤器唯一的区别就是，作为筛选条件的列不会包含在返回的结果中。

⑪ 包含结束过滤器（InclusiveStopFilter）：扫描的时候，可以设置一个开始行键和一个终止行键，默认情况下，这个行键的返回是前闭后开区间，即包含起始行，但不包含终止行，如果想要同时包含起始行和终止行，那么可以使用此过滤器。

⑫ 列计数过滤器（ColumnCountGetFilter）：用来限制每行最多返回多少列，并在遇到一行的列数超过我们所设置的限制值的时候，结束扫描操作。

⑬ 全匹配过滤器（WhileMatchFilter）：这个过滤器的应用场景也很简单，如果想要在遇到某种条件数据之前的数据时，就可以使用这个过滤器；当遇到不符合设定条件的数据时，整个扫描也就结束了。

## 【任务实施】

### 一、表的管理

① sw-desktop 客户端启动 Hbase Shell，操作如下：

```
hadoop@sw-desktop:~$ hbase shell
hbase(main):001:0>
```

② 创建 score 表，操作如下：

```
hbase(main):001:0> create 'score','name','class','course'
0 row(s) in 1.8570 seconds
=> Hbase::Table - score
```

③ 查看表，操作如下：

```
hbase(main):002:0> list
TABLE
score
1 row(s) in 0.0900 seconds
=> ["score"]
```

④ 查看表结构，操作如下：

```
hbase(main):003:0> describe 'score'
Table score is ENABLED
score
COLUMN FAMILIES DESCRIPTION
```

```
    {NAME => 'class', BLOOMFILTER => 'ROW', VERSIONS => '1', IN_MEMORY => 'false',
KEEP_DEL
    ETED_CELLS => 'FALSE', DATA_BLOCK_ENCODING => 'NONE', TTL => 'FOREVER',
COMPRESSION =>
    'NONE', MIN_VERSIONS => '0', BLOCKCACHE => 'true', BLOCKSIZE => '65536',
REPLICATION_SC
    OPE => '0'}
    {NAME => 'name', BLOOMFILTER => 'ROW', VERSIONS => '1', IN_MEMORY => 'false',
KEEP_DELE
    TED_CELLS => 'FALSE', DATA_BLOCK_ENCODING => 'NONE', TTL => 'FOREVER',
COMPRESSION => '
    NONE', MIN_VERSIONS => '0', BLOCKCACHE => 'true', BLOCKSIZE => '65536',
REPLICATION_SCO
    PE => '0'}
    {NAME => 'course', BLOOMFILTER => 'ROW', VERSIONS => '1', IN_MEMORY => 'false',
KEEP_DEL
    ETED_CELLS => 'FALSE', DATA_BLOCK_ENCODING => 'NONE', TTL => 'FOREVER',
COMPRESSION =>
    'NONE', MIN_VERSIONS => '0', BLOCKCACHE => 'true', BLOCKSIZE => '65536',
REPLICATION_SC
    OPE => '0'}
    3 row(s) in 0.2070 seconds
```

## 二、数据操作

① 添加记录，操作如下：

```
    hbase(main):004:0> put 'score','610213','name:','Tom'
    0 row(s) in 0.4310 seconds
    hbase(main):005:0> put 'score','610213','class:class','163Cloud'
    0 row(s) in 0.0210 seconds
    hbase(main):006:0> put 'score','610213','course:java','85'
    0 row(s) in 0.0340 seconds
    hbase(main):007:0> put 'score','610213','course:python','79'
    0 row(s) in 0.0450 seconds
    hbase(main):008:0> put 'score','610215','name','John'
    0 row(s) in 0.0140 seconds
    hbase(main):009:0> put 'score','610215','class:class','173BigData'
    0 row(s) in 0.0110 seconds
    hbase(main):010:0> put 'score','610215','course:java','70'
    0 row(s) in 0.0140 seconds
    hbase(main):011:0> put 'score','610215','course:java','80'
    0 row(s) in 0.0120 seconds
```

```
hbase(main):012:0> put 'score','610215','course:python','86'
0 row(s) in 0.0070 seconds
```

② 读表记录，操作如下：

```
hbase(main):013:0> get 'score','610215'
COLUMN              CELL
 class:class        timestamp=1486456047670, value=173BigData
 name:              timestamp=1486456006155, value=John
 course:java        timestamp=1486456206791, value=80
 course:python      timestamp=1486456224168, value=86
4 row(s) in 0.0500 seconds

hbase(main):014:0> get 'score','610215','course'
COLUMN              CELL
 course:java        timestamp=1486456206791, value=80
 course:python      timestamp=1486456224168, value=86
2 row(s) in 0.0160 seconds

hbase(main):015:0> get 'score','610215','course:java'
COLUMN              CELL
 course:java        timestamp=1486456206791, value=80
1 row(s) in 0.0320 seconds
```

③ 扫描记录，操作如下：

```
hbase(main):016:0> scan 'score'
ROW                 COLUMN+CELL
 610213             column=class:class, timestamp=1486455866244, value=163Cloud
 610213             column=name:, timestamp=1486455816564, value=Tom
 610213             column=course:java, timestamp=1486455895937, value=85
 610213             column=course:python, timestamp=1486455938119, value=79
 610215             column=class:class, timestamp=1486456047670, value=173BigData
 610215             column=name:, timestamp=1486456006155, value=John
 610215             column=course:java, timestamp=1486456206791, value=80
 610215             column=course:python, timestamp=1486456224168, value=86
2 row(s) in 0.0390 seconds

hbase(main):017:0> scan 'score',{COLUMNS=>'course'}
ROW                 COLUMN+CELL
 610213             column=course:java, timestamp=1486455895937, value=85
 610213             column=course:python, timestamp=1486455938119, value=79
 610215             column=course:java, timestamp=1486456206791, value=80
 610215             column=course:python, timestamp=1486456224168, value=86
2 row(s) in 0.0230 seconds
```

```
hbase(main):018:0> scan 'score',{COLUMN=>'course:java'}
ROW                COLUMN+CELL
 610213            column=course:java, timestamp=1486455895937, value=85
 610215            column=course:java, timestamp=1486456206791, value=80
2 row(s) in 0.0350 seconds
```

④ FILTER 过滤扫描记录，操作如下：

- 扫描值是 John 的记录。

```
hbase(main):019:0> scan 'score',FILTER=>"ValueFilter(=,'binary:John')"
ROW                COLUMN+CELL
 610215            column=name:, timestamp=1486456006155, value=John
1 row(s) in 0.0120 seconds
```

- 扫描值包含 To 的记录。

```
hbase(main):020:0> scan 'score',FILTER=>"ValueFilter(=,'substring:To')"
ROW                COLUMN+CELL
 610213            column=name:, timestamp=1486455816564, value=Tom
1 row(s) in 0.0520 seconds
```

- 扫描列 class 的值包含 Clou 的记录。

```
hbase(main):021:0> scan 'score',FILTER=>"ColumnPrefixFilter('class') AND ValueFilter(=,'substring:Clou')"
ROW                COLUMN+CELL
 610213            column=class:class, timestamp=1486455866244, value=163Cloud
1 row(s) in 0.0850 seconds
```

- 扫描 Rowkey 为 610 开头的记录。

```
hbase(main):022:0> scan 'score',FILTER=>"PrefixFilter('610')"
ROW                COLUMN+CELL
 610213            column=class:class, timestamp=1486455866244, value=163Cloud
 610213            column=name:, timestamp=1486455816564, value=Tom
 610213            column=course:java, timestamp=1486455895937, value=85
 610213            column=course:python, timestamp=1486455938119, value=79
 610215            column=class:class, timestamp=1486456047670, value=173BigData
 610215            column=name:, timestamp=1486456006155, value=John
 610215            column=course:java, timestamp=1486456206791, value=80
 610215            column=course:python, timestamp=1486456224168, value=86
2 row(s) in 0.0200 seconds
```

- 只拿出 key 中的第一个 column 的第一个 version 并且只要 key 的记录。

```
hbase(main):023:0> scan 'score',FILTER=>"FirstKeyOnlyFilter() AND KeyOnlyFilter()"
ROW                COLUMN+CELL
 610213            column=class:class, timestamp=1486455866244, value=
 610215            column=class:class, timestamp=1486456047670, value=
2 row(s) in 0.3200 seconds
```

FirstKeyOnlyFilter：一个 rowkey 可以有多个 version，同一个 rowkey 的同一个 column 也会有多个的值，只拿出 key 中的第一个 column 的第一个 version。

KeyOnlyFilter：只要 key，不要 value。

- 扫描从 610213 开始到 610215 结束的记录。

```
hbase(main):024:0> scan 'score',{STARTROW=>'610213',STOPROW=>'610215'}
ROW              COLUMN+CELL
 610213          column=class:class, timestamp=1486455866244, value=163Cloud
 610213          column=name:, timestamp=1486455816564, value=Tom
 610213          column=course:java, timestamp=1486455895937, value=85
 610213          column=course:python, timestamp=1486455938119, value=79
1 row(s) in 0.0630 seconds
```

- 扫描列族 name 中含有 To 的记录。

```
hbase(main):025:0> scan 'score',{COLUMNS=>['name'],FILTER=>"(ValueFilter(=,'substring:To'))"}
ROW              COLUMN+CELL
 610213          column=name:, timestamp=1486455816564, value=Tom
1 row(s) in 0.0090 seconds
```

- 扫描列族 course 中成绩大于等于 85 的记录。

```
hbase(main):026:0> scan 'score',{COLUMNS=>['course'],FILTER=>"(ValueFilter(>=,'binary:85'))"}
ROW              COLUMN+CELL
 610213          column=course:java, timestamp=1486455895937, value=85
 610215          column=course:python, timestamp=1486456224168, value=86
2 row(s) in 0.0500 seconds
```

⑤ 删除记录，操作如下：

```
hbase(main):027:0> delete 'score','610213','course:python'
0 row(s) in 0.0100 seconds
hbase(main):028:0> get 'score','610213'
COLUMN           CELL
 class:class     timestamp=1486455866244, value=163Cloud
 name:           timestamp=1486455816564, value=Tom
 course:java     timestamp=1486455895937, value=85
3 row(s) in 0.0130 seconds
hbase(main):029:0> delete 'score','610215','class:class',1486456047670
0 row(s) in 0.0440 seconds
hbase(main):030:0> get 'score','610215'
COLUMN           CELL
 name:           timestamp=1486456006155, value=John
 course:java     timestamp=1486456206791, value=80
 sourse:python   timestamp=1486456224168, value=86
3 row(s) in 0.0160 seconds
```

### 三、表的修改与删除

① 增加列族，操作如下：

```
hbase(main):031:0> alter 'score',NAME=>'address'
Updating all regions with the new schema...
0/1 regions updated.
1/1 regions updated.
Done.
0 row(s) in 3.6190 seconds
```

② 删除列族，操作如下：

```
hbase(main):032:0> alter 'score',NAME=>'address',METHOD=>'delete'
Updating all regions with the new schema...
1/1 regions updated.
Done.
0 row(s) in 2.2660 seconds
```

③ 删除表，操作如下：

```
hbase(main):033:0> disable 'score'
0 row(s) in 2.3880 seconds
hbase(main):034:0> drop 'score'
0 row(s) in 2.3370 seconds
```

## 任务 3  HBase 编程

### 【任务概述】

编写 HBase 查询程序，使用 Get 查找 610215 的班级（class），使用 Scan 查找 610213 所有列；编写数据操作程序，完成 HBase 表的创建、数据的插入、查询等。

### 【支撑知识】

HBase 数据操作访问可以通过 HTableInterface 或 HTableInterface 的 HTable 类来完成，两者都支持 HBase 的主要操作，HBase 提供几个 Java API 接口，方便编程调用。

（1）HbaseConfiguration

关系：org.apache.hadoop.hbase.HBaseConfiguration

作用：通过此类可以对 HBase 进行配置。

（2）HBaseAdmin

关系：org.apache.hadoop.hbase.client.HBaseAdmin

作用：提供一个接口来管理 HBase 数据库中的表信息。它提供创建表、删除表等方法。

（3）HTableDescriptor

关系：org.apache.hadoop.hbase.client.HTableDescriptor

作用：包含了表的名字及其对应列族。

HTableDescriptor 提供的方法如下：

```
void addFamily(HColumnDescriptor)                    //添加一个列族
HColumnDescriptor removeFamily(byte[] column)        //移除一个列族
byte[] getName()                                     //获取表的名字
byte[] getValue(byte[] key)                          //获取属性的值
void setValue(String key,String value)               //设置属性的值
```

（4）HColumnDescriptor

关系：org.apache.hadoop.hbase.client.HColumnDescriptor

作用：维护关于列的信息。

HColumnDescriptor 提供的方法如下：

```
byte[] getName()                                     //获取列族的名字
byte[] getValue()                                    //获取对应属性的值
void setValue(String key,String value)               //设置对应属性的值
```

（5）HTable

关系：org.apache.hadoop.hbase.client.HTable

作用：用户与 HBase 表进行通信。此方法对于更新操作来说是非线程安全的，如果启动多个线程尝试与单个 HTable 实例进行通信，那么写缓冲器可能会崩溃。

（6）Put

关系：org.apache.hadoop.hbase.client.Put

作用：用于对单个行执行添加操作。

（7）Get

关系：org.apache.hadoop.hbase.client.Get

作用：用于获取单个行的相关信息。

（8）Result

关系：org.apache.hadoop.hbase.client.Result

作用：存储 Get 或 Scan 操作后获取的单行值。

（9）ResultScanner

关系：Interface

作用：客户端获取值的接口。

## 【任务实施】

### 一、查询程序

① 运行 Eclipse，执行菜单栏中的"File"→"New"→"Java Project"命令，创建"SearchScore"项目，如图 5-10 所示。

② 在"SearchScore"上单击鼠标右键，在弹出的快捷菜单中选择"Build Path"→"Configure Build Path..."命令，在弹出的对话框中选择"Libraries"选项卡，单击"Add External JARs..."按钮，添加"Libraries"下所有 jar 包，如图 5-11 所示。

③ 在"SearchScore"→"src"上单击鼠标右键，在弹出的快捷菜单中选择"New"→"Class"命令，新建 SearchScore 类，如图 5-12 所示。

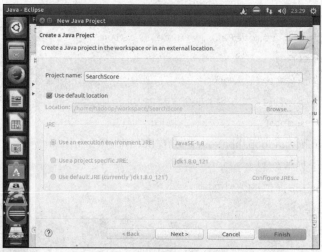

图 5-10 创建 "SearchScore" 项目

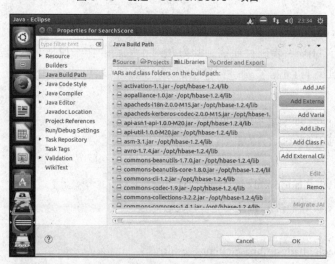

图 5-11 添加 HBase Lib 的 jar 包

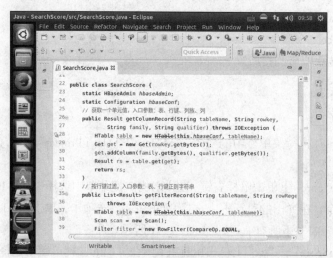

图 5-12 SearchScore 类

SearchScore.java 的代码如下：
```java
import java.io.IOException;
import java.util.ArrayList;
import java.util.Iterator;
import java.util.List;
import org.apache.hadoop.conf.Configuration;
import org.apache.hadoop.hbase.Cell;
import org.apache.hadoop.hbase.CellUtil;
import org.apache.hadoop.hbase.HBaseConfiguration;
import org.apache.hadoop.hbase.client.Get;
import org.apache.hadoop.hbase.client.HBaseAdmin;
import org.apache.hadoop.hbase.client.HTable;
import org.apache.hadoop.hbase.client.Result;
import org.apache.hadoop.hbase.client.ResultScanner;
import org.apache.hadoop.hbase.client.Scan;
import org.apache.hadoop.hbase.filter.CompareFilter.CompareOp;
import org.apache.hadoop.hbase.filter.Filter;
import org.apache.hadoop.hbase.filter.RegexStringComparator;
import org.apache.hadoop.hbase.filter.RowFilter;
import org.apache.hadoop.hbase.util.Bytes;
public class SearchScore {
  static HBaseAdmin hbaseAdmin;
  static Configuration hbaseConf;
  // 获取一个单元值，入口参数：表、行键、列族、列
  public Result getColumnRecord(String tableName, String rowkey,
      String family, String qualifier) throws IOException {
    HTable table = new HTable(this.hbaseConf, tableName);
    Get get = new Get(rowkey.getBytes());
    get.addColumn(family.getBytes(), qualifier.getBytes());
    Result rs = table.get(get);
    return rs;
  }
  // 按行键过滤，入口参数：表、行键正则字符串
  public List<Result> getFilterRecord(String tableName, String rowRegexString)
      throws IOException {
    HTable table = new HTable(this.hbaseConf, tableName);
    Scan scan = new Scan();
    Filter filter = new RowFilter(CompareOp.EQUAL,
        new RegexStringComparator(rowRegexString));
    scan.setFilter(filter);
    ResultScanner scanner = table.getScanner(scan);
```

```java
    List<Result> list = new ArrayList<Result>();
    for (Result r : scanner) {
      list.add(r);
    }
    scanner.close();
    return list;
  }
  public static void main(String[] args) throws IOException {
    Configuration conf = new Configuration();
    conf.set("hbase.zookeeper.quorum", "master,slave1,slave2");
    hbaseConf = HBaseConfiguration.create(conf);
    hbaseAdmin = new HBaseAdmin(hbaseConf);
    SearchScore searchScore = new SearchScore();
    Result rs = searchScore.getColumnRecord("score",
        "610215", "class", "class");
    for (Cell cell : rs.rawCells()) {
      System.out.print("Rowkey:"+ Bytes.toString(rs.getRow()));
      System.out.print(" Familiy-Quilifier:" +
        Bytes.toString(CellUtil.cloneFamily(cell)) + "-" +
        Bytes.toString(CellUtil.cloneQualifier(cell)));
      System.out.println(" Value:" +
        Bytes.toString(CellUtil.cloneValue(cell)));
    }
    System.out.println(" 610215的班级:" + Bytes.toString(rs.value()));
    List<Result> list = null;
    list = searchScore.getFilterRecord("score", "610213");
    Iterator<Result> it = list.iterator();
    while (it.hasNext()) {
      rs = it.next();
      String name = new String(rs.getValue(Bytes.toBytes("name"),
          Bytes.toBytes("")));
      String class1 = new String(rs.getValue(Bytes.toBytes("class"),
          Bytes.toBytes("class")));
      String java = new String(rs.getValue(Bytes.toBytes("course"),
          Bytes.toBytes("java")));
      String python = new String(rs.getValue(Bytes.toBytes("course"),
          Bytes.toBytes("python")));
      System.out.println("name:"+name+" class:"+class1
          +" java:"+java+" python:"+python);
    }
  }
}
```

④ 单击"Run"按钮，运行结果如图 5-13 所示。

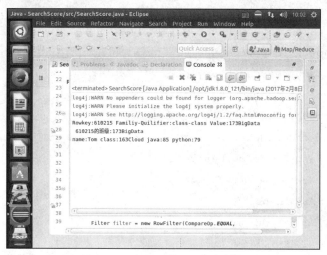

图 5-13　SearchScore 运行结果

## 二、数据操作程序

① 运行 Eclipse，执行菜单栏中的"File"→"New"→"Java Project"命令，创建 HBaseDemo 项目后，再创建 HBaseDemo 类，导入 HBase 的 jar 包，如图 5-14 所示。

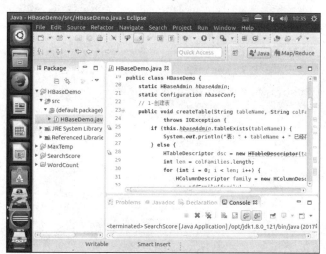

图 5-14　HBaseDemo 类

HBaseDemo.java 代码如下：

```java
import java.io.IOException;
import java.util.ArrayList;
import java.util.Iterator;
import java.util.List;
import org.apache.hadoop.conf.Configuration;
import org.apache.hadoop.hbase.HBaseConfiguration;
import org.apache.hadoop.hbase.HColumnDescriptor;
```

```java
import org.apache.hadoop.hbase.HTableDescriptor;
import org.apache.hadoop.hbase.KeyValue;
import org.apache.hadoop.hbase.client.Get;
import org.apache.hadoop.hbase.client.HBaseAdmin;
import org.apache.hadoop.hbase.client.HTable;
import org.apache.hadoop.hbase.client.Put;
import org.apache.hadoop.hbase.client.Result;
import org.apache.hadoop.hbase.client.ResultScanner;
import org.apache.hadoop.hbase.client.Scan;
public class HBaseDemo {
  static HBaseAdmin hbaseAdmin;
  static Configuration hbaseConf;
  // 1-创建表
  public void createTable(String tableName, String colFamilies[])
      throws IOException {
    if (this.hbaseAdmin.tableExists(tableName)) {
      System.out.println("表: " + tableName + " 已经存在!");
    } else {
      HTableDescriptor dsc = new HTableDescriptor(tableName);
      int len = colFamilies.length;
      for (int i = 0; i < len; i++) {
        HColumnDescriptor family = new HColumnDescriptor(colFamilies[i]);
        dsc.addFamily(family);
      }
      hbaseAdmin.createTable(dsc);
      System.out.println("创建表成功");
    }
  }
  // 2-插入一行记录
  public void insertRecord(String tableName, String rowkey, String family,
      String qualifier, String value) throws IOException {
    HTable table = new HTable(this.hbaseConf, tableName);
    Put put = new Put(rowkey.getBytes());
    put.add(family.getBytes(), qualifier.getBytes(), value.getBytes());
    table.put(put);
    System.out.println("插入行成功");
  }
  // 3-获取一行记录
  public Result getOneRecord(String tableName, String rowkey)
```

```java
        throws IOException {
    HTable table = new HTable(this.hbaseConf, tableName);
    Get get = new Get(rowkey.getBytes());
    Result rs = table.get(get);
    return rs;
}
// 4-获取所有记录
public List<Result> getAllRecord(String tableName) throws IOException {
    HTable table = new HTable(this.hbaseConf, tableName);
    Scan scan = new Scan();
    ResultScanner scanner = table.getScanner(scan);
    List<Result> list = new ArrayList<Result>();
    for (Result r : scanner) {
        list.add(r);
    }
    scanner.close();
    return list;
}
public static void main(String[] args) throws IOException {
    // 1-初始化
    Configuration conf = new Configuration();
    conf.set("hbase.zookeeper.quorum", "master,slave1,slave2");
    hbaseConf = HBaseConfiguration.create(conf);
    hbaseAdmin = new HBaseAdmin(hbaseConf);
    HBaseDemo hbaseDemo = new HBaseDemo();
    // 2-创建表
    String tableName = "demo";
    String colFamilies[] = { "article", "author" };
    hbaseDemo.createTable(tableName, colFamilies);
    // 3-插入一条记录
    hbaseDemo.insertRecord(tableName, "row1", "article", "title", "HBase");
    hbaseDemo.insertRecord(tableName, "row1", "author", "name", "Cat");
    hbaseDemo.insertRecord(tableName, "row1", "author", "nickname", "Tom");
    // 4-查询一条记录
    Result rs1 = hbaseDemo.getOneRecord(tableName, "row1");
    for (KeyValue kv : rs1.raw()) {
        System.out.println(new String(kv.getRow()));
        System.out.println(new String(kv.getFamily()));
        System.out.println(new String(kv.getQualifier()));
```

```
      System.out.println(new String(kv.getValue()));
    }
    // 5-查询整个 Table
    List<Result> list = null;
    list = hbaseDemo.getAllRecord(tableName);
    Iterator<Result> it = list.iterator();
    while (it.hasNext()) {
      Result rs2 = it.next();
      for (KeyValue kv : rs2.raw()) {
        System.out.print("row key: " + new String(kv.getRow()));
        System.out.print(" family: " + new String(kv.getFamily()));
        System.out.print(" qualifier:"
          + new String(kv.getQualifier()));
        System.out.print(" timestamp:" + kv.getTimestamp());
        System.out.println(" value: " + new String(kv.getValue()));
      }
    }
  }
}
```

② 单击"运行"按钮，运行结果如图 5-15 所示。

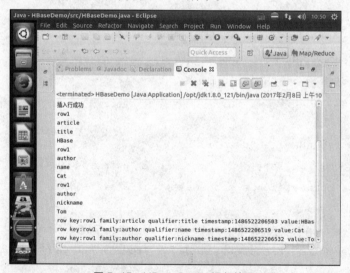

图 5-15　HBaseDemo 运行结果

## 任务 4　MapReduce 与 HBase 集成

### 【任务概述】

编写 MapReduce 程序，将运行的单词计数结果直接写入 HBase 表中；编写 MapReduce

程序，读取 HBase 的 score 表中数据，计算每位同学课程总分数。

【支撑知识】

MapReduce 的执行过程如下：

HDFS 数据 → InputFormat → Mapper → (Shuffle) → Reducer → OutputFormat

其中：

InputFormat 调用 RecordReader 完成 HDFS 数据的读入，负责数据的拆分、计算，输出数据格式为<k1,v1>；

Mapper 负责数据并行计算，输出数据格式为<k2,v2>；

Reducer 负责数据汇总计算，输出数据格式为<k3,v3>；

OutputFormat 调用 RecordWriter 完成数据的输出，结果存放在 HDFS 中。

（1）InputFormat

InputFormat 负责拆分输入数据，返回一个 RecordReader 实例，这个实例定义了键值对象类，并提供了 next() 方法来遍历输入的数据。

HBase 提供了组专用的 TableInputFormatBase 的类，该类的子类是 TableInputFormat。TableInputFormatBase 实现了大部分功能，但仍旧是抽象类，它的子类 TableInputFormat 是一个轻量级的实体类版本。这些类提供了一个扫描 HBase 全表的实现。用户需要提供一个 Scan 实例，设置起止行键、添加过滤器和指定版本数目等。TableInputFormat 将表拆分成大小合适的块，同时交给后面的 MapReduce 过程处理，如图 5-16 所示。

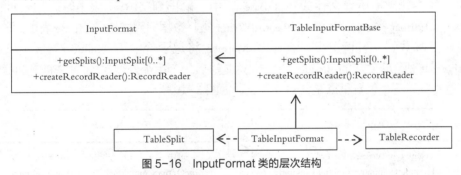

图 5-16　InputFormat 类的层次结构

（2）Mapper

Mapper 是 MapReduce 过程的第二个阶段，在这一阶段中，从 RecordReader 读到的每一个数据都由一个 map() 方法来处理。Mapper 读取一组特定的键/值对，但是可能会输出其他的类型。这个过程便于将原始数据转化为更有用的数据类型，以做进一步处理。

HBase 提供了一个 TableMapper 类，将键的类型强制转换为一个 ImmutableBytesWritable，同时将值的类型强制转换为 Result 类型，这样构成了 TableRecordReader 类返回的结果。

有一个 TableMapper 的特殊实现是 IdentityTableMapper，它也是一个展示如何将自己想要的功能添加到 HBase 提供的类中的比较好的例子。TableMapper 类没有实现任何实际的功能，它只是添加了键/值对的签名。IdentityTableMapper 只是简单地把键/值对传到下一个处理阶段而已，如图 5-17 所示。

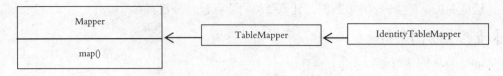

图 5-17 Mapper 类的层次结构

（3）Reducer

Reducer 阶段和类的层次结构与 Mapper 阶段十分相似，Mapper 类的输出会经过 shuffle 和 sort 处理。在 Mapper 和 Reducer 阶段的内部 shuffle 阶段中，每个不同的 Map 输出的中间结果都会被复制到 Reduce 服务器，同时 sort 阶段会将所有经过 suffled（copied）阶段处理的数据进行联合排序，这样 Reducer 得到的中间结果就是一个已排序的数据集，在这个数据集中，每一个特定的键与它所有可能的值关联在一起，如图 5-18 所示。

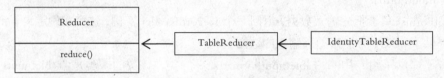

图 5-18 Reducer 类的层次结构

（4）OutputFormat

最后阶段由 OutputFormat 类处理。它的工作是将数据持久化到不同的位置。这里提供一些具体实现将结果输出到文件或者 HBase 表中。输出到 HBase 表中时，用户可以使用 TableOutputFormat 类。它使用 TableRecordWriter 类将数据写到特定的 HBase 表中。同时也需要注意一下基数。一般情况下，有许多 Mapper 将记录传输给 Reducer，但是只有一个 OutputFormat 处理它对应的 Reducer 输出。这是最后一个用于处理键/值对并将它们写到最终存储位置的类，这个位置可以是文件或者表，如图 5-19 所示。

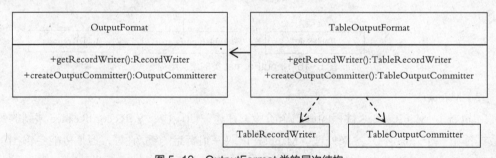

图 5-19 OutputFormat 类的层次结构

Hadoop 类需要 TableOutputCommitter 类来实现它们的功能。如果不需要 HBase 来处理数据，则它是一个空类，并没有实现实际的功能。不过其他 OutputFormat 的实现需要这个类。

当创建作业时，输出表的名字就已经被指定了。除此之外，TableOutputFormat 不会增加其他复杂性。值得注意的是，表的缓冲区自动刷写被关闭，同时缓冲区由系统内部自动管理。这对提高导入大量数据时的速度十分有帮助。

MapReduce 的支持类与 TableMapReduceUtil 类一同协作在 HBase 上执行 MapReduce 作业。它有一个静态方法能配置作业，并使作业可以使用 HBase 作为数据源或目标。

## 【任务实施】

### 一、单词计数写入 HBase 表

① 运行 Eclipse,选择"Map/Reduce Project",新建 WordIntoTable 项目,添加 Hadoop 和 HBase 的 jar 包,创建 WordIntoTable 类,如图 5-20 所示。

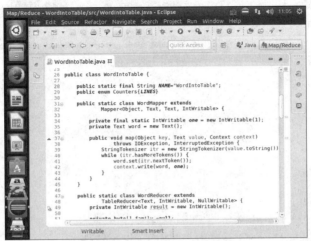

图 5-20 WordIntoTable 类

WordIntoTable.java 代码如下:
```
import java.io.IOException;
import java.util.StringTokenizer;
import org.apache.commons.cli.CommandLine;
import org.apache.commons.cli.CommandLineParser;
import org.apache.commons.cli.HelpFormatter;
import org.apache.commons.cli.Option;
import org.apache.commons.cli.Options;
import org.apache.commons.cli.PosixParser;
import org.apache.hadoop.conf.Configuration;
import org.apache.hadoop.fs.Path;
import org.apache.hadoop.hbase.HBaseConfiguration;
import org.apache.hadoop.hbase.KeyValue;
import org.apache.hadoop.hbase.client.Put;
import org.apache.hadoop.hbase.mapreduce.TableOutputFormat;
import org.apache.hadoop.hbase.mapreduce.TableReducer;
import org.apache.hadoop.hbase.util.Bytes;
import org.apache.hadoop.io.IntWritable;
import org.apache.hadoop.io.NullWritable;
import org.apache.hadoop.io.Text;
import org.apache.hadoop.mapreduce.Job;
import org.apache.hadoop.mapreduce.Mapper;
import org.apache.hadoop.mapreduce.lib.input.FileInputFormat;
import org.apache.hadoop.util.GenericOptionsParser;
```

```java
public class WordIntoTable {
  public static final String NAME="WordIntoTable";
  public enum Counters{ LINES }
    public static class WordMapper extends
      Mapper<Object, Text, Text, IntWritable> {
    private final static IntWritable one = new IntWritable(1);
    private Text word = new Text();
    public void map(Object key, Text value, Context context)
        throws IOException, InterruptedException {
      StringTokenizer itr = new StringTokenizer(value.toString());
      while (itr.hasMoreTokens()) {
        word.set(itr.nextToken());
        context.write(word, one);
      }
    }
  }
  public static class WordReducer extends
      TableReducer<Text, IntWritable, NullWritable> {
    private IntWritable result = new IntWritable();
    private byte[] family =null;
    private byte[] qualifier = null;
    protected void setup(Context context){
      String column = context.getConfiguration().get("conf.column");
      byte[][] colkey = KeyValue.parseColumn(Bytes.toBytes(column));
      family = colkey[0];
      if(colkey.length>1){
        qualifier = colkey[1];
      }
    }
    public void reduce(Text key, Iterable<IntWritable> values,
        Context context) throws IOException, InterruptedException {
      int sum = 0;
      for (IntWritable val : values) {
        sum += val.get();
      }
      try{
        byte[] rowkey= key.toString().getBytes();
        byte[] words= Bytes.toBytes(String.valueOf(sum).toString());
        Put put = new Put(rowkey);
        put.add(family,qualifier,words);
        context.write(NullWritable.get(), put);
        context.getCounter(Counters.LINES).increment(1);
      }catch(Exception e){
```

```java
      e.printStackTrace();
    }
  }
}

private static CommandLine parseArgs(String[] args){
  Options options = new Options();
  Option o = new Option("t" ,"table",true,"table to import into (must exist)");
  o.setArgName("table-name");
  o.setRequired(true);
  options.addOption(o);
  o = new Option("c","column",true,"column to store row data into");
  o.setArgName("family:qualifier");
  o.setRequired(true);
  options.addOption(o);
  o = new Option("i", "input", true,"the directory or file to read from");
  o.setArgName("path-in-HDFS");
  o.setRequired(true);
  options.addOption(o);
  options.addOption("d", "debug", false,"switch on DEBUG log level");
  CommandLineParser parser = new PosixParser();
  CommandLine cmd = null;
  try {
    cmd = parser.parse(options, args);
  } catch (Exception e) {
    System.err.println("ERROR: " + e.getMessage() + "\n");
    HelpFormatter formatter = new HelpFormatter();
    formatter.printHelp(NAME + " ", options, true);
    System.exit(-1);
  }
  return cmd;
}
public static void main(String[] args) throws Exception {
  Configuration cf = new Configuration();
  cf.set("hbase.zookeeper.quorum", "master,slave1,slave2");
  Configuration conf = HBaseConfiguration.create(cf);
  String[] otherArgs = new GenericOptionsParser(conf, initialArg())
      .getRemainingArgs();
  CommandLine cmd = parseArgs(otherArgs);
  String table = cmd.getOptionValue("t");
  String input = cmd.getOptionValue("i");
  String column = cmd.getOptionValue("c");
  conf.set("conf.column", column);
```

```java
        Job job = Job.getInstance(conf, "word into table");
        job.setJarByClass(WordIntoTable.class);
        job.setMapperClass(WordMapper.class);
        job.setReducerClass(WordReducer.class);
        job.setOutputKeyClass(Text.class);
        job.setOutputValueClass(IntWritable.class);
        job.setOutputFormatClass(TableOutputFormat.class);
        job.getConfiguration().set(TableOutputFormat.OUTPUT_TABLE, table);
        FileInputFormat.addInputPath(job, new Path(input));
        System.exit(job.waitForCompletion(true) ? 0 : 1);
    }
    private static String[] initialArg(){
        String []args = new String[6];
        args[0]="-c";
        args[1]="wordcount:result";
        args[2]="-i";
        args[3]="hdfs://master:9000/input";
        args[4]="-t";
        args[5]="test";
        return args;
    }
}
```

② 编辑并上传测试数据，操作如下：

```
hadoop@sw-desktop:~$ vi sw1.txt
Hello World
Good Hadoop
hadoop@sw-desktop:~$ vi sw2.txt
Hello Hadoop
Bye Hadoop
hadoop@sw-desktop:~$ hdfs dfs -rm -r -f /input
17/02/10 10:55:28 INFO fs.TrashPolicyDefault: Namenode trash configuration: Deletion interval = 0 minutes, Emptier interval = 0 minutes.
Deleted /input
hadoop@sw-desktop:~$ hdfs dfs -mkdir /input
hadoop@sw-desktop:~$ hdfs dfs -put sw*.txt /input
```

③ 创建 HBase 数据表，操作如下：

```
hadoop@sw-desktop:~$ hbase shell
hbase(main):001:0> create 'test','wordcount'
0 row(s) in 9.5730 seconds
=> Hbase::Table - test
```

④ 单击 Eclipse "运行" 按钮，选择 "Run on Hadoop" 选项，开始运行。

⑤ 查看运行结果，操作如下：

```
hbase(main):002:0> scan 'test'
ROW                 COLUMN+CELL
 Bye                column=wordcount:result, timestamp=1486695813124, value=1
 Good               column=wordcount:result, timestamp=1486695813124, value=1
 Hadoop             column=wordcount:result, timestamp=1486695813124, value=3
 Hello              column=wordcount:result, timestamp=1486695813124, value=2
 World              column=wordcount:result, timestamp=1486695813124, value=1
5 row(s) in 0.6420 seconds
```

## 二、读取 HBase 的 score 表，计算课程总分数

① 运行 Eclipse，选择"Map/Reduce Project"命令，新建 ScoreFromTable 项目，添加 Hadoop 和 HBase 的 jar 包，创建 ScoreFromTable 类，如图 5-21 所示。

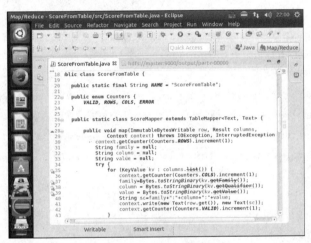

图 5-21 ScoreFromTable 类

ScoreFromTable.java 代码如下：

```
import java.io.IOException;
import org.apache.hadoop.conf.Configuration;
import org.apache.hadoop.fs.Path;
import org.apache.hadoop.hbase.HBaseConfiguration;
import org.apache.hadoop.hbase.KeyValue;
import org.apache.hadoop.hbase.client.Result;
import org.apache.hadoop.hbase.client.Scan;
import org.apache.hadoop.hbase.io.ImmutableBytesWritable;
import org.apache.hadoop.hbase.mapreduce.TableMapReduceUtil;
import org.apache.hadoop.hbase.mapreduce.TableMapper;
import org.apache.hadoop.hbase.util.Bytes;
import org.apache.hadoop.io.Text;
import org.apache.hadoop.mapreduce.Job;
import org.apache.hadoop.mapreduce.Reducer;
import org.apache.hadoop.mapreduce.lib.output.FileOutputFormat;
```

```java
public class ScoreFromTable {
  public static final String NAME = "ScoreFromTable";
  public enum Counters {
    VALID, ROWS, COLS, ERROR
  }
  public static class ScoreMapper extends TableMapper<Text, Text> {
    public void map(ImmutableBytesWritable row, Result columns,
        Context context) throws IOException, InterruptedException {
      context.getCounter(Counters.ROWS).increment(1);
      String family = null;
      String column = null;
      String value = null;
      try {
        for (KeyValue kv : columns.list()) {
          context.getCounter(Counters.COLS).increment(1);
          family=Bytes.toStringBinary(kv.getFamily());
          column = Bytes.toStringBinary(kv.getQualifier());
          value = Bytes.toStringBinary(kv.getValue());
          String sc=family+":"+column+":"+value;
          context.write(new Text(row.get()), new Text(sc));
          context.getCounter(Counters.VALID).increment(1);
        }
      } catch (Exception e) {
        e.printStackTrace();
        System.err.println("Error:" + e.getMessage() + ",Row:"
            + Bytes.toStringBinary(row.get()) + ",Value:" + value);
        context.getCounter(Counters.ERROR).increment(1);
      }
    }
  }
  public static class ScoreReducer extends
      Reducer<Text, Text, Text, Text> {
    private Text result = new Text();
    public void reduce(Text key, Iterable<Text> values, Context context)
        throws IOException, InterruptedException {
      String name=null;
      int sum=0;
      for (Text val:values) {
        String[] sp= val.toString().split(":");
        if (sp[0].equals("name"))
          name=sp[2];
        if (sp[0].equals("course"))
          sum+=Integer.valueOf(sp[2]);
```

```java
        }
        name="Name:"+name+" sum:"+ String.valueOf(sum);
        result.set(name);
        context.write(key, result);
    }
}
public static void main(String[] args) throws Exception {
    Configuration cf = new Configuration();
    cf.set("hbase.zookeeper.quorum", "master,slave1,slave2");
    Configuration conf = HBaseConfiguration.create(cf);
    String table = "score";
    String output = "hdfs://master:9000/output";
    Scan scan = new Scan();
    Job job = Job.getInstance(conf, "Score From Table");
    job.setJarByClass(ScoreFromTable.class);
    job.setMapperClass(ScoreMapper.class);
    //job.setNumReduceTasks(0);
    job.setReducerClass(ScoreReducer.class);
    job.setOutputKeyClass(Text.class);
    job.setOutputValueClass(Text.class);
    TableMapReduceUtil.initTableMapperJob(table, scan,
        ScoreMapper.class, Text.class, Text.class, job);
    FileOutputFormat.setOutputPath(job, new Path(output));
    System.exit(job.waitForCompletion(true) ? 0 : 1);
    }
  }
}
```

② 单击"运行"按钮，运行结果如图 5-22 所示。

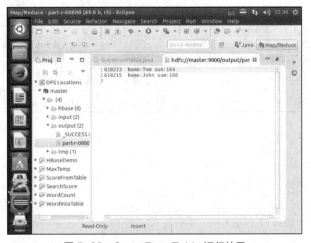

图 5-22 ScoreFromTable 运行结果

## 【同步训练】

### 一、简答题

（1）HBase 的特点是什么？

（2）HBase 是如何存储数据的？

（3）HRegionServer 的主要功能是什么？

（4）Zookeeper 的作用是什么？

### 二、操作题

（1）创建 Employee 表，在 Employee 表中插入数据，如表 5-9 所示。

表 5-9　Employee 表

| RowKey | Info | | | Address | | |
|---|---|---|---|---|---|---|
| | Company | Favorite | Age | Province | City | Town |
| Jie | SWPT | | 28 | GuangDong | ShanWei | JieShi |
| Xie | Alibaba | Tourism | 25 | ZheJiang | HangZhou | |
| Shi | SZPT | Movie | 22 | GuangDong | ShenZhen | XiLi |

（2）计算 Employee 表记录的数量，查看 Employee 表的全部记录。

（3）查看 Xie 的 City 和 Town。

（4）查看含有 GuangDong 的记录。

（5）查看列族为 Info 和列 Province 的信息。

（6）修改 Employee 表，插入列族 What，查看 Employee 表结构。

### 三、编程题

（1）编写程序，查看上题中 Employee 表中 Jie 的 City 和 Town。

（2）编写程序，完成 Company 表的创建、数据的插入和查询所有记录。Company 的信息如表 5-10 所示。

表 5-10　Company 表

| Name | | Address | | | What |
|---|---|---|---|---|---|
| ShortName | LongName | Country | Province | City | |
| WZ | Wu Zhou | China | GuangDong | ShenZhen | Cloud Computing |
| AL | Alibaba | China | ZheJiang | HangZhou | |

# PART 6 项目六
# Hive 数据仓库安装与应用

## 【项目介绍】

Hive 数据仓库也是 Hadoop 生态圈重要的成员，本项目需要完成 Hive 的安装、配置，使用 Hive CLI 客户端工具创建表、导入数据、查询数据，最后完成 Hive 程序的编写。

本项目分为以下 4 个任务：
- 任务 1　安装 Hive
- 任务 2　Hive CLI
- 任务 3　Hive 编程
- 任务 4　Hive 与 HBase 集成

## 【学习目标】

一、知识目标
- 了解 Hive 架构及组件。
- 熟悉 Hive 元数据存储模式。
- 掌握 Hive 系统的配置。
- 掌握 Hive 数据类型和常用函数。
- 掌握 Hive 创建数据库和表。
- 掌握 Hive 数据的导入和查询。
- 掌握 Hive 的基本编程。

二、能力目标
- 能够安装、配置 Hive。
- 能够使用 Hive CLI 命令创建数据库、表。
- 能够使用 Hive CLI 命令导入数据、查询数据。
- 能够使用 Java 编写 Hive 程序。

## 任务 1　安装 Hive

### 【任务概述】

Hive 需要将元数据保存到数据库，需要配置相应服务，本任务主要完成 Hive 的安装和配置、MySQL 用户授权和 hive-schema 导入 MySQL、hive metastore 服务启动和 Hive 测试。

## 【支撑知识】

### 一、Hive 简介

Hive 是基于 Hadoop 的一个数据仓库处理工具，可以将结构化的数据文件映射为一张数据库表，并提供简单的 SQL 查询功能，可以将 SQL 语句转换为 MapReduce 任务进行运行。其优点是学习成本低，可以通过类 SQL 语句快速实现简单的 MapReduce 统计，不必开发专门的 MapReduce 应用，十分适合数据仓库的统计分析。

Hive 是建立在 Hadoop 上的数据仓库基础构架。它提供了一系列的工具，可以用来进行数据提取转化加载（ETL），这是一种可以存储、查询和分析存储在 Hadoop 中的大规模数据的机制。Hive 定义了简单的类 SQL 查询语言，称为 HQL，它允许熟悉 SQL 的用户查询数据。同时，这个语言也允许熟悉 MapReduce 的开发者开发自定义的 Mapper 和 Reducer 来处理内建的 Mapper 和 Reducer 无法完成的复杂的分析工作。

Hive 构建在基于静态批处理的 Hadoop 之上，Hadoop 通常都有较高的延迟并且在作业提交和调度的时候需要大量的开销。因此，Hive 并不能在大规模数据集上实现低延迟快速的查询，例如，Hive 在几百 MB 的数据集上执行查询一般有分钟级的时间延迟。因此，Hive 并不适合那些需要低延迟的应用，例如，联机事务处理（OLTP）。Hive 查询操作过程严格遵守 Hadoop MapReduce 的作业执行模型，Hive 将用户的 HiveQL 语句通过解释器转换为 MapReduce 作业提交到 Hadoop 集群上，Hadoop 监控作业执行过程，然后返回作业执行结果给用户。Hive 并非为联机事务处理而设计，Hive 并不提供实时的查询和基于行级的数据更新操作。Hive 的最佳使用场合是大数据集的批处理作业，例如，网络日志分析。

Hive 是一种底层封装了 Hadoop 的数据仓库处理工具，使用类 SQL 的 HiveQL 语言实现数据查询，所有 Hive 的数据都存储在 Hadoop 兼容的文件系统（例如，Amazon S3、HDFS）中。Hive 在加载数据过程中不会对数据进行任何修改，只是将数据移动到 HDFS 中 Hive 设定的目录下，因此，Hive 不支持对数据的改写和添加，所有的数据都是在加载时确定的。Hive 的设计特点如下：

① 支持索引，加快数据查询速度。

② 不同的存储类型，例如，纯文本文件、HBase 中的文件。

③ 将元数据保存在关系数据库中，大大减少了在查询过程中执行语义检查的时间。

④ 可以直接使用存储在 Hadoop 文件系统中的数据。

⑤ 内置大量用户函数来操作时间、字符串和其他的数据挖掘工具，支持用户扩展 UDF 函数来完成内置函数无法实现的操作。

⑥ 类 SQL 的查询方式，将 SQL 查询转换为 MapReduce 的 Job 在 Hadoop 集群上执行。

### 二、Hive 架构

Hive 架构包括 CLI（Command Line Interface）、JDBC/ODBC、Thrift Server、Hive Web Interface（HWI）、metastore 和 Driver（Compiler、Optimizer 和 Executor）组件，如图 6-1 所示。

① Driver 组件：核心组件，整个 Hive 的核心，该组件包括 Compiler、Optimizer 和 Executor，它的作用是将我们写的 HQL 语句进行解析、编译优化，生成执行计划，然后调用底层的 MapReduce

计算框架。

② Metastore 组件：元数据服务组件，这个组件存储 Hive 的元数据，Hive 的元数据存储在关系数据库中，Hive 支持的关系数据库有 Derby、MySQL。

③ CLI：CLI 是 Command Line Interface 的缩写，即命令行接口。

④ Thrift Servers：提供 JDBC 和 ODBC 接入的能力，它用来进行可扩展且跨语言的服务开发，Hive 集成了该服务，能让不同的编程语言调用 Hive 的接口。

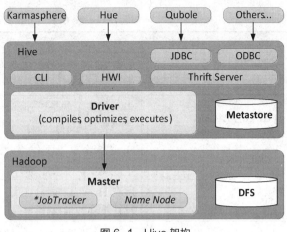

图 6-1　Hive 架构

⑤ Hive Web Interface（HWI）：Hive 客户端提供了一种通过网页的方式访问 Hive 所提供的服务。这个接口对应 Hive 的 HWI（Hive Web Interface）组件。

Hive 将通过 CLI 接入、JDBC/ODBC 接入或者 HWI 接入的相关查询，通过 Driver（Compiler、Optimizer 和 Executor）进行编译，分析优化，最后变成可执行的 MapReduce。

### 三、Metastore

Metastore 是系统目录（Catalog）用于保存 Hive 中所存储的表的元数据（Metadata）信息，它存放了表、区、列、类型、规则模型的所有信息，并且可以通过 Thrift 接口进行修改和查询。它为编译器提供高效的服务，所以，它会存放在一个传统的 RDBMS 中，利用关系模型进行管理。这个信息非常重要，所以需要备份，并且支持查询的可扩展性。

Metastore 是 Hive 被用作传统数据库（如 Oracle 和 DB2）解决方案时区别于其他类似系统的一个特征。

Metastore 包含如下几部分：

- Database 是表的名字空间，默认的数据库（database）名为"default"。
- Table 表的原数据包含的信息有列（list of columns）和它们的类型（types）、拥有者（owner）、存储空间（storage）和 SerDe 的信息。

Partition 每个分区都有自己的列（columns）、SerDe 和存储空间（storage）。

将元数据存储在 RDBMS 中，有 3 种模式可以连接到数据库。

① 单用户模式（Single User Mode）：此模式连接到一个 In-memory 的数据库 Derby，只能允许一个会话连接，一般用于单元测试，如图 6-2 所示。

② 多用户模式（Multi User Mode）：通过网络连接到一个数据库，一般使用 MySQL 作为元数据库，Hive 内部对 MySQL 提供了很好的支持，如图 6-3 所示。

使用 MySQL 作为 Hive MetaStore 的存储数据库，主要涉及的表如表 6-1 所示。

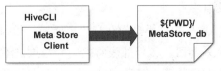

图 6-2　单用户模式

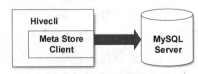

图 6-3　多用户模式

表 6-1　元数据涉及的表

| 表　名 | 说　明 | 关联键 |
| --- | --- | --- |
| TBLS | 所有 Hive 表的基本信息（表名、创建时间、所属者等） | TBL_ID,SD_ID |
| TABLE_PARAM | 表级属性（如是否外部表、表注释、最后修改时间等） | TBL_ID |
| COLUMNS | Hive 表字段信息（字段注释、字段名、字段类型、字段序号） | SD_ID |
| SDS | 所有 Hive 表、表分区所对应的 HDFS 数据目录和数据格式 | SD_ID,SERDE_ID |
| SERDE_PARAM | 序列化反序列化信息，如行分隔符、列分隔符、NULL 的表示字符等 | SERDE_ID |
| PARTITIONS | Hive 表分区信息（所属表、分区值） | PART_ID,SD_ID,TBL_ID |
| PARTITION_KEYS | Hive 分区表分区键（即分区字段） | TBL_ID |

③ 远程服务器模式（Remote Server Mode）：用于非 Java 客户端访问元数据库，在服务器端启动一个 MetaStoreServer，客户端利用 Thrift 协议通过 MetaStoreServer 访问元数据库，如图 6-4 所示。

图 6-4　远程服务器模式

### 四、数据存储

Hive 本身没有专门的数据存储格式，也没有为数据建立索引，只需要在创建表的时候告诉 Hive 数据中的列分隔符和行分隔符，Hive 就可以解析数据。所以，往 Hive 表中导入数据只是简单地将数据移动到表所在的目录中。

Hive 中主要包含以下几种数据模型：Table（表）、External Table（外部表）、Partition（分区）、Bucket（桶）。

① 表：Hive 中的表和关系型数据库中的表在概念上很类似，每个表在 HDFS 中都有相应的目录用来存储表的数据，这个目录可以通过${HIVE_HOME}/conf/hive-site.xml 配置文件中的 hive.metastore.warehouse.dir 属性来配置，这个属性默认的值是/user/hive/warehouse（这个目录在 HDFS 上），可以根据实际情况来修改这个配置。如果有一个表 sw，那么在 HDFS 中会创建/user/hive/warehouse/sw 目录（这里假定 hive.metastore.warehouse.dir 配置为/user/hive/warehouse）；sw 表所有的数据都存放在这个目录中。

② 外部表：Hive 中的外部表和表很类似，但是其数据不是放在自己表所属的目录中，而是存放到别处，这样做的好处是如果要删除这个外部表，该外部表所指向的数据是不会被删除的，它只会删除外部表对应的元数据；而如果要删除表，该表对应的所有数据包括元数据都会被删除。

③ 分区：在 Hive 中，表的每一个分区对应表下的相应目录，所有分区的数据都存储在对应的目录中。例如，sw 表有 dt 和 city 两个分区，则对应 dt=20170210,city=SW 表的目录为 /user/hive/warehouse/dt=20170210/city=SW，所有属于这个分区的数据都存放在这个目录中。

④ 桶：对指定的列计算其 Hash，根据 Hash 值切分数据，目的是并行，每一个桶对应一个文件（注意和分区的区别）。例如，将 sw 表 id 列分散至 16 个桶中，首先对 id 列的值计算 Hash，对应 Hash 值为 0 的数据存储的 HDFS 目录为 /user/hive/warehouse/sw/part-00000；而 Hash 值为 20 的数据存储的 HDFS 目录为 /user/hive/warehouse/sw/part-00020。

**五、存储模式**

Hive 由两部分组成：服务端和客户端。服务端可以装在任何节点上，可以是 NameNode 上，也可以是 DataNode 的任意一个节点上，不过在 Hadoop 的 HA 环境中应该是在两个 NameNode 中都装成 Hive 的服务，并且 hive.metastore.warehouse.dir 配置成 hdfs://****，这样其他节点安装的 Hive 就都是客户端了，并且 hive.metastore.uris 值可以指向这两个 NameNode 的 IP。多个 IP 配置的 hive-site.xml 配置属性如下：

```
<property>
  <name>hive.metastore.uris</name>
  <value>uri1,uri2,... </value>//可配置多个uri
  <description>JDBC connect string for a JDBC metastore</description>
</property>
```

Metastore 有 3 种存储模式，如图 6-5 所示。

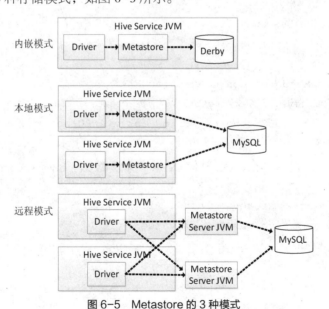

图 6-5  Metastore 的 3 种模式

① 内嵌模式：元数据保存在内嵌 Derby 中，只允许一个会话链接。hive-site.xml 配置属性如下：

```
<property>
  <name>hive.metastore.warehouse.dir</name>
  <value>/user/hive/warehouse</value>
```

```xml
    <description>location of default database for the warehouse</description>
</property>
<property>
    <name>hive.metastore.local</name>
    <value>true</value>
    <description>Use false if a production metastore server is used</description>
</property>
<property>
    <name>javax.jdo.option.ConnectionURL</name>
    <value>jdbc:derby:;databaseName=metastore_db;create=true</value>
    <description>JDBC connect string for a JDBC metastore.</description>
</property>
```

② 本地模式：本地安装 MySQL 替代 Derby 存储元数据。hive-site.xml 配置属性如下：

```xml
<property>
    <name>hive.metastore.warehouse.dir</name>
    <value>/user/hive/warehouse</value>
    <description>location of default database for the warehouse</description>
</property>
<property>
    <name>hive.metastore.local</name>
    <value>true</value>
    <description>Use false if a production metastore server is used</description>
</property>
<property>
    <name>javax.jdo.option.ConnectionURL</name>
    <value> jdbc:mysql://localhost:3306/hive?createDatabaseIfNotExist=true</value>
    <description>JDBC connect string for a JDBC metastore.</description>
</property>
```

③ 远程模式：远程安装 MySQL 替代 Derby 存储元数据。hive-site.xml 配置属性如下：

```xml
<property>
    <name>hive.metastore.warehouse.dir</name>
    <value>/user/hive/warehouse</value>
    <description>location of default database for the warehouse</description>
</property>
<property>
    <name>hive.metastore.local</name>
    <value>false</value>
    <description>Use false if a production metastore server is used</description>
</property>
<property>
    <name>javax.jdo.option.ConnectionURL</name>
```

```
    <value>jdbc:mysql://mysql_server:3306/hive?createDatabaseIfNotExist=
true</value>
    <description>JDBC connect string for a JDBC metastore.</description>
  </property>
```

用户如果不想使用默认的路径,可以配置一个不同目录来存储表数据。对于远程模式,默认存储路径是 hdfs://namenode_server/user/hive/warehouse;对于其他模式,默认路径是 file:///user/hive/warehouse。

对数据量比较小的操作,可以在本地执行,这样比提交任务到集群执行效率要快很多。配置如下参数,可以开启 Hive 的本地模式:

```
hive> set hive.exec.mode.local.auto=true;(默认为 false)
```

当一个 job 满足如下条件时才能真正使用本地模式。

① job 的输入数据大小必须小于参数:hive.exec.mode.local.auto.inputbytes.max(默认 128MB);

② job 的 map 数必须小于参数:hive.exec.mode.local.auto.tasks.max(默认 4);

③ job 的 reduce 数必须为 0 或者 1。

## 六、启动方式

Hive 启动方式有如下 3 种:

(1) Hive CLI 模式

运行命令:$ hive 或者 $ hive --service cli

作用:用于 Linux 平台命令行查询,查询语句基本与 MySQL 查询语句类似。

(2) Hive Web 界面的启动方式

运行命令:$ hive --service hwi &        (& 表示后台运行)

作用:用于通过浏览器来访问 Hive,浏览器访问地址是:http://master:9999/hwi。

(3) Hive 远程服务启动方式

运行命令:$ hive --service hiveserver2 &    (&表示后台运行)

作用:用于 Java、Python 等程序访问 hive,默认使用端口为 10000。

用 Java、Python 等程序实现通过 jdbc 等驱动的访问 Hive 就用这种起动方式了。

## 七、HiveServer 和 HiveServer2

HiveServer 和 HiveServer2 都允许远程客户端使用多种编程语言,通过 HiveServer 或 HiveServer2,客户端可以在不启动 CLI 的情况下对 Hive 中的数据进行操作。HiveServer 和 HiveServer2 都是基于 Thrift 的,但 HiveSever 有时被称为 Thrift server。由于 HiveServer 不能处理多于一个客户端的并发请求(HiveServer 使用的 Thrift 接口所导致的限制),又不能通过修改 HiveServer 的代码修正。因此,在 Hive-0.11.0 版本中重写了 HiveServer 代码,得到了 HiveServer2,进而解决了该问题。HiveServer2 支持多客户端的并发和认证,为开放 API 客户端如 JDBC、ODBC 提供更好的支持。从 hive0.15 起就不再支持 hiveserver 了,新的 hiveServer2 版本不再支持 Hive-on-MR。

可以通过 $ hive --service hiveserver2 或者 $ hiveserver2 启动 hiveserver2 服务。

hiveserver2 主要配置如下:(配置文件 hive-site.xml):

（1）配置监听端口和主机

```xml
<property>
  <name>hive.server2.thrift.port</name>
  <value>10000</value>
</property>
<property>
  <name>hive.server2.thrift.bind.host</name>
  <value>master</value>
</property>
```

（2）设置 impersonation

```xml
<property>
  <name>hive.server2.enable.doAs</name>
  <value>true</value>
</property>
```

这样 hive server 会以提交用户的身份去执行语句，如果设置为 false，则会以其 hive server daemon 的 admin user 来执行语句。

（3）hiveserver2 节点配置

```xml
<property>
  <name>hive.metastore.uris</name>
  <value>thrift://master:9083</value>
  <description>Thrift URI for the remote metastore. Used by metastore client to connect to remote metastore.</description>
</property>
```

新版本 Hiveserver2 已经不再需要 hive.metastore.local 这个配置项了（hive.metastore.uris 为空，则表示 metastore 在本地，否则就是远程），远程的话直接配置 hive.metastore.uris 即可。

（4）Zookeeper 配置

```xml
<property>
  <name>hive.support.concurrency</name>
  <value>true</value>
  <description>
  Whether Hive supports concurrency control or not.
  A ZooKeeper instance must be up and running when using zookeeper Hive lock manager </description>
</property>
<property>
  <name>hive.zookeeper.quorum</name>
  <value>master,slave1,slave2</value>
</property>
```

没有配置 hive.zookeeper.quorum 会导致无法并发执行 Hive QL 请求和数据异常。

（5）hiveserver2 的 Web UI 配置

```xml
<property>
  <name>hive.server2.webui.host</name>
  <value>0.0.0.0</value>
  <description>The host address the HiveServer2 WebUI will listen on</description>
</property>
<property>
  <name>hive.server2.webui.port</name>
  <value>10002</value>
  <description>The port the HiveServer2 WebUI will listen on. This can be set to 0 or a negative integer to disable the web UI</description>
</property>
```

Hive 2.0 以后才支持 Web UI，在以前的版本中并不支持。

# 【任务实施】

## 一、Master 节点安装 MySQL 软件

① 安装 MySQL，操作如下：

```
hadoop@master:~$ sudo apt-get install mysql-client mysql-server
```

② 设置 MySQL 参数，操作如下：

```
hadoop@master:~$ sudo vi /etc/mysql/mysql.conf.d/mysqld.cnf
```

修改参数如下：

```
bind-address = 127.0.0.1 改为 0.0.0.0
```

添加内容如下：

```
default-storage-engine = innodb
innodb_file_per_table
collation-server = utf8_general_ci
init-connect = 'SET NAMES utf8'
character-set-server = utf8
```

③ 设置开机自动启动，操作如下：

```
hadoop@master:~$ sudo systemctl enable mysql.service
Synchronizing state of mysql.service with SysV service script with /lib/systemd/systemd-sysv-install.
Executing: /lib/systemd/systemd-sysv-install enable mysql
```

④ 启动 MySQL 服务，操作如下：

```
hadoop@master:~$ sudo systemctl start mysql.service
```

⑤ 给 MySQL 用户 hive 授权，修改用户 hive 不需要 SSL，操作如下：

```
hadoop@master:~$ mysql -uroot -p123456
mysql> GRANT ALL PRIVILEGES ON *.* TO 'hive'@'%' IDENTIFIED BY '123456';
Query OK, 0 rows affected, 1 warning (0.00 sec)
mysql> GRANT ALL PRIVILEGES ON *.* TO 'hive'@'localhost' IDENTIFIED BY '123456';
```

```
Query OK, 0 rows affected, 1 warning (0.00 sec)
mysql> ALTER USER 'hive'@'%' REQUIRE none;
Query OK, 0 rows affected (0.15 sec)
mysql> exit
Bye
```

⑥ 重启 MySQL 服务，操作如下：

```
hadoop@master:~$ sudo systemctl restart mysql.service
```

## 二、Master 节点安装 Hive 软件

① 下载 Hive 软件包到 /home/hadoop 目录下，下载地址如下：

http://mirrors.aliyun.com/apache/hive/hive-2.1.1/apache-hive-2.1.1-bin.tar.gz

② 以用户 hadoop 登录 Master 节点，安装 Hive 软件，操作如下：

```
hadoop@master:~$ cd /opt
hadoop@master:/opt$ sudo tar xvzf /home/hadoop/apache-hive-2.1.1-bin.tar.gz
hadoop@master:/opt$ sudo chown -R hadoop:hadoop /opt/apache-hive-2.1.1-bin
```

## 三、Master 节点设置 Hive 参数

① Hive 配置文件改名，操作如下：

```
hadoop@master:/opt$ cd /opt/apache-hive-2.1.1-bin/conf
hadoop@...$ mv beeline-log4j2.properties.template beeline-log4j2.properties
hadoop@...$ mv hive-env.sh.template hive-env.sh
hadoop@...$ mv hive-exec-log4j2.properties.template hive-exec-log4j2.properties
hadoop@...$ mv hive-log4j2.properties.template hive-log4j2.properties
hadoop@...$ mv llap-cli-log4j2.properties.template llap-cli-log4j2.properties
hadoop@...$ mv llap-daemon-log4j2.properties.template llap-daemon-log4j2.properties
```

② 修改 hive-env.sh 文件，操作如下：

```
hadoop@master:/opt/apache-hive-2.1.1-bin/conf$ vi hive-env.sh
```

添加内容如下：

```
HADOOP_HOME=/opt/hadoop-2.7.3/
export HIVE_CONF_DIR=/opt/apache-hive-2.1.1-bin/conf/
export HIVE_AUX_JARS_PATH=/opt/apache-hive-2.1.1-bin/lib/
```

③ 新建 hive-site.xml 文件，操作如下：

```
hadoop@master:/opt/apache-hive-2.1.1-bin/conf$ vi hive-site.xml
```

内容如下：

```xml
<?xml version="1.0"?>
<?xml-stylesheet type="text/xsl" href="configuration.xsl"?>
<configuration>
  <property>
    <name>hive.metastore.warehouse.dir</name>
    <value>/hive/warehouse</value>
    <description>location of default database for the warehouse</description>
  </property>
```

```xml
<property>
  <name>javax.jdo.option.ConnectionURL</name>
  <value>jdbc:mysql://master:3306/hive?createDatabaseIfNotExist=true
      &amp;useSSL=false</value>
  <description>JDBC connect string for a JDBC metastore.</description>
</property>
<property>
  <name>javax.jdo.option.ConnectionDriverName</name>
  <value>com.mysql.jdbc.Driver</value>
  <description>Driver class name for a JDBC metastore</description>
</property>
<property>
  <name>javax.jdo.option.ConnectionUserName</name>
  <value>hive</value>
  <description>Username to use against metastore database</description>
</property>
<property>
  <name>javax.jdo.option.ConnectionPassword</name>
  <value>123456</value>
  <description>password to use against metastore database</description>
</property>
<property>
  <name>hive.querylog.location</name>
  <value>/opt/apache-hive-2.1.1-bin/logs</value>
  <description>Location of Hive run time structured log file</description>
</property>
<property>
  <name>hive.metastore.uris</name>
  <value>thrift://master:9083</value>
  <description>Thrift URI for the remote metastore. Used by metastore client to connect to remote metastore.</description>
</property>
<property>
  <name>hive.server2.webui.host</name>
  <value>0.0.0.0</value>
</property>
<property>
  <name>hive.server2.webui.port</name>
  <value>10002</value>
</property>
</configuration>
```

④ 下载"MySQL Connector/J"连接包,操作如下:

```
hadoop@master...$ cd /opt/apache-hive-2.1.1-bin/lib
```

```
hadoop@master:/opt/apache-hive-2.1.1-bin/lib$ wget http://central.maven.org/
maven2/mysql/mysql-connector-java/5.1.40/mysql-connector-java-5.1.40.jar
```

⑤ 修改环境变量，操作如下：

```
hadoop@master:/opt/apache-hive-2.1.1-bin/lib$ cd
hadoop@master:~$ vi .profile
```

添加内容如下：

```
export HIVE_HOME=/opt/apache-hive-2.1.1-bin
export PATH=$PATH:$HIVE_HOME/bin
```

⑥ 环境变量生效，操作如下：

```
hadoop@master:~$ source .profile
```

## 四、Master 节点启动 Hive 服务

① 创建数据库 hive，导入 hive-schema，操作如下：

```
hadoop@master:~$ cd /opt/apache-hive-2.1.1-bin/scripts/metastore/upgrade/ mysql
hadoop@master:/opt/apache.../mysql$ mysql -hmaster -uhive -p123456
mysql> create database hive character set latin1;
Query OK, 1 row affected (0.01 sec)
mysql> use hive;
Database changed
mysql> source hive-schema-2.1.0.mysql.sql;
mysql> exit
Bye
```

② 启动 hive metastore 服务，操作如下：

```
hadoop@master:~$ hive --service metastore &
[1] 41937
Starting Hive Metastore Server
```

③ 启动 hiveserver2 服务，操作如下：

```
hadoop@master:~$ hiveserver2 &
[2] 42425
```

## 五、客户端主机安装 Hive 软件

① 以用户 hadoop 登录客户端主机，安装 Hive 软件，客户端主机可以是任何一台 Slave 节点机或 sw-desktop 主机，操作如下：

```
hadoop@sw-desktop:~$ sudo scp -r hadoop@master:/opt/apache-hive-2.1.1-bin
/opt/apache-hive-2.1.1-bin
hadoop@master's password:
hadoop@sw-desktop:~$ sudo chown -R hadoop:hadoop /opt/apache-hive-2.1.1-bin
```

② 修改 hive-site.xml 文件，操作如下：

```
hadoop@sw-desktop:~$ vi /opt/apache-hive-2.1.1-bin/conf/hive-site.xml
```

内容如下：

```
<?xml version="1.0"?>
<?xml-stylesheet type="text/xsl" href="configuration.xsl"?>
```

```xml
<configuration>
  <property>
    <name>hive.metastore.warehouse.dir</name>
    <value>/hive/warehouse</value>
    <description>location of default database for the warehouse </description>
  </property>
  <property>
    <name>hive.querylog.location</name>
    <value>/opt/apache-hive-2.1.1-bin/logs</value>
    <description>Location of Hive run time structured log file</description>
  </property>
  <property>
    <name>hive.metastore.uris</name>
    <value>thrift://master:9083</value>
    <description>Thrift URI for the remote metastore. Used by metastore client to connect to remote metastore.</description>
  </property>
</configuration>
```

③ 修改环境变量，操作如下：

```
hadoop@sw-desktop:~$ vi .profile
```

添加内容如下：

```
export HIVE_HOME=/opt/apache-hive-2.1.1-bin
export PATH=$PATH:$HIVE_HOME/bin
```

④ 环境变量生效，操作如下：

```
hadoop@sw-desktop:~$ source .profile
```

### 六、测试 Hive

① 客户端主机测试 Hive，操作如下：

```
hadoop@sw-desktop:~$ hive
hive>
```

② 查看数据库，操作如下：

```
hive> show databases;
OK
default
Time taken: 1.243 seconds, Fetched: 1 row(s)
hive>
```

③ 打开浏览器，在地址栏中输入"http://master:10002/hiveserver2.jsp"，查看 HiveServer2 服务，如图 6-6 所示。

图 6-6　HiveServer2

## 任务 2　Hive CLI

### 【任务概述】

Hive CLI 是 Hive 的交互工具，Hive CLI 提供丰富的命令，本任务主要使用 Hive CLI 创建表和外部表、完成数据导入和查看、删除表。同时新版本也提供新的交互工具 Beeline，使用 Beeline 完成表的查询等。

### 【支撑知识】

#### 一、Hive 命令语法

Hive 命令提供命令行界面，也就是 CLI，执行命令如下：

```
$ hive --help
Usage ./hive <parameters> --service serviceName <service parameters>
Service  List:  beeline  cleardanglingscratchdir  cli  hbaseimport
hbaseschematool help hiveburninclient hiveserver2 hplsql hwi jar lineage
llapdump llap llapstatus metastore metatool orcfiledump rcfilecat schemaTool
version
 Parameters parsed:
   --auxpath : Auxillary jars
   --config : Hive configuration directory
   --service : Starts specific service/component. cli is default
 Parameters used:
   HADOOP_HOME or HADOOP_PREFIX : Hadoop install directory
   HIVE_OPT : Hive options
 For help on a particular service:
   ./hive --service serviceName --help
 Debug help:  ./hive --debug -help
```

用户可以通过--serviceName 服务名称来启用某个服务，常用服务如表 6-2 所示。

表 6-2　Hive 常用服务

| 选　项 | 名　称 | 描　述 |
| --- | --- | --- |
| cli | 命令行界面 | 用户定义表、执行查询等。这个是默认服务 |
| hiveserver | Hive Server | 监听来自其他进程的 Thrift 连接的一个守护进程 |
| hwi | Hive Web 界面 | 是一个可以执行查询语句的其他命令的简单 Web 界面 |
| jar | | hadoop jar 命令的一个扩展，这样可以执行需要 hive 环境的应用 |
| metastore | | 启动一个扩展的 Hive 元数据服务，可以供多客户端使用 |
| rcfilecat | | 一个可以打印出 RCFile 格式文件内容的工具 |

Hive CLI 是 Hive 最常用的交互方式，命令选项如下：

```
$ hive --help --service cli
usage: hive
 -d,--define <key=value>           Variable subsitution to apply to hive
```

```
  --database <databasename>         Specify the database to use
-e <quoted-query-string>            SQL from command line
-f <filename>                       SQL from files
-H,--help                           Print help information
  --hiveconf <property=value>       Use value for given property
  --hivevar <key=value>             Variable subsitution to apply to hive
                                    commands. e.g. --hivevar A=B
-i <filename>                       Initialization SQL file
-S,--silent                         Silent mode in interactive shell
-v,--verbose                        Verbose mode (echo executed SQL to the
                                    console)
```

## 二、Hive 数据类型

Hive 支持多种不同长度的整数型和浮点型数据类型，支持布尔类型，也支持无长度限制的字符串类型，Hive 支持的数据类型如表 6-3 所示。

表 6-3  数据类型

| 数据类型 | 长 度 | 例 子 |
| --- | --- | --- |
| TINYINT | 1byte 有符号整数 | 10Y |
| SMALLINT | 2byte 有符号整数 | 20S |
| INT | 4byte 有符号整数 | 30 |
| BIGINT | 8byte 有符号整数 | 40L |
| BOOLEAN | TRUE 或者 FALSE | TRUE |
| FLOAT | 单精度浮点数 | 3.14159 |
| DOUBLE | 双精度浮点数 | 3.14159 |
| DECIMAL | DECIMAL(precision,scale) | DECIMAL(10,0) |
| STRING | 字符序列。可以指定字符集，可以使用单引号或者双引号 | 'hello welcome', "a good idea" |
| VARCHAR | 1 to 65535 | VARCHAR(50) |
| CHAR | 255 | CHAR(10) |
| TIMESTAMP | 整数、浮点数或者字符串 | 1486479778533<br>1486479778533.123456789<br>'2017-02-11 15:16:33.123456789' |
| BINARY | 字节数组 | |
| STRUCT | 和 C 语言的 struct 相似，都可以通过 "." 符号访问元素内容 | STRUCT('John','Tom') |
| MAP | MAP 是一组键-值对元组集合，使用数组表示法可以访问元素 | MAP('First','Second','Third') |
| ARRAY | 数组是一组具有相同类型和名称的变量的集合。这些变量称为数组的元素，每个数组元素都有一个编号，编号从 0 开始 | ARRAY('JARRY','Tom') |

### 三、Hive 文本文件数据编码

Hive 支持逗号","分隔符和制表符"\t"分隔符,Hive 使用 field 来表示替换默认分隔符的字符。Hive 中默认的记录和字段分隔符如表 6-4 所示。

表 6-4 Hive 记录和字段默认分隔符

| 分隔符 | 描述 |
| --- | --- |
| \n | 对于文本文件来说,每一行都是一条记录,因此,换行符可以分隔记录 |
| ^A(Ctrl+A) | 用于分隔字段(列)。在 CREATE TABLE 语句中可以使用八进制编码\001 表示 |
| ^B | 用于分隔 ARRAY 或者 STRUCT 中的元素,或用于 MAP 中的键-值对之间的分隔。在 CREATE TABLE 语句中可以使用八进制编码\002 表示 |
| ^C | 用于 MAP 中键和值之间的分隔。在 CREATE TABLE 语句中可以使用八进制编码\003 表示 |

### 四、Hive 的运算符与运算函数

Hive 有 4 种类型的运算符:关系运算符、算术运算符、逻辑运算符、复杂运算符。
(1)关系运算符

关系运算符被用来比较两个操作数。Hive 的关系运算符如表 6-5 所示。

表 6-5 Hive 关系运算符

| 运算符 | 操作数 | 描述 |
| --- | --- | --- |
| A = B | 所有基本类型 | 如果表达式 A 等于表达式 B,结果为 TRUE,否则为 FALSE |
| A != B | 所有基本类型 | 如果表达式 A 不等于表达式 B,返回 TRUE,否则返回 FALSE |
| A < B | 所有基本类型 | 如果表达式 A 小于表达式 B,返回 TRUE,否则返回 FALSE |
| A <= B | 所有基本类型 | 如果表达式 A 小于或等于表达式 B,返回 TRUE,否则返回 FALSE |
| A > B | 所有基本类型 | 如果表达式 A 大于表达式 B,返回 TRUE,否则返回 FALSE |
| A >= B | 所有基本类型 | 如果表达式 A 大于或等于表达式 B,返回 TRUE,否则返回 FALSE |
| A IS NULL | 所有类型 | 如果表达式 A 的计算结果为 NULL,返回 TRUE,否则返回 FALSE |
| A IS NOT NULL | 所有类型 | 如果表达式 A 的计算结果为 NULL,返回 FALSE,否则返回 TRUE |
| A LIKE B | 字符串 | 如果字符串模式 A 匹配到 B,返回 TRUE,否则返回 FALSE |
| A RLIKE B | 字符串 | 如果 A 或 B 为 NULL,则返回 NULL;如果 A 任何子字符串匹配 Java 正则表达式 B,则返回 TRUE,否则返回 FALSE |
| A REGEXP B | 字符串 | 等同于 RLIKE |

(2)算术运算符

算术运算符支持数字类型操作数,返回数字类型。Hive 的算术运算符如表 6-6 所示。

表 6-6　Hive 算术运算符

| 运算符 | 操作数 | 描　　述 |
|---|---|---|
| A + B | 所有数字类型 | A 加 B 的结果 |
| A-B | 所有数字类型 | A 减去 B 的结果 |
| A * B | 所有数字类型 | A 乘以 B 的结果 |
| A / B | 所有数字类型 | A 除以 B 的结果 |
| A % B | 所有数字类型 | A 除以 B 产生的余数 |
| A & B | 所有数字类型 | A 和 B 的按位与结果 |
| A \| B | 所有数字类型 | A 和 B 的按位或结果 |
| A ^ B | 所有数字类型 | A 和 B 的按位异或结果 |
| ~A | 所有数字类型 | A 按位非的结果 |

（3）逻辑运算符

逻辑运算符的操作数是逻辑表达式，所有这些返回 TRUE 或 FALSE。Hive 的逻辑运算符如表 6-7 所示。

表 6-7　Hive 逻辑运算符

| 运算符 | 操作数 | 描　　述 |
|---|---|---|
| A AND B | boolean | 如果 A 和 B 都是 TRUE，为 TRUE，否则为 FALSE |
| A && B | boolean | 类似于 A AND B |
| A OR B | boolean | 如果 A 或 B 或两者都是 TRUE，为 TRUE，否则为 FALSE |
| A \|\| B | boolean | 类似于 A OR B |
| NOT A | boolean | 如果 A 是 FALSE，为 TRUE，否则为 FALSE |
| !A | boolean | 类似于 NOT A |

（4）复杂运算符

复杂的运算符提供一个表达式来接入复杂类型的元素。Hive 的复杂运算符如表 6-8 所示。

表 6-8　Hive 复杂的运算符

| 运算符 | 操作数 | 描　　述 |
|---|---|---|
| A[n] | A 是一个数组，n 为整数 | 它返回数组 A 的第 n 个元素，第一个元素的索引为 0 |
| M[key] | M 是一个 Map<K, V> 且 key 的类型为 K | 它返回对应于映射中关键字的值 |
| S.x | S 是一个结构 | 它返回 S 的 x 字段 |

（5）Hive 运算函数

Hive 内部提供了很多运算函数，如数学函数、聚合函数、表生成函数及其他内置函数。表 6-9 列举了一些常用运算函数。

表 6-9  Hive 常用运算函数

| 运算函数 | 返回类型 | 描述 |
| --- | --- | --- |
| round(double a) | BIGINT | 返回 a 四舍五入后的 BIGINT 值 |
| floor(double a) | BIGINT | 返回等于或小于 a 的最大 BIGINT 值 |
| ceil(double a) | BIGINT | 返回等于或大于 a 的最小 BIGINT 值 |
| rand(), rand(int seed) | DOUBLE | 返回一个随机数 |
| concat(string A, string B,…) | STRING | 返回 A 连接 B 产生的字符串 |
| substr(string A, int start) | STRING | 返回起始位置开始到结束的字符串 |
| substr(string A, int start, int length) | STRING | 返回从起始位置开始到给定长度的字符串 |
| upper(string A) | STRING | 将字符串 A 中的小写字母转为大写字母 |
| ucase(string A) | STRING | 与 upper() 功能相同 |
| lower(string A) | STRING | 将字符串 A 中的大写字母转为小写字母 |
| lcase(string A) | STRING | 与 lower() 功能相同 |
| trim(string A) | STRING | 删除字符串 A 两端空格 |
| ltrim(string A) | STRING | 删除字符串 A 左侧空格 |
| rtrim(string A) | STRING | 删除字符串 A 右侧空格 |
| regexp_replace(string A, string B, string C) | STRING | 用字符串 C 替换所有字符串 A 中符合正则表达式 B 匹配结果的子字符串 |
| size(Map<K.V>) | INT | 返回 Map 类型的元素数量 |
| size(Array<T>) | INT | 返回数组类型的元素数量 |
| cast(<expr> as <type>) | VALUE OF <TYPE> | 转换 expr 为指定数据类型，如果转换不成功，返回 NULL |
| from_unixtime(int unixtime [, string format]) | STRING | 转化 UNIX 时间戳到当前时区的时间格式 |
| to_date(string timestamp) | STRING | 返回时间戳字符串中的日期 |
| year(string date) | INT | 返回时间戳字符串中的年份 |
| month(string date) | INT | 返回时间戳字符串中的月份 |
| day(string date) | INT | 返回时间戳字符串中的天数 |
| get_json_object(string json_string, string path) | STRING | 解析 json 的字符串 json_string，返回 path 指定的内容。如果输入的 json 字符串无效，那么返回 NULL |
| count(*), count(expr), | IGINT | 返回检索行的总数 |
| sum(col), sum(DISTINCT col) | DOUBLE | 返回列 col 所有元素的总和 |
| avg(col), avg(DISTINCT col) | DOUBLE | 返回列 col 所有元素的平均值 |
| min(col) | DOUBLE | 返回该组中列的最小值 |
| max(col) | DOUBLE | 返回该组中列的最大值 |

## 五、数据库操作

（1）创建数据库

Hive 是一种数据库技术，可以定义数据库和表来分析结构化数据。结构化数据分析以表方式存储数据，并通过查询来分析。Default 是 Hive 默认的数据库。

CREATE DATABASE 是在 Hive 中创建数据库的语句。它的语法如下：

```
CREATE DATABASE|SCHEMA [IF NOT EXISTS] <database name>
```

例如，创建 testdb 数据库：

```
hive> CREATE DATABASE testdb;
```

或

```
hive> CREATE SCHEMA testdb;
```

（2）删除数据库

DROP DATABASE 是删除所有的表并删除数据库的语句。它的语法如下：

```
DROP DATABASE StatementDROP (DATABASE|SCHEMA) [IF EXISTS] database_name
[RESTRICT|CASCADE];
```

例如，删除 testdb 数据库：

```
hive> DROP DATABASE IF EXISTS testdb;
hive> DROP SCHEMA testdb;
```

（3）创建表

CREATE TABLE 是在 Hive 中创建表的语句。它的语法如下：

```
CREATE [TEMPORARY] [EXTERNAL] TABLE [IF NOT EXISTS] [db_name.] table_name
[(col_name data_type [COMMENT col_comment], ...)]
[COMMENT table_comment]
[ROW FORMAT row_format]
[STORED AS file_format]
```

例如，创建 employee 表，详细信息如表 6-10 所示。

表 6-10　employee 表信息

| No | 字段名称 | 数据类型 |
| --- | --- | --- |
| 1 | id | int |
| 2 | name | String |
| 3 | salary | float |
| 4 | designation | string |

```
hive> CREATE TABLE IF NOT EXISTS employee (id int, name String,
   > salary float, destination String)
   > COMMENT 'Employee details'
   > ROW FORMAT DELIMITED
   > FIELDS TERMINATED BY '\t'
   > LINES TERMINATED BY '\n'
   > STORED AS TEXTFILE;
```

（4）装入数据

一般来说，在 SQL 中创建表后，就可以使用 INSERT 语句插入数据。但在 Hive 中，可以使用 LOAD DATA 语句插入数据。有两种方法用来加载数据：一种是从本地文件系统加载数据，另一种是从 Hadoop 文件系统加载数据。

LOAD DATA 语句的语法如下：

```
LOAD DATA [LOCAL] INPATH 'filepath' [OVERWRITE] INTO TABLE tablename
[PARTITION (partcol1=val1, partcol2=val2 ...)]
```

其中：

LOCAL 是标识符指定本地路径，是可选的。

OVERWRITE 是可选的，覆盖表中的数据。

PARTITION 是可选的。

例如，/opt/employee.txt 装入 employee 表。

employee.txt 文件内容如下：

```
1201   Gopal          45000       Technical manager
1202   Manisha        45000       Proof reader
1203   Masthanvali    40000       Technical writer
1204   Kiran          40000       Hr Admin
1205   Kranthi        30000       Op Admin
hive> LOAD DATA LOCAL INPATH '/opt/employee.txt'
    > OVERWRITE INTO TABLE employee;
```

（5）修改表

ALTER TABLE 在 Hive 中用来修改表。它的语法如下：

```
ALTER TABLE name RENAME TO new_name
ALTER TABLE name ADD COLUMNS (col_spec[, col_spec ...])
ALTER TABLE name DROP [COLUMN] column_name
ALTER TABLE name CHANGE column_name new_name new_type
ALTER TABLE name REPLACE COLUMNS (col_spec[, col_spec ...])
```

例如，employee 改名为 emp：

```
hive> ALTER TABLE employee RENAME TO emp;
```

（6）删除表

DROP TABLE 在 Hive 中用来删除表。它的语法如下：

```
DROP TABLE [IF EXISTS] table_name;
```

例如，删除 employee 表：

```
hive> DROP TABLE IF EXISTS employee;
```

（7）Hive 分区

按照数据表将某一列或多列分为多个区。使用分区很容易对数据进行部分查询。表或分区细分成桶，以提供额外的结构，因此查询数据更高效。桶的工作是基于表的一些列的散列函数值。

例如，表 emp（id、name、dept、yoj）包含雇员数据（yoj 即加盟年份）。假设需要检索所有在 2015 年加盟的雇员详细信息，需要搜索整个表。但是，如果雇员数据按年份分区并将其分别存储在一个单独的文件中，则可以减少查询处理时间。下面的示例演示如何分区。

下面文件包含 emp 数据表数据：/opt/emp/file.txt

| id | name | dept | yoj |
|----|------|------|------|
| 1 | gopal | TP | 2015 |
| 2 | kiran | HR | 2015 |
| 3 | kaleel | SC | 2016 |
| 4 | Prasanth | SC | 2016 |

上面的数据划分成两个文件：

/opt/emp/file1.txt

| id | name | dept | yoj |
|----|------|------|------|
| 1 | gopal | TP | 2015 |
| 2 | kiran | HR | 2015 |

/opt/emp/file2.txt

| id | name | dept | yoj |
|----|------|------|------|
| 3 | kaleel | SC | 2016 |
| 4 | Prasanth | SC | 2016 |

① 添加分区。添加分区的语法如下：

```
ALTER TABLE table_name ADD [IF NOT EXISTS] PARTITION partition_spec
[LOCATION 'location1'] partition_spec [LOCATION 'location2'] ...;
partition_spec:
: (p_column = p_col_value, p_column = p_col_value, ...)
```

例如：

```
hive> ALTER TABLE emp
    > ADD PARTITION (year='2016')
    > location '/2015/part2015';
```

② 重命名分区。重命名分区的语法如下：

```
ALTER TABLE table_name PARTITION partition_spec RENAME TO PARTITION partition_spec;
```

例如：

```
hive> ALTER TABLE emp PARTITION (year='2016')
    > RENAME TO PARTITION (Yoj='2016');
```

③ 删除分区。删除分区的语法如下：

```
ALTER TABLE table_name DROP [IF EXISTS] PARTITION partition_spec, PARTITION partition_spec,...;
```

例如：

```
hive> ALTER TABLE emp DROP [IF EXISTS]
    > PARTITION (year='2016');
```

### 六、数据查询

Hive 查询语言是一种类似于 SQL 的语言,可以将结构化的数据文件映射为一张数据库表,并提供完整的 SQL 查询功能,可以将 SQL 语句转换为 MapReduce 任务进行运行,通过自己的 SQL 去查询分析需要的内容。SELECT 的语法如下:

```
SELECT [ALL | DISTINCT] select_expr, select_expr, ...
FROM table_reference
[WHERE where_condition]
[GROUP BY col_list]
[HAVING having_condition]
[CLUSTER BY col_list | [DISTRIBUTE BY col_list]
[ORDER BY col_list]]
[LIMIT number];
```

(1) SELECT…WHERE

例如,检索薪水超过 30000 元的员工详细信息:

```
hive> SELECT * FROM employee WHERE salary>30000;
```

(2) SELECT…ORDER BY

例如,按部门排序查询的员工详细信息:

```
hive> SELECT * FROM employee ORDER BY dept;
```

(3) SELECT…GROUP BY

例如,查询每个部门的员工数量:

```
hive> SELECT dept,count(*) FROM employee GROUP BY dept;
```

(4) SELECT…JOIN

JOIN 子句将相关表进行联接,类似于 SQL JOIN。不同类型的联接如下:

- JOIN。
- LEFT OUTER JOIN。
- RIGHT OUTER JOIN。
- FULL OUTER JOIN。

例如:

表 customers(id, name, age, address, salary)

表 orders(oid, date, customer_id, amount)

① JOIN 子句用于合并和检索来自多个表中的记录。JOIN 和 SQL 的 OUTER JOIN 类似。连接条件是使用主键和表的外键。

例如,查询执行 customers 和 orders 表的 JOIN:

```
hive> SELECT C.id, C.name, C.age, O.amount
    > FROM customers C JOIN orders O
    > ON (C.id = O.customer_id);
```

② LEFT OUTER JOIN 返回包含左表的全部行,以及右表中全部匹配的行(右表缺少匹配时使用 NULL 值填补)

例如,查询执行 customers 和 orders 表的 LEFT OUTER JOIN:

```
hive> SELECT C.id, C.name, C.age, O.amount
    > FROM customers C
    > LEFT OUTER JOIN orders O
    > ON (C.id = O.customer_id);
```

③ RIGHT OUTER JOIN 返回包含左表中全部匹配的行（左表缺少匹配时使用 NULL 值填补），以及右表的全部行。

例如，查询执行 customers 和 orders 表的 RIGHT OUTER JOIN：

```
hive> SELECT C.id, C.name, C.age, O.amount
    > FROM customers C
    >RIGHT OUTER JOIN orders O
    > ON (C.id = O.customer_id);
```

④ FULL OUTER JOIN 返回左表和右表的所有行，左右表缺少匹配时使用 NULL 值填补。

例如，查询执行 customers 和 orders 表的 RIGHT OUTER JOIN：

```
hive> SELECT C.id, C.name, C.age, O.amount
    > FROM customers C
    > FULL OUTER JOIN orders O
    > ON (C.id = O.customer_id);
```

## 七、Beeline 命令

Beeline 需要连接 HiveServer2，是基于 JDBC 客户端的 SQLLINE CLI。Beeline 部分帮助信息如下：

| 命令 | 说明 |
| --- | --- |
| !autocommit | 设置自动提交事务开（on）或关（off） |
| !close | 关闭当前数据库连接 |
| !closeall | 关闭所有当前打开的连接 |
| !columns | 列出所有指定表的所有列 |
| !commit | 提交当前事务（autocommit 设为 off） |
| !connect | 打开一个新连接 |
| !dbinfo | 获取数据库元数据信息 |
| !describe | 表结构信息 |
| !dropall | 移除当前数据库所有表 |
| !exportedkeys | 列出指定表所有输出键 |
| !go | 选择当前连接 |
| !help | 显示命令语法概要 |
| !history | 显示历史命令 |
| !importedkeys | 列出指定表所有输入键 |
| !indexes | 列出指定表所有索引 |
| !isolation | 为连接设置事务隔离 |
| !list | 列出当前连接 |

| 命令 | 说明 |
| --- | --- |
| !manual | 显示 Beeline 手册 |
| !metadata | 获得元数据信息 |
| !primarykeys | 列出指定表所有主键 |
| !procedures | 列出所有存储过程 |
| !quit | 退出 |
| !reconnect | 重新连接数据库 |
| !rollback | 回滚当前事务（autocommit 设为 off） |
| !run | 运行脚本文件 |
| !save | 保存当前变量和别名 |
| !scan | 扫描已安装的 JDBC 驱动 |
| !script | 保存脚本到文件中 |
| !set | 设置 beeline 变量 |
| !sh | 执行 shell 命令 |
| !sql | 执行 SQL 命令 |
| !tables | 列出数据库的所有表 |
| !typeinfo | 显示当前连接类型 |
| !verbose | 设置 verbose 模式为 on |

# 【任务实施】

## 一、基本命令

① 基本数据。

a. score 表信息如表 6-11 所示。

表 6-11  score 表详细信息

| No | 字段名称 | 数据类型 |
| --- | --- | --- |
| 1 | sno | int |
| 2 | name | string |
| 3 | java | decimal(10,2) |
| 4 | python | decimal(10,2) |

b. /home/hadoop/score.txt 文件信息如下：

```
610213    Tom     85    79
610215    John    80    85
610222    Marry   75    87
```

② 创建 score 表，操作如下：

```
Hive> CREATE TABLE score (
    sno int, name String,
    java decimal(10,2),
    python decimal(10,2)
    )
```

```
        ROW FORMAT DELIMITED
        FIELDS TERMINATED BY '\t'
        LINES TERMINATED BY '\n'
        STORED AS TEXTFILE;
OK
Time taken: 0.202 seconds
```

③ 导入数据，操作如下：

```
hive> load data local inpath 'score.txt' overwrite into table score;
Loading data to table default.score
OK
Time taken: 2.094 seconds
```

④ 查询 score 表，操作如下：

```
hive> select * from score;
OK
610213    Tom     85.00    79.00
610215    John    80.00    85.00
610222    Marry   75.00    87.00
Time taken: 1.167 seconds, Fetched: 3 row(s)
```

⑤ 查看 HDFS 数据，操作如下：

```
hive> dfs -ls /hive/warehouse/score;
Found 1 items
-rwxr-xr-x   3 hadoop supergroup         54 2017-02-12 11:27 /hive/warehouse/score/score.txt
hive> dfs -cat /hive/warehouse/score/score.txt;
610213    Tom     85    79
610215    John    80    85
610222    Marry   75    87
```

⑥ 查看元数据信息，操作如下：

```
hadoop@master:~$ mysql -hmaster -uhive -p123456
mysql> use hive;
Database changed
mysql> select TBL_ID,OWNER,TBL_NAME,TBL_TYPE,SD_ID from TBLS;
+--------+--------+----------+---------------+-------+
| TBL_ID | OWNER  | TBL_NAME | TBL_TYPE      | SD_ID |
+--------+--------+----------+---------------+-------+
|      1 | hadoop | score    | MANAGED_TABLE |     1 |
+--------+--------+----------+---------------+-------+
1 row in set (0.00 sec)
mysql> select * from DBS;
+-------+------------------+-----------------------+------+------------+------------+
| DB_ID | DESC             | DB_LOCATION_URI       | NAME | OWNER_NAME | OWNER_TYPE |
```

```
+----+------------------+---------------------------------------+---------+--------+------+
|  1 | Default Hive database | hdfs://master:9000/hive/ warehouse | default | public | ROLE |
+----+------------------+---------------------------------------+---------+--------+------+
1 row in set (0.00 sec)
mysql>
```

⑦ 删除表，操作如下：

```
hive> drop table score;
OK
Time taken: 0.457 seconds
```

⑧ 查看 HDFS，操作如下：

```
hive> dfs -ls /hive/warehouse;
hive>
```

HDFS 上的数据已经删除，这说明删除表时，HDFS 上的数据一起删除了。

## 二、外部表

① 基本数据如下：

```
hadoop@sw-desktop:~$ more test_ext.txt
1,master
2,slave1
3,slave2
```

② 上传数据到 HDFS 中，操作如下：

```
hive> dfs -mkdir /hive/warehouse/test_ext;
hive> dfs -put test_ext.txt /hive/warehouse/test_ext;
hive>
```

③ 创建外部表，操作如下：

```
hive> create external table test_external (
    id int,
    name string
    )
    row format delimited
    fields terminated by ','
    location '/hive/warehouse/test_ext';
OK
Time taken: 0.132 seconds
```

④ 查询 test_external 表，操作如下：

```
hive> select * from test_external;
OK
1    master
2    slave1
3    slave2
Time taken: 2.409 seconds, Fetched: 3 row(s)
hive>
```

⑤ 删除 test_external 表，操作如下：
```
hive> drop table test_external;
OK
Time taken: 0.366 seconds
```
⑥ 查看 HDFS，操作如下：
```
hive> dfs -ls /hive/warehouse/test_ext;
Found 1 items
-rw-r--r--   3 hadoop supergroup         27 2017-02-12 11:33 /hive/warehouse/test_ext/test_ext.txt
hive>
```
test_external 表已经删除，在 HDFS 中 test_ext.txt 文件并没有删除。说明外部表删除时，HDFS 上的数据没有删除，这一点与表删除不一样。

### 三、Beeline 命令

① 设置用户 hadoop 的访问权限，修改 hadoop 配置文件 core-site.xml，加入如下配置：
```
<property>
    <name>hadoop.proxyuser.hadoop.hosts</name>
    <value>*</value>
</property>
<property>
    <name>hadoop.proxyuser.hadoop.groups</name>
    <value>*</value>
</property>
```
集群节点都要修改，最后重启 HDFS，另外 /tmp 目权限也要设置。操作如下：
```
hadoop@master:~$ hdfs dfs -chmod -R 777 /tmp
```
② 连接 HiveServer2，操作如下：
```
hadoop@sw-desktop:~$ beeline
beeline> ?
beeline> !connect jdbc:hive2://master:10000 hive 123456
……
Connecting to jdbc:hive2://master:10000
Connected to: Apache Hive (version 2.1.1)
Driver: Hive JDBC (version 2.1.1)
17/02/12 16:32:32 [main]: WARN jdbc.HiveConnection: Request to set autoCommit to false; Hive does not support autoCommit=false.
Transaction isolation: TRANSACTION_REPEATABLE_READ
0: jdbc:hive2://master:10000>
```
③ 显示数据库，操作如下：
```
0: jdbc:hive2://master:10000> show databases;
+----------------+--+
| database_name  |
+----------------+--+
```

```
| default         |
+-----------------+--+
1 row selected (7.387 seconds)
```

④ 创建 test_exteranal 外部表，操作如下：

```
0: jdbc:hive2://master:10000> create external table test_external (
                              id int,
                              name string
                              )
                              row format delimited
                              fields terminated by ','
                              location '/hive/warehouse/test_ext';
No rows affected (2.408 seconds)
```

⑤ 查询 test_exteranal 表，操作如下：

```
0: jdbc:hive2://master:10000> select * from test_external;
+--------------------+--------------------+--+
| test_external.id   | test_external.name |
+--------------------+--------------------+--+
| 1                  | master             |
| 2                  | slave1             |
| 3                  | slave2             |
+--------------------+--------------------+--+
3 rows selected (4.997 seconds)
```

⑥ 关闭连接并退出，操作如下：

```
0: jdbc:hive2://master:10000> !closeall
Closing: 0: jdbc:hive2://master:10000
beeline> !quit
hadoop@sw-desktop:~$
```

## 任务 3  Hive 编程

### 【任务概述】

编写删除 test_external 表的程序；编写 HiveTable 表的创建、数据导入、数据查询程序。

### 【任务实施】

#### 一、删除表程序

① drop_test_external.java 代码如下：

```java
import java.sql.SQLException;
import java.sql.Connection;
import java.sql.DriverManager;
import java.sql.Statement;
public class drop_test_external{
```

```java
    private static String driverName = "org.apache.hive.jdbc.HiveDriver";
    private static String url = "jdbc:hive2://master:10000/default";
    private static String user = "hive";
    private static String password = "123456";
    private static String sql = "DROP TABLE IF EXISTS test_external";
    public static void main(String[] args) throws SQLException {
      try {
        // Register driver and create driver instance
        Class.forName(driverName);
        // get connection
        Connection conn = DriverManager.getConnection(url, user, password);
        // create statement
        Statement stmt = conn.createStatement();
        // execute statement
        stmt.executeUpdate(sql);
        System.out.println("Drop table successful.");
        conn.close();
      } catch (Exception e) {
        e.printStackTrace();
      }
    }
}
```

② 编写 drop_test-external.sh 脚本，内容如下：

```bash
#!/bin/bash
HADOOP_HOME=/opt/hadoop-2.7.3
HIVE_HOME=/opt/apache-hive-2.1.1-bin
CLASSPATH=.:$HIVE_HOME/conf:$(hadoop classpath)
for i in ${HIVE_HOME}/lib/*.jar ; do
    CLASSPATH=$CLASSPATH:$i
done
java -cp $CLASSPATH drop_test_external
```

③ 编译、运行脚本，操作如下：

```
hadoop@sw-desktop:~$ javac drop_test_external.java
hadoop@sw-desktop:~$ sh drop_test_external.sh
17/02/12 17:14:33 INFO jdbc.Utils: Supplied authorities: master:10000
17/02/12 17:14:33 INFO jdbc.Utils: Resolved authority: master:10000
Drop table successful.
```

## 二、数据操作程序

① hive_jdbc.java 代码如下：

```java
import java.sql.SQLException;
import java.sql.Connection;
```

```java
import java.sql.ResultSet;
import java.sql.Statement;
import java.sql.DriverManager;
public class hive_jdbc {
  private static String driverName = "org.apache.hive.jdbc.HiveDriver";
  private static String url = "jdbc:hive2://master:10000/default";
  private static String user = "hive";
  private static String passwd = "123456";
  private static String sql = "";
  private static ResultSet res;
  public static void main(String[] args) throws SQLException {
    try {
      Class.forName(driverName);
    } catch (ClassNotFoundException e) {
      e.printStackTrace();
      System.exit(1);
    }
    Connection con = DriverManager.getConnection(url,user,passwd);
    Statement stmt = con.createStatement();
    String tableName = "test_jdbc";
    stmt.execute("drop table if exists " + tableName);
    stmt.execute("create table " + tableName + " (key int, value string)");
    // show tables
    String sql = "show tables '" + tableName + "'";
    System.out.println("Running: " + sql);
    ResultSet res = stmt.executeQuery(sql);
    if (res.next()) {
      System.out.println(res.getString(1));
    }
    // describe table
    sql = "describe " + tableName;
    System.out.println("Running: " + sql);
    res = stmt.executeQuery(sql);
    while (res.next()) {
      System.out.println(res.getString(1) + "\t" + res.getString(2));
    }
    String filepath = "hdfs://master:9000/tmp/test.txt";
    sql = "load data inpath '" + filepath + "' into table " + tableName;
    System.out.println("Running: " + sql);
    stmt.execute(sql);
    // select * query
    sql = "select * from " + tableName;
```

```java
    System.out.println("Running: " + sql);
    res = stmt.executeQuery(sql);
    while (res.next()) {
      System.out.println(String.valueOf(res.getInt(1)) + "\t"
        + res.getString(2));
    }
  }
}
```

② 编写 puttxt.sh 脚本，内容如下：

```bash
#!/bin/bash
echo -e '1\x01Tom' > /tmp/test.txt
echo -e '2\x01Jarry' >> /tmp/test.txt
hdfs dfs -put /tmp/test.txt /tmp
```

③ 运行脚本，操作如下：

```
hadoop@sw-desktop:~$ chmod +x puttxt.sh
hadoop@sw-desktop:~$ ./puttxt.sh
```

④ 编写 hive_jdbc.sh 脚本，内容如下：

```bash
#!/bin/bash
HADOOP_HOME=/opt/hadoop-2.7.3
HIVE_HOME=/opt/apache-hive-2.1.1-bin
CLASSPATH=.:$HIVE_HOME/conf:$(hadoop classpath)
for i in ${HIVE_HOME}/lib/*.jar ; do
    CLASSPATH=$CLASSPATH:$i
Done
java -cp $CLASSPATH hive_jdbc
```

⑤ 编译、运行脚本，操作如下：

```
hadoop@sw-desktop:~$ javac hive_jdbc.java
hadoop@sw-desktop:~$ sh hive_jdbc.sh
17/02/12 21:40:46 INFO jdbc.Utils: Supplied authorities: master:10000
17/02/12 21:40:46 INFO jdbc.Utils: Resolved authority: master:10000
Running: show tables 'test_jdbc'
test_jdbc
Running: describe test_jdbc
Key     int
Value   string
Running: load data inpath 'hdfs://master:9000/tmp/test.txt' into table test_jdbc
Running: select * from test_jdbc
1       Tom
2       Jarry
```

# 任务 4  Hive 与 HBase 集成

## 【任务概述】

Hive 提供类似于 SQL 的处理语言 HiveQL，允许用户查询存储在 Hadoop 中的半结构化数据。Hive 0.6.0 及之后版本提供了 HBase 的支持，用户可以直接定义将 Hive 表存储为 HBase 表，并按需要映射列值。本任务创建一个 Hive 表映射到 HBase 中的 score 表，再联合创建一个 Hive 表和一个 HBase 表，并做测试。

## 【任务实施】

### 一、创建外部表

① 在 Hive CLI 创建外部表，操作如下：

```
hive> create external table hive_score (
    key string,
    name map<string,string>,
    class string,
    java string,
    python string
    )
    stored by 'org.apache.hadoop.hive.hbase.HBaseStorageHandler'
    with serdeproperties
    ("hbase.columns.mapping" =
    ":key,name:,class:class,course:java,course:python")
    tblproperties("hbase.table.name" = "score");
OK
Time taken: 1.642 seconds
```

② HiveQL 查询，操作如下：

```
hive> select * from hive_score;
OK
610213   {"":"Tom"}    163Cloud    85   79
610215   {"":"John"}   173BigData  80   86
Time taken: 0.986 seconds, Fetched: 2 row(s)
```

### 二、联合创建表

① 联合创建表，操作如下：

```
hive> create table hive_hbase(key int, value string)
    stored by 'org.apache.hadoop.hive.hbase.HBaseStorageHandler'
    with serdeproperties ("hbase.columns.mapping" = ":key,cf1:val")
    tblproperties ("hbase.table.name" = "xyz");
OK
Time taken: 13.511 seconds
```

② 从 test_external 导出数据到 hive_hbase，操作如下：

```
hive> insert overwrite table hive_hbase select * from test_external;
OK
Time taken: 132.979 seconds
```

③ 查看 hive_hbase 记录，操作如下：

```
hive> select * from hive_hbase;
OK
1    master
2    slave1
3    slave2
Time taken: 2.087 seconds, Fetched: 3 row(s)
```

④ 查看 HBase 表 xyz 记录，操作如下：

```
hbase(main):018:0> scan 'xyz'
ROW                  COLUMN+CELL
 1                   column=cf1:val, timestamp=1486912164051, value=master
 2                   column=cf1:val, timestamp=1486912164051, value=slave1
 3                   column=cf1:val, timestamp=1486912164051, value=slave2
3 row(s) in 0.0480 seconds
```

## 【同步训练】

一、简答题

（1）Hive 有什么特点？

（2）Metastore 是什么？它包含哪些内容？

（3）表和外部表有什么不同？

（4）HiveServer 和 HiveServer2 有什么不同？

二、操作题

（1）创建 employee 表，表信息如表 6-10 所示。

（2）导入数据到 employee 表中。

（3）查询 name 以 Ma 开头的记录。

（4）修改 employee 表，添加一列 address。

三、编程题

编写 Java 程序，查询 Hive 的 employee 表中的所有记录。

# 项目七 Pig 数据分析

## 【项目介绍】

Pig 为复杂的海量数据并行计算提供一个简单的操作和编程接口。本项目主要完成 Pig 的安装、Pig 命令的使用,完成数据的过滤、数据集的运算,完成 Pig 自定义过滤函数、自定义运算函数和自定义装载函数的编写和使用。

本项目分以下两个任务:
- 任务1  Pig 安装及使用
- 任务2  Pig 高级编程

## 【学习目标】

### 一、知识目标

- 了解 Pig 的运行模式。
- 掌握 Pig 的数据类型和函数。
- 熟悉 Pig 命令和语句。
- 掌握 Pig Latin 的编写。
- 掌握 Pig 数据集的运算。
- 掌握 Pig 自定义函数的编写和使用。

### 二、能力目标

- 能够安装 Pig 系统。
- 能够使用 Pig 命令处理数据。
- 能够使用 Pig 进行数据集的运算。
- 能够编写 Pig 自定义函数。
- 能够使用自定义函数处理数据。

## 任务1  Pig 安装及使用

### 【任务概述】

本任务需要安装 Pig 软件,使用 Pig 命令操作 HDFS 数据,完成 HDFS 数据基本计算、数据过滤、数据生成、数据集运算等。

## 【支撑知识】

### 一、Pig 简介

Apache Pig 是一个高级过程语言，适合使用 Hadoop 和 MapReduce 平台来查询大型半结构化数据集。Pig 在 MapReduce 的基础上创建了更简单的过程语言抽象，为 Hadoop 应用程序提供了一种更加接近结构化查询语言（SQL）的接口。

用 MapReduce 进行数据分析，当业务比较复杂的时候，使用 MapReduce 将会是一件很复杂的事情，如需要对数据进行很多预处理或转换，以便能够适应 MapReduce 的处理模式。另一方面，编写 MapReduce 程序，发布及运行作业都将是一件比较耗时的事情。Pig 的出现很好地弥补了上述不足。Pig 能够让你专心于数据及业务本身，而不是纠结于数据的格式转换，以及 MapReduce 程序的编写。从本质是上来说，当使用 Pig 进行处理时，Pig 本身会在后台生成一系列 MapReduce 操作来执行任务，但是这个过程对用户来说是透明的。

Pig 具有 3 个特性：

① 易编程。Pig Latin 程序由一系列的"操作"或"变换"构成，实际上通过"操作"将 MapRecude 程序变成数据流，使得实现简单的和并行要求高的数据分析任务变得非常容易，在它所提供的 Pig Latin 控制台上，可以用几行 Pig Latin 代码轻松完成 TB 级的数据集处理任务。

② 自动优化。系统会对编写的 Pig Latin 代码自动进行优化，程序员就可以省去优化过程，不必关心效率问题，将大量的时间专注于分析语义方面。

③ 扩展性好。程序员可以按照自己的需求编写自定义函数，其载入（Load）、存储（Store）、过滤（Filter）、连接（Join）过程均可定制。

### 二、Pig 运行模式

Pig 是作为一个客户端应用程序运行的，即使准备在 Hadoop 集群上运行 Pig，也不需要在集群上额外安装。Pig 从工作站上发出作业，并和 HDFS 进行交互。Pig 有两种执行类型或模式（Mode）：本地模式（Local Mode）和 MapReduce 模式（MapReduce Mode）。

（1）本地模式

在本地模式下，Pig 运行在单个 JVM 中，访问本地文件系统。该模式只适应用于试用 Pig 处理小规模数据集。

执行类型可用"-x"或者"-exectype"选项进行设置。如果要使用本地模式运行，那么就将该选项设置为 local。例如：

```
$ pig -x local
```

（2）MapReduce 模式

在 MapReduce 模式下，Pig 将查询翻译为 MapReduce 作业，然后在 Hadoop 集群上执行。集群可以是伪分布的，也可以是全分布的。如果要用 Pig 处理大规模数据集，应该使用全分布集群上的 MapReduce 模式。

### 三、Pig Latin

Pig 不再需要编写单独的 MapReduce 应用程序，可以使用 Pig Latin 语言编写一个脚本，在集群中自动并行处理与分发该脚本。

（1）结构

一个 Pig Latin 程序由一组语句构成。一个语句可以理解为一个操作或一个命令。例如，GROUP 操作是这样一条语句：grouped_records =GROUP records BY year;。

另一个例子是列出 Hadoop 文件系统中文件的命令：ls /。

如前面的 GROUP 语句，一条语句通常以分号结束。实际上，那是一条必须用分号表示结束的语句，不能省略了分号。而另一方面，ls 命令可以不以分号结束。一般的规则如下：在 Grunt 中交互使用的语句或命令不需要表示结束的分号。这包括交互式的 Hadoop 命令，以及用于诊断的操作，如 DESCRIBE。加上表示结束的分号也不会出错。

必须用分号表示结束的语句可以写成多行以便于阅读：

```
records =LOAD ,input/ncdc/micro-tab/sample.txt,
AS (year:chararray, temperature: int, quality:int);
```

Pig Latin 有两种注释方法。双减号表示单行注释。Pig Latin 解释器会忽略从第一个减号开始到行尾的所有内容。例如：

```
-- My program
DUMP A; -- What,s in A?
```

C 语言风格的注释更灵活。这是因为它使用/*和*/符号表示注释开始和结束。这样，注释既可以跨多行，也可以内嵌在某一行内。

Pig Latin 有一个关键词列表。其中的单词在 Pig Latin 中有特殊含义，不能用作标识符。这些单词包括操作（LOAD、ILLUSTRATE）、命令（cat、ls）、表达式（matches、FLATTEN）及函数（DIFF,MAX）等。

Pig Latin 的大小写敏感性采用混合的规则。操作和命令是大小写无关的，而别名和函数名是大小写敏感的。

（2）语句

在 Pig Latin 程序执行时，每个命令按次序进行解析。如果遇到句法错误或其他（语义）错误，解释器会终止运行，并显示错误消息。解释器会给每个关系操作建立一个逻辑计划。逻辑计划构成了 Pig Latin 程序的核心。解释器把为一个语句创建的逻辑计划加到目前为止已经解释完的程序的逻辑计划上，然后继续处理下一条语句。

特别需要注意的是，在整个程序逻辑计划没有构造完成前，Pig 并不处理数据。让 Pig 开始执行的是 DUMP 语句。此时，逻辑计划被编译成物理计划并执行。

Pig 的物理计划是一系列的 MapReduce 作业。在本地模式下，这些作业在本地 JVM 中运行；而在 MapReduce 模式下，它们在 Hadoop 集群上运行。

表 7-1 概述了能够作为 Pig 逻辑计划一部分的关系操作。有些语句并不加到逻辑计划中。例如，诊断操作 DESCRIBE、EXPLAIN 及 ILLUSTRATE。这些操作是用来让用户能够与逻辑计划进行交互以进行调试的，如表 7-2 所示。DUMP 也是一种诊断操作，它只能用于与很小的结果集进行交互调试，或与 LIMIT 结合使用，来获得某个较大的关系的一小部分行。STORE 语句应该在输出包含较多行的时候使用，这是因为 STORE 语句把结果存入文件而不是在控制台显示。

表 7-1  Pig Latin 关系操作

| 类型 | 操作 | 描述 |
|---|---|---|
| 加载与存储 | LOAD | 将数据从文件系统加载，存入关系 |
|  | STORE | 将一个关系存放到文件系统 |
|  | DUMP | 将关系打印到控制台 |
| 过滤 | FILTER | 从关系中过滤掉不需要的行 |
|  | DISTINCT | 从关系中删除重复的行 |
|  | FOREACH...GENERATE | 在关系中增加或删除字段 |
|  | STREAM | 使用外部的程序对关系进行变换 |
|  | SAMPLE | 从关系中随机取样 |
| 分组与连接 | JOIN | 连接两个或者多个关系 |
|  | COGROUP | 在两个或多个关系中对数据进行分组 |
|  | GROUP | 在一个关系中对数据进行分组 |
|  | CROSS | 获取两个或多个关系的乘积 |
| 排序 | ORDER | 根据一个或多个字段对关系进行排序 |
|  | LIMIT | 将关系的元组个数限定在一定数量内 |
| 合并与分割 | UNION | 合并两个或多个关系 |
|  | SPLIT | 把某一个关系切分成多个关系 |

表 7-2  Pig Latin 诊断操作

| 操作 | 描述 |
|---|---|
| DESCRIBE | 显示关系的模式 |
| EXPLAIN | 显示逻辑和物理计划 |
| ILLUSTRATE | 使用生成的输入子集显示逻辑计划的试运行结果 |

为了在 Pig 脚本中使用用户自定义函数，Pig Latin 提供了 REGISTER 和 DEFINE 语句，如表 7-3 所示。这些命令处理关系不会被加入逻辑计划，而是立即执行。

表 7-3  Pig Latin 的 UDF 语句

| 语句 | 描述 |
|---|---|
| REGISTER | 在 Pig 运行的环境中注册一个 JAR 文件 |
| DEFINE | 为 UDF、流式脚本或命令规范新建别名 |

Pig 提供了与 Hadoop 文件系统和 MapReduce 进行交互的命令及其他一些工具命令，如表 7-4 所示。与 Hadoop 文件系统进行交互的命令对在 Pig 处理前和处理后移动数据非常有用。文件系统相关命令可以对任何 Hadoop 文件系统的文件或目录进行操作。

表 7-4  Pig Latin 命令描述

| Hadoo Filesystem | 描述 |
|---|---|
| cat | 显示一个或多个文件的内容 |
| cd | 改变当前目录 |

| Hadoo Filesystem | 描 述 |
|---|---|
| copyFromLocal | 把一个本地文件或目录复制到 Hadoop 文件系统 |
| copyToLocal | 将一个文件或目录从 Hadoop 文件系统复制到本地文件系统 |
| cp | 把一个文件或目录复制到另一个目录 |
| fs | 访问 Hadoop 文件系统外壳程序 |
| ls | 显示文件列表信息 |
| mkdir | 创建新目录 |
| mv | 将一个文件或目录移动到另一个目录 |
| pwd | 显示当前工作目录的路径 |
| rm | 删除一个文件或目录 |
| rmf | 强制删除文件或目录（即使文件或目录不存在也不会失败） |
| Hadoop MapReduce | 描 述 |
| kill | 终止某个 MapReduce 作业 |
| exec | 在一个新的 Grunt 外壳程序中以批处理模式运行一个脚本 |
| help | 显示可用的命令选项 |
| quit | 退出解释器 |
| run | 在当前 Grunt 外壳程序中运行脚本 |
| set | 设置 Pig 选项 |

（3）数据类型

Pig 有 4 种数值类型：int、long、float 和 double。它们和 Java 中对应的数值类型相同。此外，Pig 还有 bytearray 类型，这类似于表示二进制对象的 Java 的 byte 数组。chararray 类似于用 UTF-16 格式表示文本数据的 java.lang.String。chararray 也可以加载或存储 UTF-8 格式的数据。Pig 没有任何一种数据类型对应于 Java 的 boolean、byte、short 或 char。Java 的这些数据类型都能方便地使用 Pig 的 int 类型（对数值类型）或 chararray 类型（对 char）表示。数值、文本与二进制类型都是原子类型。Pig Latin 有 3 种用于表示嵌套结构的复杂类型："元组（tuple）""包（bag）"和"映射（map）"。表 7-5 列出了 Pig Latin 的所有数据类型。

在 Pig Latin 中，如果使用 SQL 定义 null 或未定义，Pig 会用 null 替代，在输出到屏幕（或使用 STORE 存储）时，null 被显示（或存储）为一个空位。

表 7-5  Pig Latin 数据类型

| 简单数据类型 | 描 述 | 文字示例 |
|---|---|---|
| int | 32 位有符号整数 | 10 |
| long | 64 位有符号整数 | 10L、10l |
| float | 32 位浮点数 | 10.5F、10.5E2F、10.5f、10.5e2f |
| double | 64 位浮点数 | 10.5、10.5E2、10.5e2 |
| 数组类型 | 描 述 | 文字示例 |
| chararray | UTF-8 格式的字符数组 | hello world |

| 数组类型 | 描述 | 文字示例 |
|---|---|---|
| bytearray | 字节数组（blob） | |

| 复杂类型 | 描述 | 文字示例 |
|---|---|---|
| tuple | 任何类型的字段序列 | (19，2) |
| bag | 元组的无序多重集合（允许重复元组） | {(19，2), (18，1)} |
| map | 一组键-值对。键必须是字符数组，值可以是任何类型的数据 | [open#apache] |

另外，Pig Latin 还提供了操作运算符和功能函数，Pig 可以使用丰富的表达式类型，如表 7-6 和表 7-7 所示。

表 7-6　Pig Latin 运算符

| 操作符 | 说明 | 示例 |
|---|---|---|
| + | 加法运算符 | addition = FOREACH tables GENERATE dataItem+1; |
| − | 减法运算符 | substruction = FOREACH tables GENERATE dataItem−1; |
| * | 乘法运算符 | multiplication = FOREACH tables GENERATE dataItem10; |
| / | 除法运算符 | division = FOREACH tables GENERATE dataItem/10; |
| % | 求模运算 | modulo = FOREACH tables GENERATE dataItem%10; |
| ?: | 条件运算符 | X = FOREACH A GENERATE f2, (f2==1?1:0); |
| AND | 布尔运算符 | PIG Latin 不支持布尔类型，但支持布尔运算 |
| OR | 布尔运算符 | 在条件选择中，可以使用这些运算符 |
| NOT | 布尔运算符 | X = FILTER A BY (f1==8) OR (NOT (f2+f3 > f1)); |
| == | 对比运算符 | X = FILTER A BY (f1 == 8); X = FILTER A BY (f2 == 'apache'); |
| != | 对比运算符 | X = FILTER A BY (f1 != 8); |
| < | 对比运算符 | X = FILTER A BY (f1 < 8); |
| > | 对比运算符 | X = FILTER A BY (f1 > 8); |
| <= | 对比运算符 | X = FILTER A BY (f1 <= 8); |
| >= | 对比运算符 | X = FILTER A BY (f1 >= 8); |
| MATCHES | 正则表达式 | X = FILTER A BY (f1 MATCHES 'apache'); |
| IS NULL | Null Operators | X = FILTER A BY f1 is null; |
| IS NOT NULL | Null Operators | X = FILTER A BY f1 is not null; |
| + | 正负符号 | A = LOAD 'data' as (x, y, z); |
| − | 正负符号 | B = FOREACH A GENERATE −x, y; |

表 7-7　Pig Latin 功能函数

| 类别 | 描述 |
|---|---|
| AVG | 计算包中项的平均值 |
| CONCAT | 把两个字节数组或字符数组连接成一个 |

续表

| 类 别 | 描 述 |
|---|---|
| COUNT | 计算一个包中非空值的项的个数 |
| COUNTSTAR | 计算一个包的项的个数,包括空值 |
| DIFF | 计算两个包的差。如果两个参数不是包,它们相同时,则返回一个包含这两个参数的包;否则返回一个空的包 |
| MAX | 计算一个包中项的最大值 |
| MIN | 计算一个包中项的最小值 |
| SIZE | 计算一个类型的大小。数值型的大小总是1。对于字符数组,它返回字符的个数;对于字节数组,它返回字节的个数;对于容器(container,包括元组、包、映射),它返回其中项的个数 |
| SUM | 计算一个包中项的值的总和 |
| TOKENIZE | 对一个字符数组进行标记解析,并把结果词放入一个包 |
| IsEmpty | 判断一个包或映射是否为空 |
| PigStorage | 用字段分隔文本格式加载或存储关系。每一行被分为字段后(用一个可设置的分隔符<默认为一个制表符>分隔),分别对应于元组的各个字段。这是不指定加载/存储方式时的默认存储函数 |
| BinStorage | 从二进制文件加载一个关系或把关系存储到二进制文件中。该函数使用基于 Hadoop Writable 对象的 Pig 内部格式 |
| BinaryStorage | 从二进制文件加载只包含一个类型为 bytearray 的字段的元组到关系,或以这种格式存储一个关系。bytearray 中的字节逐字存放。该函数与 Pig 的流式处理结合使用 |
| TextLoader | 从纯文本格式加载一个关系。每一行对应于一个元组。每个元组只包含一个字段,即该行文本 |
| PigDump | 用元组的 toString() 形式存储关系。每行一个元组。这个函数对设计很有帮助 |
| HBaseStorage | 从 HBase 加载数据 |

如表 7-8 所示,Hadoop 提供了模式匹配语法,需要注意的是,这些正则字符的含义由 Pig 下面的 HDFS 决定,如果用户从 Linux Shell 命令行运行 Pig Latin 命令,那么用户还需要对这些正则字符进行转义以防止正则字符被拆分。匹配运算符相关语法如表 7-7 所示。

表 7-8  Hadoop 提供的正则匹配语法

| 正则字符 | 描 述 |
|---|---|
| ? | 匹配任何单个字符 |
| * | 匹配零个或多个字符 |
| [abc] | 匹配字符集和{a, b, c}所包含的任意一个字符 |
| [a-z] | 匹配指定范围 a~z 内的任意字符 |
| [^abc] | 匹配任何不在指定范围内的任意字符 |
| \c | 移除(转义)字符 c 所表达的特殊含义 |
| {ab, cd} | 配置字符串集合{ab, cd}中的任一字符串 |

## 四、Pig 命令语法

（1）从文件导入数据

```
tmp_table = LOAD 'F' USING PigStorage('\t') AS (age:int,options:chararray);
```

（2）查询整张表

```
DUMP tmp_table;
```

（3）查询前 10 行

```
tmp_table_limit = LIMIT tmp_table 10;
DUMP tmp_table_limit;
```

（4）查询某些列

```
tmp_table_user = FOREACH tmp_table GENERATE user;
DUMP tmp_table_user;
```

（5）给列取别名

```
tmp_column_alias = FOREACH tmp_table GENERATE user AS name,age AS user_age;
DUMP tmp_column_alias;
```

（6）排序

```
tmp_table_order = ORDER tmp_table BY age ASC;
DUMP tmp_table_order;
```

（7）条件查询

```
tmp_table_where = FILTER tmp_table by age > 20;
DUMP tmp_table_where;
```

（8）内连接 Inner Join

```
tmp_table_inner_join = JOIN tmp_table BY age,tmp_table2 BY age;
DUMP tmp_table_inner_join;
```

（9）左连接 Left Join

```
tmp_table_left_join = JOIN tmp_table BY age LEFT OUTER,tmp_table2 BY age;
DUMP tmp_table_left_join;
```

（10）右连接 Right Join

```
tmp_table_right_join = JOIN tmp_table BY age RIGHT OUTER,tmp_table2 BY age;
DUMP tmp_table_right_join;
```

（11）全连接 Full Join

```
tmp_table_full_join = JOIN tmp_table BY age FULL OUTER,tmp_table2 BY age;
DUMP tmp_table_full_join;
```

（12）同时对多张表交叉查询.

```
tmp_table_cross = CROSS tmp_table,tmp_table2;
DUMP tmp_table_cross;
```

（13）分组 GROUP BY

```
tmp_table_group = GROUP tmp_table BY is_male;
DUMP tmp_table_group;
```

（14）分组并统计

```
tmp_table_group_count = GROUP tmp_table BY is_male;
```

```
tmp_table_group_count = FOREACH tmp_table_group_count GENERATE group,COUNT($1);
DUMP tmp_table_group_count;
```

（15）查询去重 DISTINCT

```
tmp_table_distinct = FOREACH tmp_table GENERATE is_male;
tmp_table_distinct = DISTINCT tmp_table_distinct;
DUMP  tmp_table_distinct;
```

## 【任务实施】

### 一、客户端主机安装 Pig 软件

① 下载 Pig 软件包到/home/hadoop 目录下，下载地址如下：

```
http://mirrors.aliyun.com/apache/pig/pig-0.16.0/pig-0.16.0.tar.gz
```

② 安装 Pig 软件，操作如下：

```
hadoop@sw-desktop:~$ cd /opt
hadoop@sw-desktop:/opt$ sudo tar xvzf /home/hadoop/pig-0.16.0.tar.gz
hadoop@sw-desktop:/opt$ sudo chown -R hadoop:hadoop pig-0.16.0
```

③ 修改 Pig 参数，操作如下：

```
hadoop@sw-desktop:/opt$ cd /opt/pig-0.16.0/conf
hadoop@sw-desktop.../conf$ mv log4j.properties.template log4j.properties
hadoop@sw-desktop.../conf$ vi pig.properties
```

添加如下内容：

```
pig.logfile=/opt/pig-0.16.0/logs
log4jconf=/opt/pig-0.16.0/conf/log4j.properties
exectype=mapreduce
```

④ 修改环境变量，操作如下：

```
hadoop@sw-desktop:~$ vi /home/hadoop/.profile
```

添加如下内容：

```
export HADOOP_CONF_DIR=${HADOOP_HOME}/etc/hadoop
export PIG_HOME=/opt/pig-0.16.0
export PIG_CLASSPATH=${HADOOP_HOME}/etc/hadoop/
export PATH=$PATH:$PIG_HOME/bin
```

⑤ 环境变量生效，操作如下：

```
hadoop@sw-desktop:~$ source /home/hadoop/.profile
```

### 二、基本操作

① 基本数据如下：

文件 A.txt 的内容：            文件 B.txt 的内容：
　0,1,2　　　　　　　　　　　　0,5,2
　1,3,4　　　　　　　　　　　　1,7,8


文件 TP.txt 的内容：           文件 MP.txt 的内容：
　(1,2,3)　　　　　　　　　　　[Pig#Grunt]

```
            (2,3,4)                    [Apache#Hadoop]
            (2,4,5)                    [Pig#Pig Latin]
```

② 运行 Pig，操作如下：
```
hadoop@sw-desktop:~$ pig
grunt>
```

③ 创建 test 目录，上传文件到 HDFS，操作如下：
```
grunt> mkdir /test
grunt> copyFromlocal A.txt /test;
grunt> copyFromlocal B.txt /test;
grunt> copyFromlocal TP.txt /test;
grunt> copyFromlocal MP.txt /test;
```

④ 装载 A.txt 到变量 a，变量 b 为 a 的列$0+列$1，操作如下：
```
grunt> a = load '/test/A.txt' using PigStorage(',') as (c1:int,c2:double,c3:float);
grunt> b = foreach a generate $0+$1 as b1;
grunt> dump b;
(1.0)
(4.0)
grunt> describe b;
b: {b1: double}
```

⑤ 变量 c 为 b 的 b1 列减去 1，操作如下：
```
grunt> c = foreach b generate b1-1;
grunt> dump c;
(0.0)
(3.0)
```

⑥ 变量 d 为 a 的第 1 列，如果是 0 则输出（c1,c2），如果不是 0 则输出（c1,c3），操作如下：
```
grunt> d = foreach a generate c1,($0==0?$1:$2);
grunt> dump d;
(0,1.0)
(1,4.0)
```

⑦ 变量 f 为 a 的 c1>0 并且 c2>1 的输出，操作如下：
```
grunt> f = filter a by c1>0 and c2>1;
grunt> dump f;
(1,3.0,4.0)
```

⑧ 装载 Tuple 数据 TP.txt 到变量 tp，变量 g 为 tp 产生的输出，操作如下：
```
grunt> tp = load '/test/TP.txt' as t:tuple(c1:int,c2:int,c3:int);
grunt> describe tp;
tp: {t: (c1: int,c2: int,c3: int)}
grunt> dump tp;
((1,2,3))
((2,3,4))
```

```
((2,4,5))
grunt> g = foreach tp generate t.c1,t.c2,t.c3;
grunt> describe g;
g: {c1: int,c2: int,c3: int}
grunt> dump g;
(1,2,3)
(2,3,4)
(2,4,5)
```

⑨ 对 g 进行分组，输出 Bag 数据到变量 bg，操作如下：

```
grunt> bg = group g by c1;
grunt> describe bg;
bg: {group: int,g: {(c1: int,c2: int,c3: int)}}
grunt> dump bg;
(1,{(1,2,3)})
(2,{(2,4,5),(2,3,4)})
grunt> illustrate bg;
-----------------------------------------------
| tp    | t:tuple(c1:int,c2:int,c3:int)       |
-----------------------------------------------
|       | (2, 4, 5)                           |
|       | (2, 3, 4)                           |
-----------------------------------------------

-----------------------------------------------
| g     | c1:int   | c2:int   | c3:int   |
-----------------------------------------------
|       | 2        | 4        | 5        |
|       | 2        | 3        | 4        |
-----------------------------------------------

-------------------------------------------------------------------------
| bg    | group:int   | g:bag{:tuple(c1:int,c2:int,c3:int)}             |
-------------------------------------------------------------------------
|       | 2           | {(2, 4, 5), (2, 3, 4)}                          |
-------------------------------------------------------------------------
grunt> x = foreach bg generate g.c1;
grunt> dump x;
({(1)})
({(2),(2)})
```

⑩ 装载 Map 数据 MP.txt 到变量 mp，变量 h 为 mp 产生的输出，操作如下：

```
grunt> mp = load '/test/MP.txt' as (m:map[]);
grunt> describe mp;
mp: {m: map[]}
```

```
grunt> dump mp;
([Pig#Grunt])
([Apache#Hadop])
([Pig#Pig Latin])
grunt> h = foreach mp generate m#'Pig';
grunt> describe h;
h: {bytearray}
grunt> dump h;
(Grunt)
()
(Pig Latin)
```

### 三、数据集运算

① 加载数据，操作如下：

```
grunt> a = load '/test/A.txt' using PigStorage(',') as (a1:int, a2:int, a3:int);
grunt> b = load '/test/B.txt' using PigStorage(',') as (b1:int, b2:int, b3:int);
```

② a 与 b 并集，操作如下：

```
grunt> c = union a, b;
grunt> dump c;
(0,1,2)
(1,3,4)
(0,5,2)
(1,7,8)
```

③ 将 c 分割为 d 和 e，其中 d 的第一列数据值为 0，e 的第一列的数据值为 1（$0 表示数据集的第一列），操作如下：

```
grunt> split c into d if $0 == 0, e if $0 == 1;
grunt> dump d;
(0,1,2)
(0,5,2)
grunt> dump e;
(1,3,4)
(1,7,8)
```

④ 选择 c 中的一部分数据，操作如下：

```
grunt> f = filter c by $1 > 3;
grunt> dump f;
(0,5,2)
(1,7,8)
```

⑤ 对数据进行分组，操作如下：

```
grunt> g = group c by $2;
grunt> dump g;
```

```
(2,{(0,1,2),(0,5,2)})
(4,{(1,3,4)})
(8,{(1,7,8)})
```

⑥ 将所有的元素集合到一起，操作如下：

```
grunt> h = group c all;
grunt> dump h;
(all,{(1,3,4),(0,1,2),(1,7,8),(0,5,2)})
```

⑦ 查看 h 中元素个数，操作如下：

```
grunt> i = foreach h generate COUNT($1);
grunt> dump i;
(4)
```

⑧ 连表查询，条件是 a.$2 == b.$2，操作如下：

```
grunt> j = join a by $2, b by $2;
grunt> dump j;
(0,1,2,0,5,2)
```

⑨ 变量 k 为 c 的 $1 和 $1 * $2 的输出，操作如下：

```
grunt> k = foreach c generate $1, $1 * $2;
grunt> dump k;
(5,10)
(7,56)
(1,2)
(3,12)
```

## 任务 2  Pig 高级编程

### 【任务概述】

本任务主要使用 Pig 查找气象数据中每年最高气温，编写 Pig 自定义过滤函数过滤无效温度，编写 Pig 自定义运算函数、自定义加载函数、自定义函数和客户端程序。

### 【任务实施】

一、查询气温

① 气温数据文件 temperature.txt 的内容如下：

```
1990 21
1990 18
1991 21
1992 30
1992 999
1990 23
```

其中，999 表示无用或缺失数据。

② 查找每年最高气温。

a. Grunt方式运行：

```
grunt> copyFromLocal temperature.txt /test
grunt> records = load '/test/temperature.txt' USING PigStorage(' ')
       as (year: chararray,temperature:int);
grunt> valid_records = filter records by temperature!=999;
grunt> grouped_records = group valid_records by year;
grunt> max_temperature = foreach grouped_records
       generate group,MAX(valid_records.temperature);
grunt> dump max_temperature;
(1990,23)
(1991,21)
(1992,30)
```

b. 脚本方式运行：

```
--编写脚本--
hadoop@sw-desktop:~$ vi max_temp.pig
records = load '/test/temperature.txt' USING PigStorage(' ')
  as (year: chararray,temperature:int);
valid_records = filter records by temperature!=999;
grouped_records = group valid_records by year;
max_temperature = foreach grouped_records
  generate group,MAX(valid_records.temperature);
dump max_temperature;
--运行脚本--
hadoop@sw-desktop:~$ pig max_temp.pig
(1990,23)
(1991,21)
(1992,30)
```

## 二、编写用户自定义函数

（1）编写自定义过滤函数

① 打开Ecplise工具，新建"Map/Reduce Project"项目，项目名为"TempFilter"，并为项目创建"IsValidTemp"类，添加"pig-0.16.0-core-h2.jar"包，如图7-1所示。

IsValidTemperature.java 代码如下：

```java
import java.io.IOException;
import org.apache.pig.FilterFunc;
import org.apache.pig.backend.executionengine.ExecException;
import org.apache.pig.data.Tuple;
public class IsValidTemp extends FilterFunc {
    @Override
    public Boolean exec(Tuple tuple) throws IOException {
        if(tuple ==null ||tuple.size()==0)
```

```
            return false;
        try {
            Object obj=tuple.get(0);
            if (obj==null)
                return false;
            int temperature=(Integer)obj;
            return temperature!=999;
        }catch(ExecException e) {
            throw new IOException(e);
        }
    }
}
```

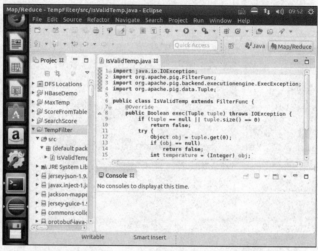

图 7-1 IsValidTemp 类

② 右击"TempFilter"命令，在列表框中选择"Export"选项后弹出对话框，选择"java"→"JAR file"命令，在"JAR file:"文本框中输入文本"IsValidTemp.jar"，导出 jar 包，如图 7-2 所示。

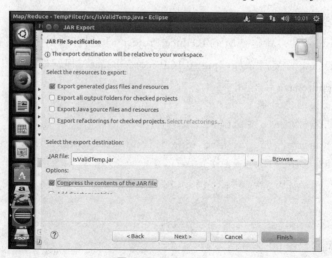

图 7-2 导出 jar 包

③ 运行自定义过滤函数包，操作如下：

```
grunt> copyFromLocal /home/hadoop/workspace/IsValidTemp.jar /test
grunt> register hdfs://master:9000/test/IsValidTemp.jar;
grunt> records = load '/test/temperature.txt' USING PigStorage(' ')
    as (year: chararray,temperature:int);
grunt> valid_records = filter records by IsValidTemp(temperature);
grunt> dump valid_records;
(1990,21)
(1990,18)
(1991,21)
(1992,30)
(1990,23)
```

（2）编写自定义运算函数

① 打开 Ecplise 工具，新建"Map/Reduce Project"项目，项目命名为"EvalTemp"，并为项目创建"EvalTemp"类，添加"pig-0.16.0-core-h2.jar"包，如图 7-3 所示。

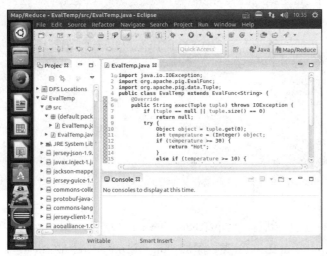

图 7-3　EvalTemp 类

EvalTemp.java 代码如下：

```
import java.io.IOException;
import org.apache.pig.EvalFunc;
import org.apache.pig.data.Tuple;
public class EvalTemp extends EvalFunc<String> {
    @Override
    public String exec(Tuple tuple) throws IOException {
        if (tuple == null || tuple.size() == 0)
            return null;
        try {
            Object object = tuple.get(0);
            int temperature = (Integer)object;
```

```
            if (temperature >= 30){
                return "Hot";
            }
            else if(temperature >=10){
                return "Moderate";
            }
            else {
                return "Cool";
            }
        } catch(Exception e) {
            throw new IOException(e);
        }
    }
}
```

② 导出 EvalTemp.jar 包，运行自定义运算函数包，操作如下：

```
grunt> copyFromLocal /home/hadoop/workspace/EvalTemp.jar /test
grunt> register hdfs://master:9000/test/EvalTemp.jar;
grunt> result = foreach valid_records generate year,temperature,
       EvalTemp(temperature);
grunt> dump result;
(1990,21,Moderate)
(1990,18,Moderate)
(1991,21,Moderate)
(1992,30,Hot)
(1990,23,Moderate)
```

（3）自定义加载函数

① 打开 Eclipse 工具，新建 "Map/Reduce Project" 项目，项目命名为 "WordCountLoad"，并为项目创建 "WordCountLoad" 类，添加 "pig-0.16.0-core-h2.jar" 包，如图 7-4 所示。

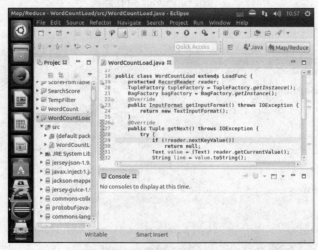

图 7-4　WordCountLoad 类

WordCountLoad.java 代码如下：

```java
import java.io.IOException;
import java.util.List;
import java.util.ArrayList;
import org.apache.hadoop.io.Text;
import org.apache.hadoop.mapreduce.InputFormat;
import org.apache.hadoop.mapreduce.Job;
import org.apache.hadoop.mapreduce.RecordReader;
import org.apache.hadoop.mapreduce.lib.input.FileInputFormat;
import org.apache.hadoop.mapreduce.lib.input.TextInputFormat;
import org.apache.pig.LoadFunc;
import org.apache.pig.backend.executionengine.ExecException;
import org.apache.pig.backend.hadoop.executionengine.mapReduceLayer.PigSplit;
import org.apache.pig.data.BagFactory;
import org.apache.pig.data.DataBag;
import org.apache.pig.data.Tuple;
import org.apache.pig.data.TupleFactory;
public class WordCountLoad extends LoadFunc {
    protected RecordReader reader;
    TupleFactory tupleFactory = TupleFactory.getInstance();
    BagFactory bagFactory = BagFactory.getInstance();
    @Override
    public InputFormat getInputFormat() throws IOException {
        return new TextInputFormat();
    }
    @Override
    public Tuple getNext() throws IOException {
        try {
            if (!reader.nextKeyValue()) return null;
            Text value = (Text)reader.getCurrentValue();
            String line = value.toString();
            String[] words = line.split("\\s+");
            List<Tuple> tuples = new ArrayList<Tuple>();
            Tuple tuple = null;
            for (String word : words) {
                tuple= tupleFactory.newTuple();
                tuple.append(word);
                tuples.add(tuple);
            }
            DataBag bag = bagFactory.newDefaultBag(tuples);
            Tuple result = tupleFactory.newTuple(bag);
```

```java
            return result;
        } catch (InterruptedException e) {
            throw new ExecException(e);
        }
    }
    @Override
    public void prepareToRead(RecordReader reader,PigSplit arg1) throws IOException
    {
        this.reader = reader;
    }
    @Override
    public void setLocation(String location, Job job) throws IOException {
        FileInputFormat.setInputPaths(job,location);
    }
}
```

② 导出 WordCountLoad.jar 包，运行自定义加载函数包，操作如下：

```
--sw1.txt 和 sw2.txt--
grunt> cat /input/sw1.txt
Hello World
Good Hadoop
grunt> cat /input/sw2.txt
Hello Hadoop
Bye Hadoop
--注册自定义加载包--
grunt> copyFromLocal /home/hadoop/workspace/WordCountLoad.jar /test
grunt> register hdfs://master:9000/test/WordCountLoad.jar;
--运行加载包--
grunt> records = load 'hdfs://master:9000/input' USING WordCountLoad()
    as (words:bag{word: (w:chararray)});
grunt> dump records;
({(Hello),(World)})
({(Good),(Hadoop)})
({(Hello),(Hadoop)})
({(Bye),(Hadoop)})
--计数--
grunt> flatten_records = foreach records generate flatten($0);
grunt> grouped_records= group flatten_records by words::w;
grunt> result= foreach grouped_records generate group,COUNT(flatten_records);
grunt> final_result= order result by $1 desc,$0;
--输出结果--
```

```
grunt> dump final_result;
(Hadoop,3)
(Hello,2)
(Bye,1)
(Good,1)
(World,1)
```

（4）编写自定义函数和客户端程序

① 打开 Eclipse 工具，新建"Map/Reduce Project"项目"WordPig"，新建"WordUpper"类，添加"pig-0.16.0-core-h2.jar"包，如图 7-5 所示。

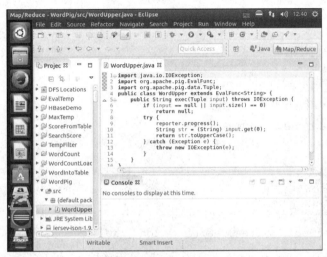

图 7-5　WordUpper 类

WordUpper.java 代码如下：

```java
import java.io.IOException;
import org.apache.pig.EvalFunc;
import org.apache.pig.data.Tuple;
public class WordUpper extends EvalFunc<String> {
    public String exec(Tuple input) throws IOException {
        if (input == null || input.size() == 0)
            return null;
        try{
            reporter.progress();
            String str = (String)input.get(0);
            return str.toUpperCase();
        }catch(Exception e){
         throw new IOException(e);
        }
    }
}
```

② 导出 WordUpper.jar 包，上传到 HDFS，测试运行，操作如下：

```
grunt> copyFromLocal /home/hadoop/workspace/WordUpper.jar /test
grunt> register hdfs://master:9000/test/WordUpper.jar;
grunt> records = load '/input/sw1.txt' USING PigStorage('\n') as (line:chararray);
grunt> dump records;
(Hello World)
(Good Hadoop)
grunt> result = foreach records generate WordUpper(line);
grunt> dump result;
(HELLO WORLD)
(GOOD HADOOP)
```

③ 编写客户端程序 WordClient.java。

WordClient.java 代码如下：

```java
import org.apache.pig.PigServer;
public class WordClient {
    public static void main(String[] args) throws Exception{
        PigServer pigServer=new PigServer("mapreduce");
        pigServer.registerJar("hdfs://master:9000/test/WordUpper.jar");
        pigServer.registerQuery("A = load 'hdfs://master:9000/input/sw1.txt'
           using PigStorage('\\n') as (line:chararray);");
        pigServer.registerQuery("B = foreach A generate WordUpper(line);");
        pigServer.store("B", "hdfs://master:9000/output/sw1");
    }
}
```

④ 编写脚本，操作如下：

```bash
#!/bin/bash
HADOOP_HOME=/opt/hadoop-2.7.3
PIG_HOME=/opt/pig-0.16.0
CLASSPATH=.:$PIG_HOME/conf:$(hadoop classpath)
for i in ${PIG_HOME}/lib/*.jar ; do
    CLASSPATH=$CLASSPATH:$i
done
CLASSPATH=$CLASSPATH:$PIG_HOME/pig-0.16.0-core-h2.jar
java -cp $CLASSPATH WordClient
```

⑤ 运行脚本，查看结果，操作如下：

```
hadoop@sw-desktop:~$ javac -cp /opt/pig-0.16.0/pig-0.16.0-core-h2.jar WordClient.java
hadoop@sw-desktop:~$ sh w.sh
```

```
hadoop@sw-desktop:~$ hdfs dfs -cat /output/sw1/part-m-00000
HELLO WORLD
GOOD HADOOP
```

## 【同步训练】

### 一、操作题

有两个数据文件：

员工（部门号,员工号,月销售额）employee.txt：

D01,E0101,80

D01,E0102,59

D01,E0103,75

D02,E0201,82

D02,E0202,81

D02,E0203,78

D03,E0301,56

D03,E0302,91

D03,E0306,83

部门（部门号,主管）dept.txt：

D01,Zhang

D02,Sun

D03,Wang

D04,Yang

（1）使用 Pig 上传 employee.txt 和 dept.txt 到 HDFS 上。

（2）使用 Pig 加载 employee.txt 到 A，加载 dept.txt 到 B。

（3）员工表筛选部门为 D01 的员工。

（4）员工表和部门表按部门号相等关联。

（5）员工表按月销售额进行排序。

（6）统计各部门月销售大于 80（万元）的员工人数。

### 二、编程题

（1）编写 Pig 自定义过滤函数，过滤出员工号为"E0203"的员工，并使用 Pig 命令调用该过滤函数查找该位员工。

（2）编写 Pig 自定义运算函数，如果员工月销售额小于 60（万元），业绩为"差"，大于等于 60（万元）且小于 80（万元），业绩为"一般"，大于等于 80（万元）且小于 90（万元），业绩为"良好"，大于等于 90（万元），业绩为"优秀"。并使用 Pig 命令调用该运算函数，输出结果（员工号、月销售额、业绩）。

# 项目八 Sqoop 数据迁移

## 【项目介绍】

Sqoop 是一个数据转移工具，本项目主要使用 Sqoop 工具完成 MySQL 与 HDFS 之间的数据转移、MySQL 与 Hive 之间及 MySQL 与 HBase 之间的数据转移。

本项目分以下两个任务：
- 任务 1　Sqoop 安装及 MySQL 与 HDFS 数据迁移
- 任务 2　MySQL 与 Hive/HBase 数据转移

## 【学习目标】

### 一、知识目标

- 掌握 Sqoop 的相关功能。
- 掌握 Sqoop Import 操作。
- 掌握 Sqoop Export 操作。

### 二、能力目标

- 能够使用 Sqoop Import 导入 MySQL 数据到 HDFS。
- 能够使用 Sqoop Export 导出 HDFS 数据到 MySQL。
- 能够使用 Sqoop Import 导入 MySQL 数据到 Hive。
- 能够使用 Sqoop Import 导入 MySQL 数据到 HBase。

## 任务 1　Sqoop 安装及 MySQL 与 HDFS 数据迁移

### 【任务概述】

本任务需要完成 Sqoop 的安装，使用 Sqoop Import 和 Sqoop Export 完成 MySQL 和 HDFS 之间数据的转移，并检查其结果。

## 【支撑知识】

### 一、Sqoop 简介

Sqoop 是一个用来将 Hadoop 和关系型数据库中的数据相互转移的工具,可以将一个关系型数据库(如 MySQL、Oracle、Postgres 等)中的数据导入到 HDFS 中,也可以将 HDFS 的数据导出到关系型数据库中,如图 8-1 所示。

图 8-1 Sqoop 操作

运行 Sqoop help 能够获取 Sqoop 帮助信息,如下所示:

```
$ sqoop help
usage: sqoop COMMAND [ARGS]
Available commands:
  codegen            Generate code to interact with database records
  create-hive-table  Import a table definition into Hive
  eval               Evaluate a SQL statement and display the results
  export             Export an HDFS directory to a database table
  help               List available commands
  import             Import a table from a database to HDFS
  import-all-tables  Import tables from a database to HDFS
  import-mainframe   Import datasets from a mainframe server to HDFS
  job                Work with saved jobs
  list-databases     List available databases on a server
  list-tables        List available tables in a database
  merge              Merge results of incremental imports
  metastore          Run a standalone Sqoop metastore
  version            Display version information
See 'sqoop help COMMAND' for information on a specific command.
```

### 二、Sqoop 的基本命令

Sqoop 提供一系列命令,包括导入操作(Import)、导出操作(Export)、导入所有表(Import-All-Tables)、列出所有数据库实例(List-Databases)和列出数据库实例中所有表(List-Tables)等,使用帮助命令获取 Sqoop 工具的详细命令参数,如下所示:

```
$ sqoop help import
usage: sqoop import [GENERIC-ARGS] [TOOL-ARGS]
Common arguments:
  --connect <jdbc-uri>                         Specify JDBC connect
                                               string
  --connection-manager <class-name>            Specify connection manager
                                               class name
  --connection-param-file <properties-file>    Specify connection
```

```
    --driver <class-name>              parameters file
                                       Manually specify JDBC
                                       driver class to use
    --hadoop-home <hdir>               Override
                                       $HADOOP_MAPRED_HOME_ARG
    --hadoop-mapred-home <dir>         Override
                                       $HADOOP_MAPRED_HOME_ARG
    --help                             Print usage instructions
    ...
```

下面列举一些 Sqoop 的常用命令。

（1）列出 MySQL 的所有数据库

sqoop list-databases --connect jdbc:mysql://IP:3306/ --username 用户 --password 密码

（2）连接 MySQL 并列出数据库中的表

sqoop list-tables --connect jdbc:mysql://IP:3306/数据库 --username 用户 --password 密码

（3）将 MySQL 的表数据导入 HDFS

sqoop import --connect jdbc:mysql://IP:3306/数据库 --username 用户 --password 密码 --table 表 --num-mappers Map 任务数 --target-dir HDFS 存放位置

参数说明：

--num-mappers 或 -m 选项指定 Map 任务个数，--target-dir 选项指定 HDFS 存放位置。

（4）HDFS 导出到 MySQL 表中

sqoop exprot --connect jdbc:mysql://IP:3306/数据库 --username 用户 --password 密码 --table 表 --num-mappers Map 任务数 --export-dir HDFS 存放位置

（5）将 MySQL 的表结构复制到 Hive 中

sqoop create-Hive-table --connect jdbc:mysql:/IP:3306/ --username 用户 --password 密码 --table 表 --hive-table hive 表 --fields-terminated-by "\0001" --lines-terminated-by "\n"

参数说明：

--fields-terminated-by "\0001"是设置每列之间的分隔符，"\0001"是 ASCII 码中的 1，它也是 Hive 的默认行内分隔符，而 Sqoop 的默认行内分隔符为","，--lines-terminated-by "\n" 设置的是每行之间的分隔符，此处为换行符，也是默认的分隔符。

（6）将数据从 MySQL 导入文件到 Hive 表中

sqoop import --connect jdbc:mysql:/IP:3306/ --username 用户 --password 密码 --table 表 --hive-import --hive-table hive 表 --num-mappers Map 任务数 --fields-terminated-by "\0001";

参数说明：

--fields-terminated-by "\0001" 需同创建 Hive 表时保持一致。

（7）将 Hive 中的表数据导入 MySQL 表中

sqoop export --connect jdbc:mysql://IP:3306/数据库 --username 用户 --password 密码 --table 表 --export-dir HDFS 文件 --input-fields-terminated-by '\0001'

（8）使用--query 语句将数据从 MySQL 导入文件到 Hive 表中

sqoop import --append --connect jdbc:mysql://IP:3306/数据库 --username 用户

--password 密码 --query "select ... from ..." --num-mappers Map 任务数 --target-dir HDFS 存放位置 --fields-terminated-by ","

（9）使用 --columns --where 语句将数据从 MySQL 导入 Hive 表中

sqoop import --append --connect jdbc:mysql://IP:3306/数据库 --username --用户 --password 密码 --table 表 --columns "列名" --where "条件" --num-mappers Map 任务数 --target-dir HDFS 存放位置 --fields-terminated-by ","

【任务实施】

### 一、客户端主机安装 Sqoop 软件

① 下载 Sqoop 软件包到 /home/hadoop 目录下，下载地址如下：

https://mirrors.aliyun.com/apache/sqoop/1.4.6/sqoop-1.4.6.bin__hadoop-2.0.4-alpha.tar.gz

② 安装 Sqoop 软件，操作如下：

```
hadoop@sw-desktop:~$ cd /opt
hadoop@...$ sudo tar xvzf /home/hadoop/sqoop-1.4.6.bin__hadoop-2.0.4-alpha.tar.gz
hadoop@...$ sudo chown -R hadoop:hadoop sqoop-1.4.6.bin__hadoop-2.0.4-alpha
```

③ 修改 Sqoop 配置参数，操作如下：

```
hadoop@...$ cd /opt/sqoop-1.4.6.bin__hadoop-2.0.4-alpha/conf
hadoop@...$ vi sqoop-env.sh
```

添加如下内容：

```
export HADOOP_COMMON_HOME=/opt/hadoop-2.7.3
export HADOOP_MAPRED_HOME=/opt/hadoop-2.7.3
export HBASE_HOME=/opt/hbase-1.2.4
export HIVE_HOME=/opt/apache-hive-2.1.1-bin
export HCAT_HOME=/opt/apache-hive-2.1.1-bin/hcatalog
export ZOOCFGDIR=/opt/zookeeper-3.4.9
```

④ 修改环境变量，操作如下：

```
hadoop@...$ vi /home/hadoop/.profile
```

添加如下内容：

```
export SQOOP_HOME=/opt/sqoop-1.4.6.bin__hadoop-2.0.4-alpha
export PATH=$PATH:$SQOOP_HOME/bin
```

⑤ 环境变量生效，操作如下：

```
hadoop@...$ source /home/hadoop/.profile
```

⑥ 复制 mysql-connector-java-5.1.40.jar，操作如下：

```
hadoop@...$ cp /home/hadoop/mysql-connector-java-5.1.40.jar \
/opt/sqoop-1.4.6.bin__hadoop-2.0.4-alpha/lib
```

### 二、安装 MySQL-Client 软件包

① 在客户端主机上安装 MySQL-Client，操作如下：

```
hadoop@sw-desktop:~$ sudo apt-get install mysql-client
```

② MySQL 用户授权，操作如下：

```
hadoop@master:~$ mysql -uroot -p123456
mysql> GRANT ALL PRIVILEGES ON *.* TO 'sqoop'@'%' IDENTIFIED BY '123456';
Query OK, 0 rows affected, 1 warning (0.30 sec)
mysql> GRANT ALL PRIVILEGES ON *.* TO 'sqoop'@'localhost' IDENTIFIED BY '123456';
Query OK, 0 rows affected, 1 warning (0.00 sec)
```

③ 创建数据，操作如下：

```
hadoop@sw-desktop:~$ mysql -hmaster -usqoop -p123456
mysql> create database sqoop;
Query OK, 1 row affected (0.06 sec)
mysql> use sqoop;
Database changed
mysql> create table dept(id int,name varchar(20),primary key(id));
Query OK, 0 rows affected (0.50 sec)
mysql> insert into dept values(610213,'云计算技术与应用');
Query OK, 1 row affected (0.26 sec)
mysql> insert into dept values(610215,'大数据技术与应用');
Query OK, 1 row affected (0.02 sec)
mysql> insert into dept values(590108,'软件技术');
Query OK, 1 row affected (0.01 sec)
mysql> select * from dept;
+--------+--------------------------+
| id     | name                     |
+--------+--------------------------+
| 590108 | 软件技术                 |
| 610213 | 云计算技术与应用         |
| 610215 | 大数据技术与应用         |
+--------+--------------------------+
3 rows in set (0.01 sec)
```

## 三、MySQL 与 HDFS 数据迁移

① 查看 MySQL 数据库，操作如下：

```
hadoop@sw-desktop:~$ sqoop list-databases \
--connect jdbc:mysql://master:3306/ \
--username sqoop --password 123456
…
information_schema
hive
```

```
mysql
performance_schema
sqoop
sys
```

② 查看 MySQL 表,操作如下:

```
hadoop@sw-desktop:~$ sqoop list-tables \
--connect jdbc:mysql://master:3306/ sqoop \
--username sqoop --password 123456
…
dept
```

③ MySQL 表导入 HDFS,操作如下:

```
hadoop@sw-desktop:~$ sqoop import --connect jdbc:mysql://master:3306/sqoop \
--username sqoop --password 123456 --table dept -m 1 --target-dir /user/dept
```

④ 查看 HDFS,操作如下:

```
hadoop@sw-desktop:~$ hdfs dfs -ls /user/dept
Found 2 items
-rw-r--r--   3 hadoop supergroup          0 2017-03-11 16:03 /user/dept/_SUCCESS
-rw-r--r--   3 hadoop supergroup         84 2017-03-11 16:03 /user/dept/part-m-00000
hadoop@sw-desktop:~$ hdfs dfs -cat /user/dept/part-m-00000
590108,软件技术
610213,云计算技术与应用
610215,大数据技术与应用
```

⑤ 清空 MySQL 的 dept 表,操作如下:

```
mysql> use sqoop;
Database changed
mysql> truncate dept;
Query OK, 0 rows affected (0.29 sec)
```

⑥ 数据从 HDFS 导出到 MySQL 表,操作如下:

```
hadoop@sw-desktop:~$ sqoop export --connect jdbc:mysql://master:3306/sqoop \
--username sqoop --password 123456 --table dept -m 1 --export-dir /user/dept
```

⑦ 查询 dept 表,操作如下:

```
mysql> select * from dept;
+--------+----------------------+
| id     | name                 |
+--------+----------------------+
| 590108 | 软件技术             |
| 610213 | 云计算技术与应用     |
| 610215 | 大数据技术与应用     |
+--------+----------------------+
3 rows in set (0.00 sec)
```

⑧ Sqoop Import 增量导入 HDFS 上，操作如下：

```
mysql> insert into dept values(590101,'计算机应用技术');
Query OK, 1 row affected (0.38 sec)

hadoop@sw-desktop:~$ sqoop import --connect jdbc:mysql://master:3306/sqoop \
--username sqoop --password 123456 --table dept -m 1 --target-dir /user/dept \
--incremental append --check-column id

hadoop@sw-desktop:~$ hdfs dfs -ls /user/dept
Found 3 items
-rw-r--r--   3 hadoop supergroup          0 2017-03-11 16:03 /user/dept/_SUCCESS
-rw-r--r--   3 hadoop supergroup         84 2017-03-11 16:03 /user/dept/part-m-00000
-rw-r--r--   3 hadoop supergroup        113 2017-03-11 16:56 /user/dept/part-m-00001
hadoop@sw-desktop:~$ hdfs dfs -cat /user/dept/part-m-00001
590101,计算机应用技术
590108,软件技术
610213,云计算技术与应用
610215,大数据技术与应用
```

注：如果 append 模式增量导入指定 --last-value 选项值，则只导入新插入的数据。如果更新表，可以使用 lastmodified 模式增量导入，操作如下：

```
hadoop@sw-desktop:~$ sqoop import \
--connect jdbc:mysql://master:3306/sqoop \
--username sqoop --password 123456 --table dept -m 1 \
--target-dir /user/dept --incremental lastmodified \
--check-column name --last-value '计算机应用' --append
```

## 任务 2　MySQL 与 Hive/HBase 数据转移

### 【任务概述】

Hive 和 HBase 都是 Hadoop 生态系统成员，Hive 是数据仓库，HBase 是分布式数据库。本任务主要完成 MySQL 和 Hive 之间，以及 MySQL 和 HBase 之间的数据转移。

### 【任务实施】

#### 一、MySQL 与 Hive 之间的数据转移

① MySQL 的 dept 表导入 Hive，操作如下：

```
hadoop@sw-desktop:~$ sqoop import --connect jdbc:mysql://master:3306/sqoop \
--username sqoop --password 123456 --table dept -m 1 --hive-import
```

参数说明：--hive-import 指定导入 Hive，--hive-table 指定 Hive 表名，--hive-overwrite 可以覆盖原 Hive 表，--create-hive-table 可以将原来 MySQL 表结构复制到 Hive 表中。

② 查看 Hive 表数据，操作如下：

```
hive> show tables;
OK
dept
Time taken: 0.422 seconds, Fetched: 1 row(s)
hive> select * from dept;
OK
590101    计算机应用技术
590108    软件技术
610213    云计算技术与应用
610215    大数据技术与应用
Time taken: 3.494 seconds, Fetched: 4 row(s)
```

③ 清空 MySQL 的 dept 表，操作如下：

```
mysql> use sqoop;
Database changed
mysql> truncate dept;
Query OK, 0 rows affected (0.92 sec)
```

④ Hive 表数据导出到 MySQL，操作如下：

```
hadoop@sw-desktop:~$ sqoop export --connect jdbc:mysql://master:3306/sqoop \
--username sqoop --password 123456 --table dept -m 1 \
--export-dir /hive/warehouse/dept --input-fields-terminated-by '\0001'
```

⑤ 查询 dept 表，操作如下：

```
mysql> select * from dept;
+--------+--------------------------+
| id     | name                     |
+--------+--------------------------+
| 590101 | 计算机应用技术           |
| 590108 | 软件技术                 |
| 610213 | 云计算技术与应用         |
| 610215 | 大数据技术与应用         |
+--------+--------------------------+
4 rows in set (0.00 sec)
```

## 二、MySQL 与 HBase 之间的数据转移

① 创建 HBase 表，操作如下：

```
hbase(main):001:0> create 'hbase_dept','col_family'
0 row(s) in 16.1080 seconds

=> Hbase::Table - hbase_dept
hbase(main):002:0>
```

② 将 MySQL 的 dept 表导入 HBase 中，操作如下：

```
hadoop@sw-desktop:~$ sqoop import --connect jdbc:mysql://master:3306/sqoop \
--username sqoop --password 123456 --table dept --hbase-create-table \
--hbase-table hbase_dept --column-family col_family --hbase-row-key id
```

③ 查看 HBase 的 dept 记录，操作如下：

```
hbase(main):003:0> scan 'hbase_dept'
ROW                COLUMN+CELL
 590101   column=col_family:name, timestamp=1489226988879, value=计算机应用技术
 590108   column=col_family:name, timestamp=1489226988879, value=软件技术
 610213   column=col_family:name, timestamp=1489226975527, value=云计算技术与应用
 610215   column=col_family:name, timestamp=1489226975527, value=大数据技术与应用
4 row(s) in 0.1840 seconds
```

注：由于 HBase 编码问题，汉字无法直接显示，上述汉字为解码结果。

## 【同步训练】

一、简答题

（1）Sqoop 是什么？
（2）Sqoop 支持的功能有哪些？

二、操作题

（1）安装 MySQL 客户端。
（2）安装 Sqoop 工具。
（3）创建 MySQL 表并输入数据，使用 Sqoop 工具把该数据导入 HDFS。

# PART 9 项目九 Spark 部署及数据分析

## 【项目介绍】

Apache Spark 是专为大规模数据处理而设计的快速通用的计算引擎,部署 Spark 集群是本项目的重点,运用 Spark 函数进行数据转化和分析,在 Spark 集群环境下使用 Scala 语言、Java 语言和 Python 语言完成单词计数的编写和运行,最后使用 Spark Python 完成 K-Means 聚类计算。

本项目分以下 3 个任务:
- 任务 1　Spark 部署
- 任务 2　Spark 数据分析
- 任务 3　Spark 编程

## 【学习目标】

### 一、知识目标

- 了解 Spark 的性能特点。
- 了解伯克利数据栈及各功能模块。
- 了解 Spark 架构及基本组件。
- 了解 Scala 语言。
- 掌握 Spark 平台的部署。
- 掌握 Spark 数据集的运算。
- 掌握 RDD 及 RDD 算子。
- 掌握 Spark 程序的编写。
- 掌握 K-Means 聚类分析算法。

### 二、能力目标

- 能够部署 Spark 集群。
- 能够使用 Scala 命令行进行数据处理。
- 能够使用 Spark 运算函数。
- 能够使用蒙特卡罗方法计算圆周率 $\pi$。

- 能够使用 Scala-IDE 工具编写 Spark 程序。
- 能够使用 Spark-submit 运行程序。
- 能够使用 Python 完成聚类计算。

## 任务 1　Spark 部署

### 【任务概述】

Spark 集群环境需要安装 Java 和 Scala，配置 Spark 环境参数，完成 Spark 集群部署，启动 Spark 服务，并测试 Spark 运行情况。

### 【支撑知识】

#### 一、Spark 简介

Spark 是 UC Berkeley AMP Lab（加州大学伯克利分校的 AMP 实验室）所开源的类 Hadoop MapReduce 的通用并行框架，Spark 拥有 Hadoop MapReduce 所具有的优点；但不同于 MapReduce 的是 Job 中间输出结果可以保存在内存中，从而不再需要读/写 HDFS，因此，Spark 能更好地适用于数据挖掘与机器学习等需要迭代的 MapReduce 的算法。

Spark 是一种与 Hadoop 相似的开源集群计算环境，但是两者之间还存在一些不同之处，这些有用的不同之处使 Spark 在某些工作负载方面表现得更加优越，换句话说，Spark 启用了内存分布数据集，除了能够提供交互式查询外，还可以优化迭代工作负载。

Spark 是在 Scala 语言中实现的，它将 Scala 用作其应用程序框架。与 Hadoop 不同，Spark 和 Scala 能够紧密集成，其中的 Scala 可以像操作本地集合对象一样轻松地操作分布式数据集。

尽管创建 Spark 是为了支持分布式数据集上的迭代作业，但是实际上它是对 Hadoop 的补充，可以在 Hadoop 文件系统中并行运行。通过名为 Mesos 的第三方集群框架可以支持此行为。Spark 由加州大学伯克利分校 AMP 实验室（Algorithms Machines People Lab）开发，可用来构建大型的、低延迟的数据分析应用程序。

Spark 的性能特点如下：

（1）快速处理能力

随着实时大数据应用越来越多，Hadoop 作为离线的高吞吐、低响应框架已不能满足这类需求。Hadoop MapReduce 的 Job 将中间输出和结果存储在 HDFS 中，读/写 HDFS 造成磁盘 I/O 成为瓶颈。Spark 允许将中间输出和结果存储在内存中，节省了大量的磁盘 I/O。同时 Spark 自身的 DAG 执行引擎也支持数据在内存中的计算。Spark 官网声称性能比 Hadoop 快 100 倍，即便是内存不足需要磁盘 I/O，其速度也是 Hadoop 的 10 倍以上。

（2）易于使用

Spark 现在支持 Java、Scala、Python 和 R 等语言编写应用程序，大大降低了门槛。Spark 提供 80 多个高等级操作符，允许在 Scala、Python、R 的 Shell 中进行交互式查询。

（3）支持流式计算

与 MapReduce 只能处理离线数据相比，Spark 还支持实时的流计算。Spark 依赖 Spark Streaming 对数据进行实时处理，其流式处理能力要强于 Storm。

（4）可用性高

Spark 自身实现了 Standalone 部署模式，此模式下的 Master 可以有多个，解决了单点故障问题。此模式完全可以使用其他集群管理器替换，如 YARN、Mesos、EC2 等。

（5）丰富的数据源支持

Spark 除了可以访问操作系统自身的文件系统和 HDFS，还可以访问 Cassandra、HBase、Hive、Tachyon，以及任何 Hadoop 的数据源。这极大地方便了已经使用 HDFS、HBase 的用户顺利迁移到 Spark。

## 二、伯克利数据分析栈

Spark 已经发展成为包含众多子项目的大数据计算平台。伯克利将 Spark 的整个生态系统称为伯克利数据分析栈（Berkeley Data Analytics Stack，BDAS），如图 9-1 所示。其核心框架是 Spark，同时 BDAS 涵盖支持结构化数据 SQL 查询与分析的查询引擎 Spark SQL 和 Shark，提供机器学习功能的系统 MLbase 及底层的分布式机器学习库 MLlib、并行图计算框架 GraphX、流计算框架 Spark Streaming、采样近似计算查询引擎 BlinkDB、内存分布式文件系统 Tachyon、资源管理框架 Mesos 等子项目。这些子项目在 Spark 上层提供了更高层、更丰富的计算范式。

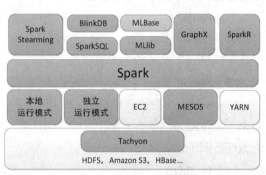

图 9-1　伯克利数据分析栈

（1）Spark

Spark 是整个 BDAS 的核心组件，是一个大数据分布式编程框架，不仅实现了 MapReduce 的算子 Map 函数和 Reduce 函数及计算模型，还提供更为丰富的算子，如 filter、join、groupByKey 等。Spark 将分布式数据抽象为弹性分布式数据集（Resilient Distributed Datasets，RDD），实现了应用任务调度、RPC、序列化和压缩，并为运行在其上的上层组件提供 API。其底层采用 Scala 这种函数式语言书写而成，并且所提供的 API 深度借鉴 Scala 函数式的编程思想，提供与 Scala 类似的编程接口。Spark 将数据在分布式环境下分区，然后将作业转化为有向无环图（Directed Acyclic Graph，DAG），并分阶段进行 DAG 的调度和任务的分布式并行处理。

（2）Shark

Shark 是构建在 Spark 和 Hive 基础之上的数据仓库。目前，Shark 已经完成学术使命，终止开发，但其架构和原理仍具有借鉴意义。它提供了能够查询 Hive 中所存储数据的一套 SQL 接口，兼容现有的 Hive QL 语法。这样，熟悉 Hive QL 或者 SQL 的用户可以基于 Shark 进行快速的 Ad-Hoc、Reporting 等类型的 SQL 查询。Shark 底层复用 Hive 的解析器、优化器，以及元数据存储和序列化接口。Shark 会将 Hive QL 编译转化为一组 Spark 任务，进行分布式运算。

（3）Spark SQL

Spark SQL 提供在大数据上的 SQL 查询功能，类似于 Shark 在整个生态系统的角色，它们可以统称为 SQL on Spark。之前，Shark 的查询编译和优化器依赖于 Hive，使得 Shark 不得不维护一套 Hive 分支，而 Spark SQL 使用 Cataly 作为查询解析和优化器，并在底层使用 Spark 作为执行引擎实现 SQL 的 Operator。用户可以在 Spark 上直接书写 SQL，相当于为 Spark 扩充了一套 SQL 算子，这无疑更加丰富了 Spark 的算子和功能，同时 Spark SQL 不断兼容不同的持久化存储（如 HDFS、Hive 等），为其发展奠定了广阔的基础。

（4）Spark Streaming

Spark Streaming 通过将流数据按指定时间片累积为 RDD，然后将每个 RDD 进行批处理，进而实现大规模的流数据处理。其吞吐量能够超越现有主流流处理框架 Storm，并提供丰富的 API 用于流数据计算。

（5）GraphX

GraphX 基于 BSP 模型，在 Spark 之上封装类似 Pregel 的接口，进行大规模同步全局的图计算，尤其是当用户进行多轮迭代时，基于 Spark 内存计算的优势尤为明显。

（6）Tachyon

Tachyon 是一个分布式内存文件系统，可以理解为内存中的 HDFS。为了提供更高的性能，将数据存储剥离 Java Heap。用户可以基于 Tachyon 实现 RDD 或者文件的跨应用共享，并提供高容错机制，保证数据的可靠性。

（7）Mesos

Mesos 是一个资源管理框架，提供类似于 YARN 的功能。用户可以在其中插件式地运行 Spark、MapReduce、Tez 等计算框架的任务。Mesos 会对资源和任务进行隔离，并实现高效的资源任务调度。

（8）BlinkDB

BlinkDB 是一个用于在海量数据上进行交互式 SQL 的近似查询引擎。它允许用户通过在查询准确性和查询响应时间之间做出权衡，完成近似查询。其数据的精度被控制在允许的误差范围内。为了达到这个目标，BlinkDB 的核心思想如下：通过一个自适应优化框架，随着时间的推移，从原始数据建立并维护一组多维样本；通过一个动态样本选择策略，选择一个适当大小的示例，然后基于查询的准确性和响应时间满足用户查询需求。

（9）MLBase/MLlib

MLBase 是 Spark 生态系统中专注于机器学习的组件，它的目标是让机器学习的门槛更低，让一些可能并不了解机器学习的用户能够方便地使用 MLBase。MLBase 定义了 4 个边界：MLRuntime、MLlib、MLI 和 ML Optimizer。

① MLRuntime 是由 Spark Core 提供的分布式内存计算框架，运行由 Optimizer 优化过的算法进行数据的计算并输出分析结果。

② MLlib 是 Spark 实现的一些常见的机器学习算法和实用程序，包括分类、回归、聚类、协同过滤、降维及底层优化。该算法可以进行可扩充。

③ MLI 是一个进行特征抽取和高级 ML 编程抽象算法实现的 API 或平台。

④ MLOptimizer 会选择它认为最适合的已经在内部实现好了的机器学习算法和相关参数，来处理用户输入的数据，并返回模型或其他帮助分析的结果。

MLBase 的核心是其优化器（ML Optimizer），它可以把声明式的任务转化成复杂的学习计划，最终产出最优的模型和计算结果。

（10）SparkR

R 语言是遵循 GNU 协议的一款开源、免费的软件，广泛应用于统计计算和统计制图，但是它只能单机运行。为了能够使用 R 语言分析大规模分布式的数据，伯克利分校 AMP 实验室开发了 SparkR，并在 Spark 1.4 版本中加入了该组件。通过 SparkR 可以分析大规模的数据集，并通过 R Shell 交互式地在 SparkR 上运行作业。SparkR 特性如下：
- 提供了 Spark 中 RDD 的 API，用户可以在集群上通过 R Shell 交互性地运行 Spark 任务。
- 支持序化闭包功能，可以将用户定义函数中所引用到的变量自动序化发送到集群中其他的机器上。

SparkR 还可以很容易地调用 R 开发包，只需要在集群上执行操作前用 includePackage 读取 R 开发包就可以了。

### 三、Spark 架构

Spark 架构采用了分布式计算中的 Master-Slave 模型。Master 是对应集群中的含有 Master 进程的节点，Slave 是集群中含有 Worker 进程的节点。Master 作为整个集群的控制器，负责整个集群的正常运行；Worker 相当于计算节点，接收主节点命令与进行状态汇报；Executor 负责任务的执行；Client 作为用户的客户端负责提交应用，Driver 负责控制一个应用的执行，如图 9-2 所示。

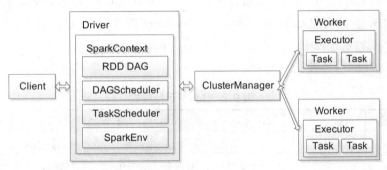

图 9-2　Spark 架构图

Spark 集群部署后，需要在主节点和从节点分别启动 Master 进程和 Worker 进程，对整个集群进行控制。在一个 Spark 应用的执行过程中，Driver 和 Worker 是两个重要角色。Driver 程序是应用逻辑执行的起点，负责作业的调度，即 Task 任务的分发，而多个 Worker 用来管理计算节点和创建 Executor 并行处理任务。在执行阶段，Driver 会将 Task 和 Task 所依赖的 file 和 jar 序列化后传递给对应的 Worker 机器，同时 Executor 对相应数据分区的任务进行处理。

Spark 基本组件及功能如表 9-1 所示。

表 9-1　Spark 基本组件及功能

| 组　　件 | 功　　能 |
| --- | --- |
| ClusterManager | 在 Standalone 模式中即为 Master（主节点），控制整个集群，监控 Worker。在 YARN 模式中为资源管理器 |
| Worker | 从节点，负责控制计算节点，启动 Executor 或 Driver。在 YARN 模式中为 NodeManager，负责计算节点的控制 |
| Driver | 运行 Application 的 main() 函数并创建 SparkContext |
| Executor | 执行器，在 Worker Node 上执行任务的组件，用于启动线程池运行任务。每个 Application 拥有独立的一组 Executors |

续表

| 组件 | 功能 | |
|---|---|---|
| SparkContext | 整个应用的上下文，控制应用的生命周期 | |
| RDD | Spark 的基本计算单元，一组 RDD 可形成执行的有向无环图（RDD Graph） | |
| DAG Scheduler | 根据作业（Job）构建基于 Stage 的 DAG，并提交 Stage 给 TaskScheduler | |
| TaskScheduler | 将任务（Task）分发给 Executor 执行 | |
| SparkEnv | 线程级别的上下文，存储运行时的重要组件的引用；SparkEnv 内创建并包含如下一些重要组件的引用 | |
| | MapOutPutTracker | 负责 Shuffle 元信息的存储 |
| | BroadcastManager | 负责广播变量的控制与元信息的存储 |
| | BlockManager | 负责存储管理、创建和查找块 |
| | MetricsSystem | 监控运行时性能指标信息 |
| | SparkConf | 负责存储配置信息 |

Spark 的整体流程如下：Client 提交应用，Master 找到一个 Worker 启动 Driver，Driver 向 Master 或者资源管理器申请资源，之后将应用转化为 RDD Graph，再由 DAGScheduler 将 RDD Graph 转化为 Stage 的有向无环图提交给 TaskScheduler，由 TaskScheduler 提交任务给 Executor 执行。在任务执行的过程中，其他组件协同工作，确保整个应用顺利执行。

Spark 运行模式有多种，如表 9-2 所示。

表 9-2　Spark 运行模式

| 运行环境 | 模式 | 描述 |
|---|---|---|
| Local | 本地模式 | 常用于本地开发测试，本地还分 local 单线程和 local-cluster 多线程 |
| Standalone | 集群模式 | 典型的 Master/Slave 模式，Master 有单点故障；Spark 支持 Zookeeper 来实现 HA |
| On yarn | 集群模式 | 运行在 Yarn 资源管理器框架之上，由 Yarn 负责资源管理，Spark 负责任务调度和计算 |
| On mesos | 集群模式 | 运行在 Mesos 资源管理器框架之上，由 Mesos 负责资源管理，Spark 负责任务调度和计算 |
| On cloud | 集群模式 | 比如 AWS 的 EC2，使用这个模式能很方便地访问 Amazon 的 S3；Spark 支持多种分布式存储系统：HDFS 和 S3 |

Spark 常用术语如表 9-3 所示。

表 9-3　Spark 常用术语

| 术语 | 描述 |
|---|---|
| Application | Spark 的应用程序，包括一个 Driver Program 和若干 Executor |
| SparkContext | Spark 应用程序入口，负责调度各个运算资源，协调各个 Worker Node 上的 Executor |
| Driver Program | 运行 Application 的 main()函数并且创建 SparkContext |
| Executor | 是 Application 运行在 Worker Node 上的一个进程，该进程负责运行 Task，并且负责将数据存在内存或者磁盘上。每个 Application 都会申请各自的 Executor 来处理任务 |

续表

| 术语 | 描述 |
|---|---|
| Cluster Manager | 在集群上获取资源的外部服务（如 Standalone、Mesos、Yarn） |
| Worker Node | 集群中任何可以运行 Application 代码的节点，运行一个或多个 Executor 进程 |
| Task | 运行在 Executor 上的工作单元 |
| Job | SparkContext 提交的具体 Action 操作，常和 Action 对应 |
| Stage | 每个 Job 会被拆分很多组 Task，每组任务被称为 Stage，也称为 TaskSet |
| RDD | 是 Resilient Distributed Datasets 的简称，中文为弹性分布式数据集；是 Spark 最核心的模块和类 |
| DAGScheduler | 根据 Job 构建基于 Stage 的 DAG，并提交 Stage 给 TaskScheduler |
| TaskScheduler | 将 TaskSet 提交给 Worker Node 集群运行并返回结果 |
| Transformation | 是 Spark API 的一种类型，Transformation 返回值还是一个 RDD，所有的 Transformation 采用的都是懒策略，如果只是将 Transformation 提交，是不会执行计算的 |
| Action | 是 Spark API 的一种类型，Action 返回值不是一个 RDD，而是一个 Scala 集合；只有在 Action 被提交的时候计算才被触发 |

### 四、Scala 语言

Scala 是 Scalable Language 的简写，是一门多范式的编程语言。联邦理工学院洛桑（EPFL）的 Martin Odersky 于 2001 年基于 Funnel 的工作开始设计 Scala。Funnel 是把函数式编程思想和 Petri 网相结合的一种编程语言。Scala 类似于 Java 的编程语言，设计初衷是实现可伸缩的语言，并集成面向对象编程和函数式编程的各种特性。

（1）面向对象特性

Scala 是一种纯面向对象的语言，每个值都是对象。对象的数据类型及行为由类和特质描述。类抽象机制的扩展有两种途径：一种途径是子类继承，另一种途径是灵活的混入机制。这两种途径能避免多重继承的种种问题。

（2）函数式编程

Scala 也是一种函数式语言，其函数也能作为值来使用。Scala 提供了轻量级的语法用以定义匿名函数，支持高阶函数，允许嵌套多层函数，并支持柯里化（Currying）。Scala 的 case class 及其内置的模式匹配相当于函数式编程语言中常用的代数类型。

更进一步，程序员可以利用 Scala 的模式匹配，编写类似正则表达式的代码处理 XML 数据。

（3）静态类型

Scala 具备类型系统，通过编译时检查，保证代码的安全性和一致性。类型系统具体支持以下特性：

- 泛型类。
- 协变和逆变。
- 标注。
- 类型参数的上下限约束。
- 把类别和抽象类型作为对象成员。

- 复合类型。
- 引用自己时显式指定类型。
- 视图。
- 多态方法。

（4）扩展性

Scala 的设计秉承一项事实，即在实践中，某个领域特定的应用程序开发往往需要特定于该领域的语言扩展。Scala 提供了许多独特的语言机制，可以以库的形式轻易无缝添加新的语言结构。

- 任何方法可用作前缀或后缀操作符。
- 可以根据预期类型自动构造闭包。

（5）并发性

Scala 使用 Actor 作为其并发模型，Actor 是类似线程的实体，通过邮箱发收消息。Actor 可以复用线程，因此，可以在程序中使用数百万个 Actor，而线程只能创建数千个。在 2.10 之后的版本中，使用 Akka 作为其默认 Actor 实现。

Spark 主要的编程语言是 Scala，选择 Scala 是因为它的简洁性（Scala 可以很方便地在交互式下使用）和性能（JVM 上的静态强类型语言）。

## 【任务实施】

### 一、Master 节点安装软件

① 下载 Scala 和 Spark 软件包到 /home/hadoop 目录下，下载地址如下：

```
http://www.scala-lang.org/download/
http://mirrors.aliyun.com/apache/spark/spark-2.1.0/spark-2.1.0-bin-hadoop2.7.tgz
```

② 安装软件，操作如下：

```
hadoop@master:~$ cd /opt
hadoop@master:/opt$ sudo tar xvzf /home/hadoop/scala-2.12.1.tgz
hadoop@master:/opt$ sudo chown -R hadoop:hadoop /opt/scala-2.12.1
hadoop@master:/opt$ sudo tar xvzf /home/hadoop/spark-2.1.0-bin-hadoop2.7.tgz
hadoop@master:/opt$ sudo chown -R hadoop:hadoop /opt/spark-2.1.0-bin-hadoop2.7
```

### 二、Master 节点设置 Spark 参数

① 新建 spark-env.sh 文件，操作如下：

```
hadoop@master:~$ cd /opt/spark-2.1.0-bin-hadoop2.7/conf
hadoop@master:/opt/.../conf$ vi spark-env.sh
```

内容如下：

```
export JAVA_HOME=/opt/jdk1.8.0_121
export HADOOP_HOME=/opt/hadoop-2.7.3
export HADOOP_CONF_DIR=$HADOOP_HOME/etc/hadoop
export SCALA_HOME=/opt/scala-2.12.1
export SPARK_HOME=/opt/spark-2.1.0-bin-hadoop2.7
export SPARK_MASTER_IP=master
export SPARK_WORKER_MEMORY=2g
```

② 新建 slaves 文件，操作如下：
```
hadoop@master:/opt/.../conf$ vi slaves
```
内容如下：
```
slave1
slave2
```
③ 修改环境变量，操作如下：
```
hadoop@master:~$ vi /home/hadoop/.profile
```
添加如下内容：
```
export HADOOP_CONF_DIR=$HADOOP_HOME/etc/hadoop
export SCALA_HOME=/opt/scala-2.12.1
export SPARK_HOME=/opt/spark-2.1.0-bin-hadoop2.7
export PATH=$PATH:$SCALA_HOME/bin:$SPARK_HOME/bin
```
④ 使环境变量生效，操作如下：
```
hadoop@master:~$ source /home/hadoop/.profile
```

### 三、Slave 节点安装软件

① 以用户 hadoop 登录 Slave1 节点安装软件，操作如下：
```
hadoop@slave1:~$ sudo scp -r hadoop@master:/opt/scala-2.12.1 /opt
hadoop@slave1:~$ sudo scp -r hadoop@master:/opt/spark-2.1.0-bin-hadoop2.7 /opt
hadoop@slave1:~$ sudo chown -R hadoop:hadoop /opt/scala-2.12.1
hadoop@slave1:~$ sudo chown -R hadoop:hadoop /opt/spark-2.1.0-bin-hadoop2.7
```
② 以用户 hadoop 登录 Slave2 节点安装软件，操作如下：
```
hadoop@slave2:~$ sudo scp -r hadoop@master:/opt/scala-2.12.1 /opt
hadoop@slave2:~$ sudo scp -r hadoop@master:/opt/spark-2.1.0-bin-hadoop2.7 /opt
hadoop@slave2:~$ sudo chown -R hadoop:hadoop /opt/scala-2.12.1
hadoop@slave2:~$ sudo chown -R hadoop:hadoop /opt/spark-2.1.0-bin-hadoop2.7
```
③ 修改 Slave1 节点和 Slave2 节点环境变量，操作如下：
```
hadoop@slave...$ vi /home/hadoop/.profile
```
添加如下内容：
```
export HADOOP_CONF_DIR=$HADOOP_HOME/etc/hadoop
export SCALA_HOME=/opt/scala-2.12.1
export SPARK_HOME=/opt/spark-2.1.0-bin-hadoop2.7
export PATH=$PATH:$SCALA_HOME/bin:$SPARK_HOME/bin
```
④ Slave1 节点和 Slave2 节点环境变量生效，操作如下：
```
hadoop@slave...$ source /home/hadoop/.profile
```

### 四、测试 Spark

① 登录各集群节点启动 Zookeeper 服务，并查看服务状态，操作如下：
```
hadoop@...:~$ zkServer.sh start
hadoop@...:~$ zkServer.sh status
```

② 在 Master 节点上启动 Hadoop 服务，操作如下：

hadoop@master:~$ start-dfs.sh

hadoop@master:~$ start-yarn.sh

hadoop@master:~$ mr-jobhistory-daemon.sh start historyserver

③ 在 Master 节点上启动 Spark 服务，操作如下：

hadoop@master:~$ /opt/spark-2.1.0-bin-hadoop2.7/sbin/start-all.sh

④ 查看各集群节点进程，操作如下：

● Master 节点：

hadoop@master:~$ jps

7712 Jps

6485 QuorumPeerMain

7190 ResourceManager

7495 JobHistoryServer

6745 NameNode

7004 SecondaryNameNode

7614 Master

● Slave1 节点：

hadoop@slave1:~$ jps

3312 DataNode

3473 NodeManager

3126 QuorumPeerMain

3751 Jps

3678 Worker

● Slave2 节点：

hadoop@slave2:~$ jps

3155 Jps

2531 QuorumPeerMain

3082 Worker

2716 DataNode

2877 NodeManager

⑤ 打开浏览器，在地址栏中输入"http://master:8080"，查看 Spark 集群情况，如图 9-3 所示。

⑥ 打开浏览器，在地址栏中输入"http://slave1:8081"，查看 Worker 执行情况，如图 9-4 所示。

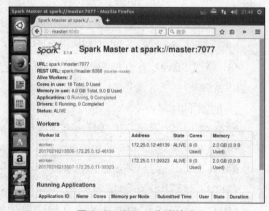

图 9-3　Spark 集群情况

⑦ 先启动 spark-shell，再打开浏览器，在地址栏中输入"http://master:4040"，查看"Spark Jobs"，如图 9-5 所示。

hadoop@master:~$ spark-shell

Setting default log level to "WARN".

```
......
Spark context Web UI available at http://172.25.0.10:4040
Spark context available as 'sc' (master = local[*], app id = local-
1496366196168).
Spark session available as 'spark'.
Welcome to
      ____              __
     / __/__  ___ _____/ /__
    _\ \/ _ \/ _ `/ __/  '_/
   /___/ .__/\_,_/_/ /_/\_\   version 2.1.0
      /_/

Using Scala version 2.11.8 (Java HotSpot(TM) 64-Bit Server VM, Java 1.8.0_121)
Type in expressions to have them evaluated.
Type :help for more information.

scala>
```

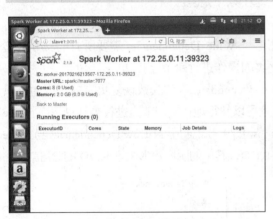

图 9-4 Spark Worker

图 9-5 Spark Jobs

## 任务 2　Spark 数据分析

### 【任务概述】

本任务主要运行测试 Spark 示例 SparkPi 和 JavaWordCount；使用 Spark 函数 map、colleect、filter、flatMap、union、join、lookup、groupByKey、sortByKey 完成数据分析。

### 【支撑知识】

#### 一、Spark 的核心 RDD

与许多专有的大数据处理平台不同，Spark 建立在统一抽象的 RDD 之上，使得它可以以基本一致的方式应对不同的大数据处理场景，包括 MapReduce、Streaming、SQL、Machine Learning 及 Graph 等；即 Matei Zaharia 所谓的"设计一个通用的编程抽象（Unified Programming Abstraction）"，这也正是 Spark 让人着迷的地方。

RDD 是一个容错的、并行的数据结构，可以让用户显式地将数据存储到磁盘和内存中，并能控制数据的分区。同时，RDD 还提供了一组丰富的操作来操作这些数据。在这些操作中，诸如 map、flatMap、filter 等转换操作实现了 monad 模式，很好地契合了 Scala 的集合操作。除此之外，RDD 还提供了诸如 join、groupBy、reduceByKey 等更为方便的操作（注意，reduceByKey 是 Action，而非 Transformation），以支持常见的数据运算。

通常来讲，针对数据处理有几种常见模型，包括 Iterative Algorithms、Relational Queries、MapReduce、Stream Processing。例如，Hadoop MapReduce 采用了 MapReduces 模型，Storm 采用了 Stream Processing 模型。RDD 混合了这 4 种模型，使得 Spark 可以应用于各种大数据处理场景。

RDD 作为数据结构，本质上是一个只读的分区记录集合。一个 RDD 可以包含多个分区，每个分区就是一个 Dataset 片段。RDD 可以相互依赖。如果 RDD 的每个分区最多只能被一个子 RDD 的一个分区使用，则称之为窄依赖（Narrow Dependency）；若多个子 RDD 分区都可以依赖，则称之为宽依赖（Wide Dependency）。不同的操作依据其特性，可能会产生不同的依赖。例如，Map 操作会产生 Narrow Dependency，而 Join 操作则产生 Wide Dependency。

Spark 之所以将依赖分为 Narrow 与 Wide，基于以下两点原因：

首先，Narrow Dependencies 可以支持在同一个 Cluster Node 上以管道形式执行多条命令，例如，在执行了 Map 后，紧接着执行 Filter。相反，Wide Dependencies 需要所有的父分区都是可用的，可能还需要调用类似 MapReduce 之类的操作进行跨节点传递。

其次，则是从失败恢复的角度考虑。Narrow Dependencies 的失败恢复更有效，因为它只需要重新计算丢失的 Parent Partition 即可，而且可以并行地在不同节点进行重计算。而 Wide Dependencies 牵涉 RDD 各级的多个 Parent Partitions。图 9-6 说明了 Narrow Dependencies 与 Wide Dependencies 之间的区别，每一个方框表示一个 RDD，其中的阴影矩形表示 RDD 的分区。

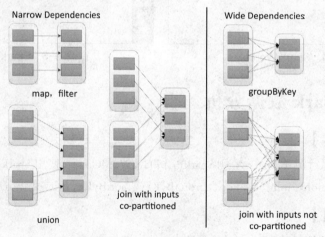

图 9-6　窄依赖与宽依赖

当用户对一个 RDD 执行 Action（如 count 或 save）操作时，调度器会根据该 RDD 的 Lineage，来构建一个由若干阶段（Stage）组成的一个 DAG（有向无环图）以执行程序，如图 9-7 所示。

实线圆角方框标识的是 RDD。阴影背景的矩形是分区，若已存于内存中，则用黑色背景标识。RDD G 上一个 Action 的执行将会以宽依赖为分区来构建各个 Stage，对各 Stage 内部的窄依赖则前后连接构成流水线。在本例中，Stage 1 的输出已经存在 RAM 中，所以，直接执行 Stage 2，然后执行 Stage 3。

每个 Stage 都包含尽可能多的连续的窄依赖型转换。各个阶段之间的分界则是宽依赖所需的 Shuffle 操作，或者是 DAG 中一个经由该分区能更快到达父 RDD 的已计算分区。之后，调度器运行多个任务来计算各个阶段所缺失的分区，直到最终得出目标 RDD。

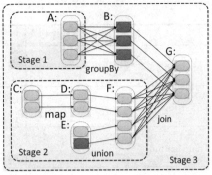

图 9-7 作业调度

调度器向各机器的任务分配采用延时调度机制并根据数据存储位置（本地性）来确定。若一个任务需要处理的某个分区刚好存储在某个节点的内存中，则该任务会分配给那个节点。否则，如果一个任务处理的某个分区，该分区含有的 RDD 提供较佳的位置（例如，一个 HDFS 文件），我们把该任务分配到这些位置。

对应宽依赖类的操作（如 Shuffle 依赖），会将中间记录物理化到保存父分区的节点上。这和 MapReduce 物化 Map 的输出类似，能简化数据的故障恢复过程。

对于执行失败的任务，只要它对应 Stage 的父类信息仍然可用，它便会在其他节点上重新执行。如果某些 Stage 变为不可用（例如，因为 Shuffle 在 Map 阶段的某个输出丢失了），则重新提交相应的任务以并行计算丢失的分区。

若某个任务执行缓慢（即"落后者"Straggler），系统则会在其他节点上执行该任务的复制，这与 MapReduce 做法类似，并取最先得到的结果作为最终的结果。

## 二、RDD 算子

RDD 支持两种类型的算子（Operation）：Transformation 算子和 Action 算子。Transformation 算子可以将已有 RDD 转换得到一个新的 RDD，而 Action 算子则是基于数据集计算，并将结果返回给驱动器（Driver）。例如，Map 是一个 Transformation 算子，它将数据集中每个元素传给一个指定的函数，并将该函数返回结果构建为一个新的 RDD；而 Reduce 是一个 Action 算子，它可以将 RDD 中的所有元素传给指定的聚合函数，并将最终的聚合结果返回给驱动器（还有一个 reduceByKey 算子，其返回的聚合结果是一个数据集）。

Spark 中所有 Transformation 算子都是懒惰的，也就是说，这些算子并不立即计算结果，而是记录下对基础数据集（如一个数据文件）的转换操作。只有等到某个 Action 算子需要计算一个结果返回给驱动器的时候，Transformation 算子所记录的操作才会被计算。这种设计使 Spark 可以运行得更加高效，例如，Map 算子创建了一个数据集，同时该数据集下一步会调用 Reduce 算子，那么 Spark 将只会返回 Reduce 的最终聚合结果（单独的一个数据）给驱动器，而不是将 Map 所产生的数据集整个返回给驱动器。

默认情况下，每次调用 Action 算子的时候，每个由 Transformation 转换得到的 RDD 都会被重新计算。然而，也可以通过调用 Persist（或者 Cache）操作来持久化一个 RDD，这意味着 Spark 将会把 RDD 的元素都保存在集群中，因此，下一次访问这些元素的速度将大大提高。同时，Spark 还支持将 RDD 元素持久化到内存或者磁盘上，甚至可以支持跨节点多副本，Spark 支持的一些常用 Transformation 算子如表 9-4 所示。

表 9-4　Transformation 算子

| 名　称 | 作　用 |
|---|---|
| map(func) | 返回一个新的分布式数据集，其中每个元素都是由源 RDD 中一个元素经 func 转换得到的 |
| filter(func) | 返回一个新的数据集，其中包含的元素来自源 RDD 中元素经 func 过滤后（func 返回 true 时才选中）的结果 |
| flatMap(func) | 类似于 Map，但每个输入元素可以映射到 0~n 个输出元素（所以，要求 func 必须返回一个 Seq 而不是单个元素） |
| mapPartitions(func) | 类似于 Map，但基于每个 RDD 分区（或者数据 block）独立运行，所以，如果 RDD 包含元素类型为 T，则 func 必须是 Iterator&lt;T&gt; =&gt; Iterator&lt;U&gt; 的映射函数 |
| mapPartitionsWithIndex(func) | 类似于 mapPartitions，只是 func 多了一个整型的分区索引值，因此，如果 RDD 包含元素类型为 T，则 func 必须是 Iterator&lt;T&gt; =&gt; Iterator&lt;U&gt; 的映射函数 |
| sample(withReplacement, fraction, seed) | 采样部分（比例取决于 fraction）数据，同时可以指定是否使用回置采样（withReplacement），以及随机数种子（seed） |
| union(otherDataset) | 返回源数据集和参数数据集（otherDataset）的并集 |
| intersection(otherDataset) | 返回源数据集和参数数据集（otherDataset）的交集 |
| distinct([numTasks])) | 返回对源数据集进行元素去重后的新数据集 |
| groupByKey([numTasks]) | 只对包含键值对的 RDD 有效，如源 RDD 包含 (K, V) 对，则该算子返回一个新的数据集包含 (K, Iterable&lt;V&gt;) 对。<br>注意：默认情况下，输出计算的并行度取决于源 RDD 的分区个数。当然，也可以通过设置可选参数 numTasks 来指定并行任务的个数 |
| reduceByKey(func, [numTasks]) | 如果源 RDD 包含元素类型 (K, V) 对，则该算子也返回包含(K, V) 对的 RDD，只不过每个 Key 对应的 Value 是经过 func 聚合后的结果，而 func 本身是一个 (V, V) =&gt; V 的映射函数。<br>另外，和 groupByKey 类似，可以通过可选参数 numTasks 指定 reduce 任务的个数 |
| aggregateByKey(zeroValue)(seqOp, combOp, [numTasks]) | 如果源 RDD 包含 (K, V) 对，则返回新 RDD 包含 (K, U) 对，其中每个 Key 对应的 Value 都是由 combOp 函数 和 一个 "0" 值 zeroValue 聚合得到的。允许聚合后 Value 类型和输入 Value 类型不同，避免了不必要的开销。和 groupByKey 类似，可以通过可选参数 numTasks 指定 reduce 任务的个数 |
| sortByKey([ascending], [numTasks]) | 如果源 RDD 包含元素类型 (K, V) 对，其中 K 可排序，则返回新的 RDD 包含 (K, V) 对，并按照 K 排序（升序还是降序取决于 ascending 参数） |
| join(otherDataset, [numTasks]) | 如果源 RDD 包含元素类型 (K, V) 且参数 RDD（otherDataset）包含元素类型(K, W)，则返回的新 RDD 中将包含内关联后 key 对应的 (K, (V, W)) 对。外关联(Outer joins)操作请参考 leftOuterJoin、rightOuterJoin 及 fullOuterJoin 算子 |

| 名称 | 作用 |
|---|---|
| cogroup(otherDataset, [numTasks]) | 如果源 RDD 包含元素类型 (K, V) 且参数 RDD（otherDataset）包含元素类型(K, W)，则返回的新 RDD 中包含 (K, (Iterable<V>, Iterable<W>))。该算子还有个别名：groupWith |
| cartesian(otherDataset) | 如果源 RDD 包含元素类型 T 且参数 RDD（otherDataset）包含元素类型 U，则返回的新 RDD 包含前两者的笛卡儿积，其元素类型为 (T, U) 对 |
| pipe(command, [envVars]) | 以 Shell 命令行管道处理 RDD 的每个分区，如 Perl 或者 bash 脚本。RDD 中每个元素都将依次写入进程的标准输入（stdin），然后按行输出到标准输出（stdout），每一行输出字符串即成为一个新的 RDD 元素 |
| coalesce(numPartitions) | 将 RDD 的分区数减少到 numPartitions。当以后大数据集被过滤成小数据集后，减少分区数，可以提升效率 |
| repartition(numPartitions) | 将 RDD 数据重新混洗（reshuffle）并随机分布到新的分区中，使数据分布更均衡，新的分区个数取决于 numPartitions。该算子总是需要通过网络混洗所有数据 |
| repartitionAndSortWithinPartitions(partitioner) | 根据 partitioner（Spark 自带有 HashPartitioner 和 RangePartitioner 等）重新分区 RDD，并且在每个结果分区中按 Key 进行排序。这是一个组合算子，功能上等价于先 repartition 再在每个分区内排序，但这个算子内部做了优化（将排序过程下推到混洗同时进行），因此性能更好 |

Spark 支持的一些常用 Action 算子如表 9-5 所示。有一些 Spark 算子会触发众所周知的混洗（Shuffle）事件。Spark 中的混洗机制是用于将数据重新分布，其结果是所有数据将在各个分区间重新分组。一般情况下，混洗需要跨执行器（Executor）或跨机器复制数据，这也是混洗操作一般都比较复杂而且开销大的原因。

表 9-5 Action 算子

| 名称 | 作用 |
|---|---|
| reduce(func) | 将 RDD 中的元素按 func 进行聚合（func 是一个 (T,T) => T 的映射函数，其中 T 为源 RDD 元素类型，并且 func 需要满足交换律和结合律以便支持并行计算） |
| collect() | 将数据集中的所有元素以数组形式返回驱动器（Driver）程序。通常用于在 RDD 进行了 filter 或其他过滤操作后，将一个足够小的数据子集返回到驱动器内存中 |
| count() | 返回数据集中的元素个数 |
| first() | 返回数据集中的首个元素（类似于 take(1)） |
| take(n) | 返回数据集中的前 n 个元素 |

续表

| 名 称 | 作 用 |
|---|---|
| takeSample(withReplacement, num, [seed]) | 返回数据集的随机采样子集，最多包含 num 个元素，withReplacement 表示是否使用回置采样，最后一个参数为可选参数 seed，随机数生成器的种子 |
| takeOrdered(n, [ordering]) | 按元素排序（可以通过 ordering 自定义排序规则）后，返回前 n 个元素 |
| saveAsTextFile(path) | 将数据集中的元素保存到指定目录下的文本文件中（或者多个文本文件），支持本地文件系统、HDFS 或者其他任何 Hadoop 支持的文件系统。保存过程中，Spark 会调用每个元素的 toString 方法，并将结果保存成文件中的一行 |
| saveAsSequenceFile(path) (Java and Scala) | 将数据集中的元素保存到指定目录下的 Hadoop Sequence 文件中，支持本地文件系统、HDFS 或者其他任何 Hadoop 支持的文件系统。适用于实现了 Writable 接口的键值对 RDD。在 Scala 中，同样也适用于能够被隐式转换为 Writable 的类型（Spark 实现了所有基本类型的隐式转换，如 Int、Double、String 等） |
| saveAsObjectFile(path) (Java and Scala) | 将 RDD 元素以 Java 序列化的格式保存成文件，保存结果文件可以使用 SparkContext.objectFile 来读取 |
| countByKey() | 只适用于包含键值对(K, V)的 RDD，并返回一个哈希表，包含 (K, Int) 对，表示每个 Key 的个数 |
| foreach(func) | 在 RDD 的每个元素上运行 func 函数。通常被用于累加操作，如更新一个累加器（Accumulator）或者和外部存储系统互操作。注意：用 foreach 操作出累加器之外的变量可能导致未定义的行为 |

### 三、Spark 代码执行

Spark 应用程序从编写到提交、执行、输出的整个过程如图 9-8 所示，图中描述的步骤如下：

① 用户使用 SparkContext 提供的 API（常用的有 textFile、sequenceFile、runJob、stop 等）编写 Driver application 程序。此外，SQLContext、HiveContext 及 StreamingContext 对 SparkContext 进行封装，并提供了 SQL、Hive 及流式计算相关的 API。

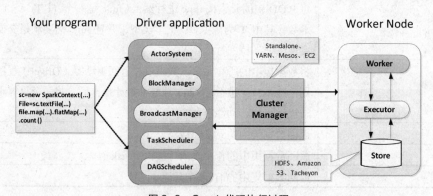

图 9-8 Spark 代码执行过程

② 使用SparkContext提交的用户应用程序,首先会使用BlockManager和BroadcastManager将任务的Hadoop配置进行广播。然后由DAGScheduler将任务转换为RDD并组织成DAG,DAG还将被划分为不同的Stage。最后由TaskScheduler借助ActorSystem将任务提交给集群管理器(Cluster Manager)。

③ 集群管理器给任务分配资源,即将具体任务分配到Worker上,Worker创建Executor来处理任务的运行。Standalone、YARN、Mesos、EC2等都可以作为Spark的集群管理器。

### 四、蒙特卡罗方法

蒙特卡罗(Monte Carlo)方法,又称随机抽样或统计试验方法,属于计算数学的一个分支,它是在20世纪40年代中期为了适应当时原子能事业的发展而发展起来的。

蒙特卡罗方法的基本思想如下:当所要求解的问题是某种事件出现的概率,或者是某个随机变量的期望值时,它们可以通过某种"试验"的方法,得到这种事件出现的频率,或者这个随机变数的平均值,并用它们作为问题的解。简而言之,就是用频率来代替概率,当实验样本足够大时,就可以得到比较精确的解结果。

蒙特卡罗方法计算圆周率π:在一个边长为2r的正方形内随机投点,该点落在此正方形的内切圆中的概率即为内切圆与正方形的面积比值,如图9-9所示。

正方形内部有一个相切的圆,内切圆面积:正方形面积=$\pi \times r^2:(2r)\times(2r) = \pi/4$。所以,$\pi = 4\times$(内切圆面积:正方形面积)= $4\times$(投在内切圆内的点:投在正方形内的点)。

取半径为$r=1$,点为$(x, y)$,($-1\leq x \leq 1$,$-1\leq y \leq 1$)。如果$x^2+y^2=1$,则点$(x, y)$在圆的边界上,如果$x^2+y^2<1$,则点$(x, y)$在圆内。

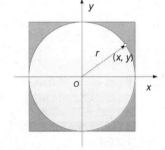

图9-9 蒙特卡罗方法计算圆周率π

### 【任务实施】

#### 一、运行Spark示例

(1) 计算圆周率π

① 运行SparkPi,操作如下:

```
hadoop@master:~$ cd /opt/spark-2.1.0-bin-hadoop2.7/
hadoop@master:/opt/...$ run-example SparkPi 10 > SparkPi.txt
```

② 查看运行结果,如下所示:

```
hadoop@master:/opt/...$ more SparkPi.txt
Pi is roughly 3.1413991413991416
```

③ SparkPi代码如下:

```
object SparkPi {
  def main(args: Array[String]) {
    val spark = SparkSession
      .builder
      .appName("Spark Pi")
      .getOrCreate()
```

```
    val slices = if (args.length > 0) args(0).toInt else 2
    val n = math.min(100000L * slices, Int.MaxValue).toInt // avoid overflow
    val count = spark.sparkContext.parallelize(1 until n, slices).map { i =>
      val x = random * 2 - 1
      val y = random * 2 - 1
      if (x*x + y*y < 1) 1 else 0
    }.reduce(_ + _)
    println("Pi is roughly " + 4.0 * count / (n - 1))
    spark.stop()
  }
}
```

Slices 表示切片（默认两个），每个切片有 10 万个点，Map 负责计算点是否投在内切圆内，Reduce 负责计算投在内切圆内的点的个数（count），count / (n − 1) = 投在内切圆内的点/投在正方形内的点。random * 2 − 1 的取值范围为（−1,1）。

（2）单词计数

① 数据准备如下：

```
hadoop@master:/opt/...$ hdfs dfs -ls /input
Found 2 items
-rw-r--r--   3 hadoop supergroup         24 2017-02-10 10:55 /input/sw1.txt
-rw-r--r--   3 hadoop supergroup         24 2017-02-10 10:55 /input/sw2.txt
hadoop@master:/opt/...$ hdfs dfs -cat /input/sw1.txt
Hello World
Good Hadoop
hadoop@master:/opt/...$ hdfs dfs -cat /input/sw2.txt
Hello Hadoop
Bye Hadoop
```

② 运行 JavaWordCount 程序，操作如下：

```
hadoop@master:/opt/...$ bin/spark-submit --master spark://master:7077 \
--class org.apache.spark.examples.JavaWordCount \
--executor-memory 2g \
examples/jars/spark-examples_2.11-2.1.0.jar \
/input > WordCount.txt
```

③ 查看运行结果，如下所示：

```
hadoop@master:/opt/...$ more WordCount.txt
Bye: 1
Hello: 2
World: 1
Good: 1
Hadoop: 3
```

④ JavaWordCount 代码如下：

```java
public final class JavaWordCount {
  private static final Pattern SPACE = Pattern.compile(" ");
  public static void main(String[] args) throws Exception {
    if (args.length < 1) {
      System.err.println("Usage: JavaWordCount <file>");
      System.exit(1);
    }
    //SparkSession 为用户提供了一个统一的切入点来使用 Spark 的各项功能，
    //SparkSession 的设计遵循了工厂设计模式（Factory Design Pattern）。
    SparkSession spark = SparkSession
      .builder()
      .appName("JavaWordCount")
      .getOrCreate();
    //创建完 SparkSession 之后，就可以使用它来读取数据，
    //textFile() 读取数据，javaRDD() 定义一个 RDD，返回每一行作为一个元素的 RDD
    JavaRDD<String> lines = spark.read().textFile(args[0]).javaRDD();
    //FlatMapFunction<String, String>第一个参数为传入的内容，
    //第二个参数为函数操作完后返回的结果类型，将每一行映射成多个单词
    JavaRDD<String> words = lines.flatMap(new FlatMapFunction<String, String>() {
      @Override
      public Iterator<String> call(String s) {
        return Arrays.asList(SPACE.split(s)).iterator();
      }
    });
    //map 操作：将单词转化成(word, 1)对
    //PairFunction<String, String, Integer>第一个参数为内容，
    //第三个参数为函数操作完后返回的结果类型
    JavaPairRDD<String, Integer> ones = words.mapToPair(
      new PairFunction<String, String, Integer>() {
        @Override
        public Tuple2<String, Integer> call(String s) {
          return new Tuple2<>(s, 1);
        }
      });
    //reduce 操作：分组并按键值添加对以产生计数
    JavaPairRDD<String, Integer> counts = ones.reduceByKey(
      new Function2<Integer, Integer, Integer>() {
        @Override
        public Integer call(Integer i1, Integer i2) {
          return i1 + i2;
```

```java
      }
    });
    //完成单词集合计算，返回整个RDD
    List<Tuple2<String, Integer>> output = counts.collect();
    for (Tuple2<?,?> tuple : output) {
      System.out.println(tuple._1() + ": " + tuple._2());
    }
    //Spark服务关闭
    spark.stop();
  }
}
```

## 二、Spark 函数运用

① 下载测试数据到 /homt/hadoop 目录下，下载地址如下：

http://statweb.stanford.edu/~tibs/ElemStatLearn/datasets/spam.data

② 上传文件到 HDFS，操作如下：

```
hadoop@master:~$ hdfs dfs -mkdir /spark
hadoop@master:~$ hdfs dfs -put spam.data /spark
hadoop@master:~$ hdfs dfs -text /spark/spam.data
0 0.64 0.64 0 0.32 0 0 0 0 0.64 0 0 0 0.32 0 1.29 1.93 0 0.96 0 0 0 0 0 0 0 0 0
 0 0 0 0 0 0 0 0 0 0 0 0 0 0 0 0 0 0.778 0 0 3.756 61 278 1
 0.21 0.28 0.5 0 0.14 0.28 0.21 0.07 0 0.94 0.21 0.79 0.65 0.21 0.14 0.14 0.07 0.28 3.47 0
 1.59 0 0.43 0.43 0 0 0 0 0 0 0 0 0 0.07 0 0 0 0 0 0 0 0 0 0 0.132 0 0.372 0.18
 0.048 5.114 101 1028 1
 0.06 0 0.71 0 1.23 0.19 0.19 0.12 0.64 0.25 0.38 0.45 0.12 0 1.75 0.06 0.06 1.03 1.36 0.3
 2 0.51 0 1.16 0.06 0 0 0 0 0 0 0 0 0 0 0.06 0 0 0.12 0 0.06 0.06 0 0 0.01 0.143
 0 0.276 0.184 0.01 9.821 485 2259 1
 0 0 0 0.63 0 0.31 0.63 0.31 0.63 0.31 0.31 0.31 0 0 0.31 0 0 3.18 0 0.31 0 0 0 0 0
 0 0 0 0 0 0 0 0 0 0 0 0 0 0 0 0 0.137 0 0.137 0 0 3.537 40 191 1
 0 0 0 0.63 0 0.31 0.63 0.31 0.63 0.31 0.31 0.31 0 0 0.31 0 0 3.18 0 0.31 0 0 0 0 0
 0 0 0 0 0 0 0 0 0 0 0 0 0 0 0 0 0.135 0 0.135 0 0 3.537 40 191 1
 0 0 0 1.85 0 0 1.85 0 0 0 0 0 0 0 0 0 0 0 0 0 0 0 0 0 0 0 0 0 0 0 0 0 0 0 0
 0 0 0 0 0 0 0.223 0 0 0 0 3 15 54 1
......
```

③ 加载文件，操作如下：

```
hadoop@master:~$ spark-shell
scala> val inFile=sc.textFile("/spark/spam.data")
inFile: org.apache.spark.rdd.RDD[String] = /spark/spam.data MapPartitionsRDD[1] at textFile at <console>:24
```

④ 显示第一行数据，操作如下：

```
scala> inFile.first()
res0: String = 0 0.64 0.64 0 0.32 0 0 0 0 0 0 0.64 0 0 0 0.32 0 1.29 1.93
0 0.96 0 0 0 0 0 0 0 0 0 0 0 0 0 0 0 0 0 0 0 0 0 0 0 0 0 0 0 0 0.778 0 0
3.756 61 278 1
```

⑤ Spark 数据集运算

● map：文本映射成双精度。

```
scala> val nums = inFile.map(x=>x.split(' ').map(_.toDouble))
nums: org.apache.spark.rdd.RDD[Array[Double]] = MapPartitionsRDD[2] at map at <console>:26
scala> nums.first()
res1: Array[Double] = Array(0.0, 0.64, 0.64, 0.0, 0.32, 0.0, 0.0, 0.0, 0.0,
0.0, 0.0, 0.64, 0.0, 0.0, 0.0, 0.32, 0.0, 1.29, 1.93, 0.0, 0.96, 0.0, 0.0, 0.0,
0.0, 0.0, 0.0, 0.0, 0.0, 0.0, 0.0, 0.0, 0.0, 0.0, 0.0, 0.0, 0.0, 0.0, 0.0, 0.0,
0.0, 0.0, 0.0, 0.0, 0.0, 0.0, 0.0, 0.0, 0.0, 0.0, 0.778, 0.0, 0.0, 3.756,
61.0, 278.0, 1.0)
```

● collect：list 转化成 rdd。

```
scala> val rdd = sc.parallelize(List(1,2,3,4,5))
rdd: org.apache.spark.rdd.RDD[Int] = ParallelCollectionRDD[3] at parallelize at <console>:24
scala> val mapRdd = rdd.map(2*_)
mapRdd: org.apache.spark.rdd.RDD[Int] = MapPartitionsRDD[4] at map at <console>:26
scala> mapRdd.collect
res2: Array[Int] = Array(2, 4, 6, 8, 10)
```

● filter：数据过滤。

```
scala> val filterRdd = mapRdd.filter(_>5)
filterRdd: org.apache.spark.rdd.RDD[Int] = MapPartitionsRDD[5] at filter at <console>:28
scala> filterRdd.collect
res3: Array[Int] = Array(6, 8, 10)
```

● flatMap：将单词转化成(key, value)对。

```
scala> val rdd = sc.textFile("/input")
rdd: org.apache.spark.rdd.RDD[String] = /input MapPartitionsRDD[7] at textFile at <console>:24
scala> rdd.cache
res4: rdd.type = /input MapPartitionsRDD[7] at textFile at <console>:24
scala> val wordCount=rdd.flatMap(_.split(' ')).map(x=>(x,1)).reduceByKey(_+_)
wordCount: org.apache.spark.rdd.RDD[(String, Int)] = ShuffledRDD[10] at reduceByKey at <console>:26
```

```
scala> wordCount.collect
res5: Array[(String, Int)] = Array((Bye,1), (Hello,2), (World,1), (Good,1),
(Hadoop,3))
scala> wordCount.saveAsTextFile("/output/w")
---查看HDFS,检查结果---
hadoop@slave1:~$ hdfs dfs -ls /output/w/
Found 3 items
-rw-r--r--   3 hadoop supergroup          0 2017-02-11 09:28 /output/w/_SUCCESS
-rw-r--r--   3 hadoop supergroup         28 2017-02-11 09:28 /output/w/part-00000
-rw-r--r--   3 hadoop supergroup         20 2017-02-11 09:28 /output/w/part-00001
hadoop@slave1:~$ hdfs dfs -cat /output/w/part-00000
(Bye,1)
(Hello,2)
(World,1)
hadoop@slave1:~$ hdfs dfs -cat /output/w/part-00001
(Good,1)
(Hadoop,3)
```

- union:联合运算。

```
scala> val rdd1 = sc.parallelize(List(('a',1),('a',2)))
rdd1: org.apache.spark.rdd.RDD[(Char, Int)] = ParallelCollectionRDD[12] at
parallelize at <console>:24
scala> val rdd2 = sc.parallelize(List(('b',1),('b',2)))
rdd2: org.apache.spark.rdd.RDD[(Char, Int)] = ParallelCollectionRDD[13] at
parallelize at <console>:24
scala> rdd1 union rdd2
res7: org.apache.spark.rdd.RDD[(Char, Int)] = UnionRDD[14] at union at
<console>:29
scala> res7.collect
res9: Array[(Char, Int)] = Array((a,1), (a,2), (b,1), (b,2))
```

- join:连接运算。

```
scala> val rdd1 = sc.parallelize(List(('a',1),('a',2),('b',3),('b',4)))
rdd1: org.apache.spark.rdd.RDD[(Char, Int)] = ParallelCollectionRDD[15] at
parallelize at <console>:24
scala> val rdd2 = sc.parallelize(List(('a',5),('a',6),('b',7),('b',8)))
rdd2: org.apache.spark.rdd.RDD[(Char, Int)] = ParallelCollectionRDD[16] at
parallelize at <console>:24
scala> val rdd3 = rdd1 join rdd2
rdd3: org.apache.spark.rdd.RDD[(Char, (Int, Int))] = MapPartitionsRDD[19]
at join at <console>:28
scala> rdd3.collect
res10: Array[(Char, (Int, Int))] = Array((a,(1,5)), (a,(1,6)), (a,(2,5)),
(a,(2,6)), (b,(3,7)), (b,(3,8)), (b,(4,7)), (b,(4,8)))
```

- lookup：按 key 查找 values。

```
scala> var rdd1 = sc.parallelize(List(('a',1),('a',2),('b',3),('b',4)))
rdd1: org.apache.spark.rdd.RDD[(Char, Int)] = ParallelCollectionRDD[20] at parallelize at <console>:24
scala> rdd1.lookup('a')
res11: Seq[Int] = WrappedArray(1, 2)
```

- groupByKey：按 key 值进行分组。

```
scala> val wc = sc.textFile("/input").flatMap(_.split(' ')).map((_,1)).groupByKey
wc: org.apache.spark.rdd.RDD[(String, Iterable[Int])] = ShuffledRDD[27] at groupByKey at <console>:24
scala> wc.collect
res12: Array[(String, Iterable[Int])] = Array((Bye,CompactBuffer(1)), (Hello,CompactBuffer(1, 1)), (World,CompactBuffer(1)), (Good,CompactBuffer(1)), (Hadoop,CompactBuffer(1, 1, 1)))
```

- sortByKey：按 key 进行排序，参数 false 表示倒序。

```
scala> val wc = sc.textFile("/input").flatMap(_.split(' ')).map((_,1)).sortByKey(false)
wc: org.apache.spark.rdd.RDD[(String, Int)] = ShuffledRDD[34] at sortByKey at <console>:24
scala> wc.collect
res13: Array[(String, Int)] = Array((World,1), (Hello,1), (Hello,1), (Hadoop,1), (Hadoop,1), (Hadoop,1), (Good,1), (Bye,1))
```

## 任务 3　Spark 编程

### 【任务概述】

Spark 支持多种开发语言，Scala IDE 开发工具支持 Scala 和 Java 项目开发，本任务使用 Scala、Java 和 Python 语言分别编写单词计数程序；最后使用 Python 语言编写聚类分析算法 K-Means。

### 【支撑知识】

#### 一、K-Means 算法

（1）K-Means 算法简介

K-Means 算法是很典型的基于距离的聚类算法，采用距离作为相似性的评价指标，即认为两个对象的距离越近，其相似度就越大。该算法认为簇是由距离靠近的对象组成的，因此，把得到紧凑且独立的簇作为最终目标。

K-Means 算法主要解决的问题如图 9-10 所示。从图中可以看出有 3 个聚类，但如何通过计算机程序找出聚类呢？

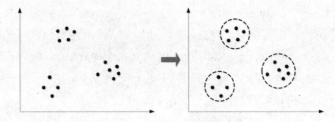

图 9-10　K-Means 聚类

（2）算法概要

图 9-11 中有 A、B、C、D、E 共 5 个点，2 个灰色点是种子点，也就是用来查找聚类中心。有 2 个种子点，即 $K=2$。

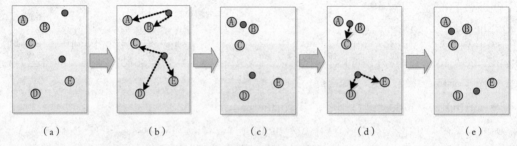

图 9-11　K-Means 算法

① K-Means 的算法如下：

a. 随机在图中取 $K$（$K=2$）个种子点（如图 9-11（a）所示）。

b. 计算全部点到 $K$ 个种子点的距离，假如点 Pi 离种子点 Si 最近，那么 Pi 属于 Si 点群（如图 9-11（b）所示，点 A、B 离上面种子点最近，C、D、E 离下面种子点最近）。

c. 重新计算聚类中心（如图 9-11（c）所示，由 A 和 B 计算新的聚类中心，由 C、D、E 计算新的聚类中心）。

d. 重复步骤 b 和 c，直到新聚类中心和原聚类中心相等或者小于指定阈值（如图 9-11（e）所示：上面灰色点是 A、B、C 的聚类中心，下面灰色点是 D、E 的聚类中心）。

② 计算点距离算法有如下三种。

a. 闵可夫斯基距离公式：

$$d_{ij} = \sqrt[\lambda]{\sum_{k=1}^{n} |x_{ik} - x_{jk}|^{\lambda}}$$

式中，$\lambda$ 可以随意取值，可以是负数，也可以是正数，或是无穷大。

b. 欧几里得距离公式：

$$d_{ij} = \sqrt{\sum_{k=1}^{n} (x_{ik} - x_{jk})^2}$$

也就是第一个公式 $\lambda=2$ 的情况。

c. 曼哈顿距离公式：

$$d_{ij} = \sum_{k=1}^{n} |x_{ik} - x_{jk}|$$

③ K-Means 主要有两个最重大的缺陷,都和初始值有关。

a. K 是事先给定的,这个 K 值的选定是非常难以估计的。

b. K-Means 算法需要用初始种子点,随机种子点影响运算结果。

c. K-Means 算法需要不断地计算调整后的聚类中心,因此,当数据量非常大时,算法的时间开销是非常大的。

④ K-Means++。

K-Means++算法可以有效地选择种子点,选择种子点的过程如下:

a. 从输入的数据点集合中随机选择一个点作为第一个种子点。

b. 对于数据集中的每一个点 $x$,计算它与最近种子点的距离 $D(x)$。

c. 选择一个新的数据点作为新的种子点,选择的原则是 $D(x)$ 较大的点,被选取作为种子点的概率较大。

d. 重复步骤 b 和 c,直到 $k$ 个种子点被选出来。

## 二、Python 及组件

(1) Python

Python 是一种面向对象的解释型计算机程序设计语言,由荷兰人 Guido van Rossum 于 1989 年发明,第一个公开发行版发行于 1991 年。Python 是纯粹的自由软件,源代码和解释器 CPython 遵循 GPL(GNU General Public License)协议。Python 语法简洁清晰,特色之一是强制用空白符(White Space)作为语句缩进。

Python 具有丰富和强大的库,常被称为胶水语言,能够把用其他语言制作的各种模块(尤其是 C/C++)很轻松地连接在一起。常见的一种应用情形是,使用 Python 快速生成程序的原型(有时甚至是程序的最终界面),然后对其中有特别要求的部分,用更合适的语言改写,如 3D 游戏中的图形渲染模块,性能要求特别高,就可以用 C/C++重写,而后封装为 Python 可以调用的扩展类库。需要注意的是,在使用扩展类库时可能需要考虑平台问题,某些可能不提供跨平台的实现。

(2) NumPy

NumPy 是 Python 的一种开源的数值计算扩展。这种工具可用来存储和处理大型矩阵,比 Python 自身的嵌套列表结构(Nested List Structure)要高效得多(该结构也可以用来表示矩阵(Matrix))。NumPy 将 Python 相当于变成一种免费的更强大的 MATLAB 系统。

(3) SciPy

SciPy 是一款方便、易于使用、专为科学和工程设计的 Python 工具包。SciPy 函数库在 NumPy 库的基础上增加了众多的数学、科学及工程计算中常用的库函数。例如,线性代数、常微分方程数值求解、信号处理、图像处理、稀疏矩阵等。

(4) MatPlotLib

MatPlotLib 是 Python 著名的绘图库,它提供了一整套和 MATLAB 相似的命令 API,适合交互式制图。也可以将它作为绘图控件,嵌入 GUI 应用程序中。它的文档相当完备,并且 Gallery 页面中有上百幅缩略图,打开之后都有源程序。因此,如果需要绘制某种类型的图,只需要在这个页面中浏览/复制/粘贴一下,就能搞定。

（5）Pandas

Pandas 是 Python 的一个数据分析包，最初由 AQR Capital Management 于 2008 年 4 月开发，并于 2009 年底开源出来，目前由专注于 Python 数据包开发的 PyData 开发 team 继续开发和维护，属于 PyData 项目的一部分。Pandas 最初被作为金融数据分析工具而开发出来，因此，Pandas 为时间序列分析提供了很好的支持。Pandas 的名称来自于面板数据（Panel Data）和 Python 数据分析（Data Analysis）。Panel Data 是经济学中关于多维数据集的一个术语，在 Pandas 中也提供了 Panel 的数据类型。

（6）PyCharm

PyCharm 是由 JetBrains 打造的一款 Python IDE，带有一整套可以帮助用户在使用 Python 语言开发时提高其效率的工具，如调试、语法高亮、Project 管理、代码跳转、智能提示、自动完成、单元测试、版本控制。此外，该 IDE 提供了一些高级功能，以用于支持 Django 框架下的专业 Web 开发。

PyCharm 支持 Google App Engine，支持 IronPython。这些功能在先进代码分析程序的支持下，使 PyCharm 成为 Python 专业开发人员和刚起步人员使用的有力工具。

## 【任务实施】

### 一、客户端搭建 Scala-IDE 开发环境

① 下载 Scala-IDE 软件包到 /home/hadoop 目录下，下载地址如下：

```
http://scala-ide.org/download/sdk.html
```

② 安装 Scala-IDE 开发软件，操作如下：

```
hadoop@sw-desktop:~$ cd /opt/
hadoop@sw-desktop:/opt$ sudo tar xvzf /home/hadoop/\
scala-SDK-4.5.0-vfinal-2.11-linux.gtk.x86_64.tar.gz
hadoop@sw-desktop:/opt$ sudo chown -R hadoop:hadoop eclipse
```

③ 从 Master 节点复制 Scala 和 Spark 软件，操作如下：

```
hadoop@sw-desktop:/opt$ sudo scp -r hadoop@master:/opt/scala-2.12.1 /opt
hadoop@sw-desktop:/opt$ sudo scp -r hadoop@master:/opt/spark-2.1.0-bin-hadoop2.7 /opt
hadoop@sw-desktop:/opt$ sudo chown -R hadoop:hadoop /opt/scala-2.12.1
hadoop@sw-desktop:/opt$ sudo chown -R hadoop:hadoop /opt/spark-2.1.0-bin-hadoop2.7
```

④ 修改环境变量，操作如下：

```
hadoop@sw-desktop:~$ vi /home/hadoop/.profile
```

添加如下内容：

```
export HADOOP_CONF_DIR=$HADOOP_HOME/etc/hadoop
export SCALA_HOME=/opt/scala-2.12.1
export SPARK_HOME=/opt/spark-2.1.0-bin-hadoop2.7
export PATH=$PATH:$SCALA_HOME/bin:$SPARK_HOME/bin
```

⑤ 环境变量生效，操作如下：

```
hadoop@sw-desktop:~$ source /home/hadoop/.profile
```

## 二、单词计数编程

（1）使用 Scala 语言实现

① 在图形窗口运行 Scala-IDE，操作如下：

```
hadoop@sw-desktop:~$ /opt/eclipse/eclipse &
```

② 选择"File"→"New"→"Scala Project"命令，弹出"New Scala Project"对话框，在"Project name"文本框中输入"ScalaWC"，如图 9-12 所示，单击"Finish"按钮，完成项目的创建。

③ 右击"ScalaWC"命令，在列表框中选择"New"→"Scala Class"命令，弹出对话框，在"Name"文本框中输入"WordCount"，单击"Finish"按钮，完成类的创建，再输入代码，如图 9-13 所示。

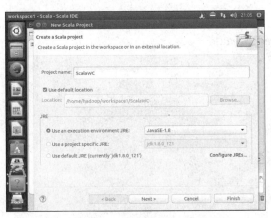

图 9-12 "New Scala Project"对话框

图 9-13 Scala Class

WordCount.scala 代码如下：

```scala
import org.apache.spark.sql.SparkSession
object WordCount {
  def main(args: Array[String]) {
    if (args.length < 1) {
      System.err.println("Usage: WordCount <file>")
      System.exit(1)
    }
    val spark = SparkSession
      .builder
      .appName("Scala WordCount")
      .getOrCreate()
    val line = spark.read.textFile(args(0)).rdd
    line.flatMap(_.split(" ")).map((_, 1))
      .reduceByKey(_ + _).collect().foreach(println)
    spark.stop()
  }
}
```

④ 在 Scala-IDE 中添加 "/opt/spark-2.1.0-bin-hadoop2.7/jars" 目录下所有的 jar 包。

⑤ 右击 "ScalaWC" 命令，在列表框中选择 "Scala" → "Set the Scala Installation" 命令，弹出对话框，选择 "Fix Scala Installation: 2.10.6(bundled)" 选项，完成设置。

⑥ 右击 "ScalaWC" 命令，在列表框中选择 "Export…" 选项，弹出对话框，选择 "Java" → "JAR File" 命令，在 "JAR file:" 文本框中输入文本 "ScalaWC.jar"，导出 jar 包。

⑦ 运行 ScalaWC.jar，操作如下：

```
hadoop@sw-desktop:~$ cd workspace1
hadoop@sw-desktop:~/workspace1$ spark-submit --master spark://master:7077\
--class WordCount \
ScalaWC.jar \
/input > WordCount.txt
```

⑧ 查看运行结果，操作如下：

```
hadoop@sw-desktop:~/workspace1$ more WordCount.txt
(Bye,1)
(Hello,2)
(World,1)
(Good,1)
(Hadoop,3)
```

（2）使用 Java 语言实现

① 在图形窗口运行 Scala-IDE，操作如下：

```
hadoop@sw-desktop:~$ /opt/eclipse/eclipse &
```

② 选择 "File" → "New" → "Other…" 命令，弹出对话框，选择 "Java Project" 命令，在 "Project Name" 文本框中输入文本 "JavaWC"，单击 "Finish" 按钮，完成项目的创建。

③ 右击 "JavaWC" 命令，在列表框中选择 "New" → "Other…" 命令，弹出对话框，选择 "Class" 选项，在 "Name" 文本框中输入文本 "WordCount"，单击 "Finish" 按钮完成类的创建，输入代码，如图 9-14 所示。

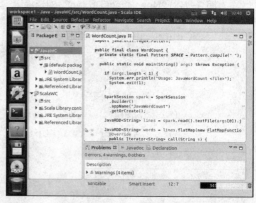

图 9-14　WordCount 类

WordCount.java 代码如下：

```
import scala.Tuple2;
import org.apache.spark.api.java.JavaPairRDD;
import org.apache.spark.api.java.JavaRDD;
import org.apache.spark.api.java.function.FlatMapFunction;
import org.apache.spark.api.java.function.Function2;
import org.apache.spark.api.java.function.PairFunction;
import org.apache.spark.sql.SparkSession;
import java.util.Arrays;
import java.util.Iterator;
```

```java
import java.util.List;
import java.util.regex.Pattern;
public final class WordCount {
  private static final Pattern SPACE = Pattern.compile(" ");
  public static void main(String[] args) throws Exception {
    if (args.length < 1) {
      System.err.println("Usage: JavaWordCount <file>");
      System.exit(1);
    }
    SparkSession spark = SparkSession
      .builder()
      .appName("JavaWordCount")
      .getOrCreate();
    JavaRDD<String> lines = spark.read().textFile(args[0]).javaRDD();
    JavaRDD<String> words = lines.flatMap(new FlatMapFunction<String, String>() {
      @Override
      public Iterator<String> call(String s) {
        return Arrays.asList(SPACE.split(s)).iterator();
      }
    });
    JavaPairRDD<String, Integer> ones = words.mapToPair(
      new PairFunction<String, String, Integer>() {
        @Override
        public Tuple2<String, Integer> call(String s) {
          return new Tuple2<>(s, 1);
        }
      });
    JavaPairRDD<String, Integer> counts = ones.reduceByKey(
      new Function2<Integer, Integer, Integer>() {
        @Override
        public Integer call(Integer i1, Integer i2) {
          return i1 + i2;
        }
      });
    List<Tuple2<String, Integer>> output = counts.collect();
    for (Tuple2<?,?> tuple : output) {
      System.out.println(tuple._1() + ": " + tuple._2());
    }
    spark.stop();
  }
}
```

④ 在 Scala-IDE 中添加 "/opt/spark-2.1.0-bin-hadoop2.7/jars" 目录下所有的 jar 包。

⑤ 右击 "JavaWC" 命令，在列表框中选择 "Export…" 命令，弹出对话框，选择 "Java" → "JAR File" 命令，在 "JAR file:" 文本框中输入文本 "JavaWC.jar"，导出 jar 包。

⑥ 运行 JavaWC.jar，操作如下：

```
hadoop@sw-desktop:~$ cd workspace1
hadoop@sw-desktop:~/workspace1$ spark-submit --master spark://master:7077 \
--class WordCount \
JavaWC.jar \
/input > WordCount.txt
```

⑦ 查看运行结果，操作如下：

```
hadoop@sw-desktop:~/workspace1$ more WordCount.txt
Bye: 1
Hello: 2
World: 1
Good: 1
Hadoop: 3
```

（3）使用 Python 语言实现

① 编写 WordCount.py，代码如下：

```python
from __future__ import print_function
import sys
from operator import add
from pyspark.sql import SparkSession
if __name__ == "__main__":
    if len(sys.argv) != 2:
        print("Usage: wordcount <file>", file=sys.stderr)
        exit(-1)
    spark = SparkSession\
        .builder\
        .appName("PythonWordCount")\
        .getOrCreate()
    lines = spark.read.text(sys.argv[1]).rdd.map(lambda r: r[0])
    counts = lines.flatMap(lambda x: x.split(' ')) \
              .map(lambda x: (x, 1)) \
              .reduceByKey(add)
    output = counts.collect()
    for (word, count) in output:
        print("%s: %i" % (word, count))
    spark.stop()
```

② 运行 WordCount.py，操作如下：

```
hadoop@sw-desktop:~/workspace1$ spark-submit --master spark://master:7077 \
--name PythonWC WordCount.py \
```

```
/input > WordCount.txt
```

③ 查看运行结果，操作如下：

```
hadoop@sw-desktop:~/workspace1$ more WordCount.txt
Bye: 1
World: 1
Good: 1
Hello: 2
Hadoop: 3
```

## 三、搭建 Python 开发环境

① 下载 Python 相关软件包，下载地址如下：

```
pip: https://pypi.python.org/pypi/pip#downloads
NumPy: http://packages.ubuntu.com/yakkety/python-numpy
SciPy: http://packages.ubuntu.com/yakkety/python-scipy
MatPlotLib: http://packages.ubuntu.com/yakkety/python-matplotlib
pandas: http://pandas.pydata.org/pandas-docs/stable/install.html
python-sklearn: http://packages.ubuntu.com/yakkety/python-sklearn
python-tk: http://packages.ubuntu.com/yakkety/python-tk
```

② 各集群节点安装 Python 相关组件，操作如下：

```
hadoop@master:~$ sudo apt-get install g++
hadoop@master:~$ sudo apt-get install gfortran
hadoop@master:~$ sudo apt-get install python-numpy
hadoop@master:~$ sudo apt-get install python-scipy
hadoop@master:~$ sudo apt-get install python-pandas
hadoop@master:~$ sudo apt-get install python-sklearn
hadoop@master:~$ sudo apt-get install python-dap python-sklearn-doc ipython
```

③ 开发客户端安装 Python 可视化组件，操作如下：

```
hadoop@sw-desktop:~$ sudo apt-get install g++
hadoop@sw-desktop:~$ sudo apt-get install gfortran
hadoop@sw-desktop:~$ sudo apt-get install python-pip
hadoop@sw-desktop:~$ sudo apt-get install python-numpy
hadoop@sw-desktop:~$ sudo apt-get install python-scipy
hadoop@sw-desktop:~$ sudo apt-get install python-matplotlib
hadoop@sw-desktop:~$ sudo apt-get install python-pandas
hadoop@sw-desktop:~$ sudo apt-get install python-sklearn
hadoop@sw-desktop:~$ sudo apt-get install python-dap python-sklearn-doc ipython
```

另外，也可以使用 pip 安装 Python 组件，操作如下：

```
hadoop@sw-desktop:~$ sudo pip install --upgrade pip
hadoop@sw-desktop:~$ sudo pip install numpy
hadoop@sw-desktop:~$ sudo pip install scipy
hadoop@sw-desktop:~$ sudo pip install matplotlib
```

```
hadoop@sw-desktop:~$ sudo pip install pandas
hadoop@sw-desktop:~$ sudo pip install sklearn
```

④ 客户端下载 PyCharm 包到/home/hadoop 目录下，下载地址如下：

http://www.jetbrains.com/pycharm/download

⑤ 客户端安装 PyCharm 开发工具，操作如下：

```
hadoop@sw-desktop:~$ cd /opt/
hadoop@.../opt$ sudo tar xzvf /home/hadoop/pycharm-community-2016.3.2.tar.gz
hadoop@.../opt$ sudo chown -R hadoop:hadoop /opt/pycharm-community-2016.3.2
```

⑥ 在客户端运行 PyCharm，操作如下：

```
hadoop@sw-desktop:~$
/opt/pycharm-community-2016.3.2/bin/pycharm.sh &
```

⑦ PyCharm 运行界面如图 9-15 所示。

## 四、K-Means 聚类算法

① 编写 kmeans.py，代码如下：

```
#coding=utf-8
from __future__ import print_function
import sys
import numpy as np
from pyspark.sql import SparkSession
# 把读入的数据都转化为 float 类型的数据
def parseVector(line):
    return np.array([float(x) for x in line.split(' ')])
# 求取该点应该分到哪个点集中去，返回的是序号
def closestPoint(p, centers):
    bestIndex = 0
    closest = float("+inf")
    for i in range(len(centers)):
        tempDist = np.sum((p - centers[i]) ** 2)
        # 找出 p 最靠近那个点
        if tempDist < closest:
            closest = tempDist
            bestIndex = i
    return bestIndex
if __name__ == "__main__":
    if len(sys.argv) != 4:
        print("Usage: kmeans <file> <k> <convergeDist>", file=sys.stderr)
        exit(-1)
    spark = SparkSession\
        .builder\
        .appName("PythonKMeans")\
```

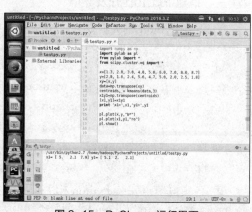

图 9-15  PyCharm 运行界面

```python
        .getOrCreate()
    lines = spark.read.text(sys.argv[1]).rdd.map(lambda r: r[0])
    # 分割数据
    data = lines.map(parseVector).cache()
    K = int(sys.argv[2])
    # 阈值
    convergeDist = float(sys.argv[3])
    # 初始化K个默认的点,False表示不应大于K,1为种子
    kPoints = data.takeSample(False, K, 1)
    tempDist = 1.0
    # 距离差大于阈值循环
    while tempDist > convergeDist:
        # 对所有数据执行map过程,最终生成的是(index, (point, 1))的rdd
        closest = data.map(
            lambda p: (closestPoint(p, kPoints), (p, 1)))
        # 执行reduce过程,把每一个K的集合的距离相加,生成(sum,num)的rdd
        pointStats = closest.reduceByKey(
            lambda p1_c1, p2_c2: (p1_c1[0] + p2_c2[0], p1_c1[1] + p2_c2[1]))
        # 生成新的聚类中心点
        newPoints = pointStats.map(
            # st[0] => K, st[1][0]距离和, st[1][0]/st[1][1]平均距离
            lambda st: (st[0], st[1][0] / st[1][1])).collect()
        # 计算一下新旧的聚类中心点的距离差的和
        tempDist = sum(np.sum((kPoints[iK] - p) ** 2) for (iK, p) in newPoints)
        for (iK, p) in newPoints:
            # 保存为聚类中心点
            kPoints[iK] = p
    print("Final centers: " + str(kPoints))
    spark.stop()
```

② 测试数据如下:

```
hadoop@master:~$ hdfs dfs -cat /test/kmeans.txt
1.30 2.00
2.00 1.60
3.00 2.40
4.00 5.60
5.00 4.70
6.00 5.00
7.00 2.00
8.00 2.50
8.70 1.80
```

③ 运行 kmeans.py,操作如下:

```
hadoop@sw-desktop:~$ spark-submit --master spark://master:7077 \
```

```
--name kmeans kmeans.py /test/kmeans.txt 3 0.001 > kmean.txt
```

④ 查看运行结果，操作如下：

```
hadoop@master:~$ more kmean.txt
Final centers: [array([ 2.1, 2. ]),array([ 7.9, 2.1]),array([ 5., 5.1])]
```

⑤ 聚类中心如图9-16所示，星点为测试数据点，圆点为聚类中心。

⑥ 实际应用一般直接调用 K-Means 函数，其主要代码如下：

```
# Loads data.
dataset = spark.read.format("libsvm").load("data/mllib/sample_kmeans_data.txt")
# Trains a k-means model.
kmeans = KMeans().setK(2).setSeed(1)
model = kmeans.fit(dataset)
# Evaluate clustering by computing Within Set Sum of Squared Errors.
wssse = model.computeCost(dataset)
print("Within Set Sum of Squared Errors = " + str(wssse))
# Shows the result.
centers = model.clusterCenters()
print("Cluster Centers: ")
for center in centers:
    print(center)
```

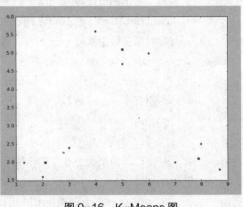

图9-16　K-Means 图

## 【同步训练】

### 一、简答题

（1）Spark 的性能特点是什么？
（2）伯克利数据分析栈的功能模块有哪些？
（3）Worker 的功能是什么？
（4）RDD 是什么？
（5）什么是宽依赖？什么是窄依赖？两者有什么不同？

### 二、操作题

（1）在 Hadoop 集群上部署 Spark。
（2）使用 Spark 工具分析气象数据，找出每年的最高气温。
（3）使用 faltMap 函数进行单词计数。

### 三、编程题

（1）使用 Scala 语言编写程序，查找气象数据每年的最高气温，并测试运行。
（2）使用 Python 语言编写程序，查找气象数据每年的最高气温，并测试运行。

# PART 10 项目十 大数据综合实例编程

## 【项目介绍】

能耗数据分析处理是 MapRecude 大数据处理的典型应用，使用 Java 完成程序的编写；使用 Mahout 完成 K-Means 聚类计算；使用 Scala/Python 完成决策树/随机森林进行模型训练和样本分类预测；使用 Apriori 和 FP-Tree 算法计算频繁项集及其支持度，进行关联分析，挖掘事务元素项之间的关联规则及其置信度。

本项目分以下 4 个任务：
- 任务 1　MapReduce 大数据处理
- 任务 2　Mahout 的 K-Means 计算
- 任务 3　决策树和随机森林的分类预测
- 任务 4　频繁项集计算与关联分析

## 【学习目标】

### 一、知识目标

- 了解 Map 输入格式。
- 了解 Map 与 Recude 之间的数据传递。
- 了解决策树和随机森林的概念。
- 了解 Apriori 和 FP-Tree 算法。
- 掌握 Map 和 Recude 模块的编写。
- 掌握 HBase API 的使用。
- 掌握 Mahout 下的 K-Means 的实现。
- 掌握使用决策树进行模型训练和样本分类预测。
- 掌握使用随机森林进行模型训练和样本分类预测。
- 掌握使用 Apriori 和 FP-Tree 计算频繁项集及其支持度。
- 掌握关联分析及置信度计算。

### 二、能力目标

- 能够使用 MapReduce 编写大数据处理程序。

- 能够使用 Mahout 进行大数据聚类运算。
- 能够使用决策树进行模型训练和样本分类预测。
- 能够使用随机森林进行模型训练和样本分类预测。
- 能够使用 Scala 编写 Apriori 和 FP-Tree 算法。
- 能够挖掘事务元素项之间的关联规则及其置信度。

## 任务 1　MapReduce 大数据处理

### 【任务概述】

某办公单位部署了能耗分项计量系统,共有两栋楼,每栋楼两个采集器,共部署了 4 个采集器,每个采集器有 6 块功能电表,共计 24 块功能电表,每个功能表采集 6 个功能项的值。采集器接收功能电表发来的数据,然后按规定格式把这些数据写入 XML 文件,发送到能耗分项计量处理系统。处理系统会把从采集器接收到的数据写到指定的 HDFS 文件系统。采集器每 10 秒发送一次数据到处理系统。4 个采集器独立工作,但是能耗分项计量大数据处理系统会把不同采集器在同一时间发送的数据作为一组数据进行存放,并使用采集器编号+数据采集时间作为文件名。

程序设计要求如下:

① MapReduce 程序从 HDFS 获取能耗数据。

② MapReduce 程序的 Map 阶段对 XML 格式的能耗数据进行解析,得到能耗基础用能数据,包括楼栋编号、采集器编号、数据采集时间、电表编号、功能项、功能项值,并把能耗基础用能数据写入 HBase 数据库。

③ MapReduce 程序的 Reduce 阶段需要计算每 10 秒(同一个采集器相邻采集时间)内的每个电表用电量的变化值,并将变化值写入 HBase 数据库中。

### 【任务分析】

#### 一、能耗数据描述

采集器接收到的数据最终会以 XML 文件格式写入指定的 HDFS 文件系统中,如图 10-1 所示。

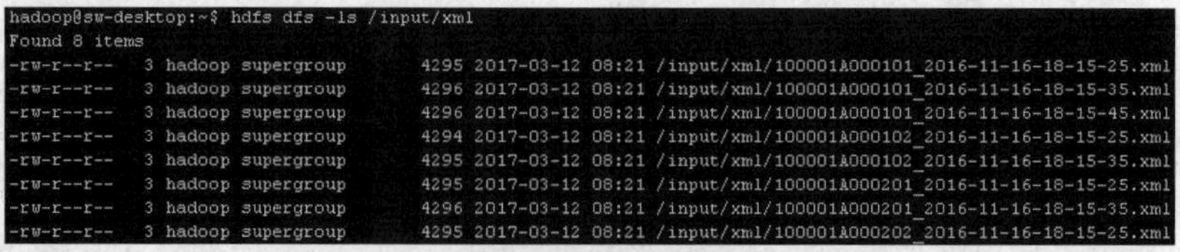

```
hadoop@sw-desktop:~$ hdfs dfs -ls /input/xml
Found 8 items
-rw-r--r--   3 hadoop supergroup   4295 2017-03-12 08:21 /input/xml/100001A000101_2016-11-16-18-15-25.xml
-rw-r--r--   3 hadoop supergroup   4296 2017-03-12 08:21 /input/xml/100001A000101_2016-11-16-18-15-35.xml
-rw-r--r--   3 hadoop supergroup   4296 2017-03-12 08:21 /input/xml/100001A000101_2016-11-16-18-15-45.xml
-rw-r--r--   3 hadoop supergroup   4294 2017-03-12 08:21 /input/xml/100001A000102_2016-11-16-18-15-25.xml
-rw-r--r--   3 hadoop supergroup   4295 2017-03-12 08:21 /input/xml/100001A000102_2016-11-16-18-15-35.xml
-rw-r--r--   3 hadoop supergroup   4295 2017-03-12 08:21 /input/xml/100001A000201_2016-11-16-18-15-25.xml
-rw-r--r--   3 hadoop supergroup   4296 2017-03-12 08:21 /input/xml/100001A000201_2016-11-16-18-15-35.xml
-rw-r--r--   3 hadoop supergroup   4295 2017-03-12 08:21 /input/xml/100001A000202_2016-11-16-18-15-25.xml
```

图 10-1　HDFS 中的能耗数据

能耗数据 XML 文件的内容格式如图 10-2 所示。

该文件由编号为 100001A000101 的采集器(<gateway_id>指明采集器编号)产生。该采集器从属于编号为 100001A0001 的楼栋(<building_id>指明楼栋编号)。每个采集器采集了多个电表的值,电表用<meter id="XXXXX">表示,电表编号由 id 定义。每个电表有多个功能项，<function id="XXXXXXP">定义了功能项编号（由 id 指定），功能项 id 的最后一位（功能标记）定义了该功能项所代表的物理测量含义。

```xml
<?xml version='1.0' encoding='UTF-8'?>
<root>
    <common>
        <building_id>100001A0001</building_id>
        <gateway_id>100001A000101</gateway_id>
        <type>peroid</type>
    </common>
    <data operation=" query/reply/report/continuous/continuous_ack">
        <sequence>2016-11-16-18-15-25</sequence>
        <parser>no</parser>
        <time>1479291325770</time>
        <total>0</total>
        <current>0</current>
        <meter id="510107D001010001">
            <function id="100001A0001000101" error="0" coding="100001A000101A00">98</function>
            <function id="100001A0001000102" error="0" coding="100001A000101A00">217</function>
            <function id="100001A0001000103" error="0" coding="100001A000101A00">128</function>
            <function id="100001A0001000104" error="0" coding="100001A000101A00">59</function>
            <function id="100001A0001000105" error="0" coding="100001A000101A00">5398</function>
            <function id="100001A0001000106" error="0" coding="100001A000101A00">1083</function>
        </meter>
        ……
    </data>
</root>
```

图 10-2　XML 文件内容

common：表示一个采集器的共有属性部分。

building_id：表示楼栋编号。

gateway_id：由 building_id 和最后两位数字编号组成，表示采集器编号。

sequence：表示采集器收集数据的时间，也就是读取电表数据的时间。

time：表示采集器收集数据的时间，同 sequence 含义一样，采用 Linux 时间戳。

meter：表示一个多功能电表，每个电表编号由 meter id 指定。

function：每个电表包括多个功能项，每个功能项由 id 区分。
各个功能项的物理测量含义字典如表 10-1 所示。

表 10-1 电表功能项字典

| 功 能 标 记 | 功 能 含 义 |
|---|---|
| 100001A0001000101 功能标记"1" | 用电量，单位 kW |
| 100001A0001000102 功能标记"2" | 瞬时电流，单位 A |
| 100001A0001000103 功能标记"3" | 其他能源，单位 A |
| 100001A0001000104 功能标记"4" | 其他能源，单位 B |
| 100001A0001000105 功能标记"5" | 其他能源，单位 C |
| 100001A0001000106 功能标记"6" | 未定义，可忽略 |

## 二、MapReduce 大数据处理程序分析

程序设计流程如图 10-3 所示。

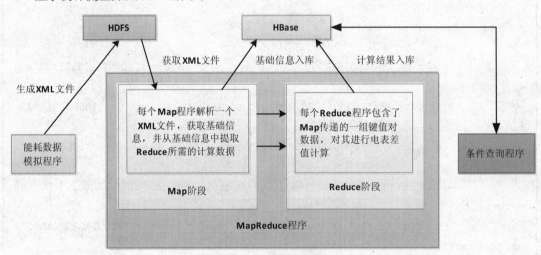

图 10-3　程序设计流程

（1）main()模块

① 完成 Job 的启动。

② XmlFileInputFormat.class 避免 XML 文件切分。

（2）map()模块

① 调用 Tools 类，读取 XML 信息。

② 调用 HBaseOps 类，把 XML 读取的信息存入 HBase 数据库。

③ 生成<Key,Value>键值对，传递给 reduce()模块进行计算。Key 格式为：采集器_电表编号，Value 格式为：采集时间_用电量。

（3）reduce()模块

① KVmap<Key,Value>保存不同批次（Job）的最后一组<Key,Value>键值对（按采集时间排序）。并把 Value 传递给下个 Job 一起计算，避免计算电表用电量的变化值缺失。

② 对 map()模块传递过来的 Value 进行拆分。

③ 计算电表用电量的变化值并存入 HBase 数据库。

（4）Tools 类

Tools 类已经被封装到 XmlTool.jar，并添加到了工程的 library 中，在给定代码框架中已经通过"import xmltool.Tools;"引入该类，Tools 提供的 API 如表 10-2 所示。

表 10-2 Tools API

| Tools API 名称 | 功能 |
| --- | --- |
| private HashMap<T, T> publicMap | 存储 XML 文件解析后的 common 数据 |
| private static HashMap<T, HashMap<T, T>> contentMap | 存储 XML 文件解析后的 meter 数据 |
| private static HashMap<T, T> funcMap | 存储 XML 文件解析后的 function 数据 |
| public void initXml(String xml) | 该方法将传入 XML 解析，并存入上面的 3 种 HashMap |
| public HashMap<T, HashMap<T, T>> getContent() | 返回 contentMap |
| public HashMap<T, T> getPublic() | 返回 publicMap |

## 【支撑知识】

### 一、能耗分项计量

能耗分项计量是指对建筑的水、电、燃气、集中供热、集中供冷等各种能耗进行监测，从而得出建筑物的总能耗量和不同能源种类、不同功能系统的能耗量。要完成能耗分项计量，需要安装大量仪表，按分钟甚至按秒对各项能源使用情况进行采集和上报。一个普通建筑需要几十块仪表，每个仪表每秒会产生多项能耗数据。而要对一个单位进行全面能耗分项计量，将需要部署上千块能耗采集仪表，所以，每个单位每秒将会产生上万条能耗数据，并且每个分项能耗数据的格式不同。由此可见，能耗分项计量是一个典型的大数据应用，具有数据量大、数据产生速度快、数据结构复杂等特点。

### 二、Map 输入格式

MapReduce 的一个输入分片（Split）就是由单个 Map 处理的输入块。每一个 Map 操作处理一个输入分片。每个分片被划分为若干个记录，每条记录就是一个键值对，Map 一个接一个地处理每条记录。InputFormat 负责产生输入分片并将它们分割成记录。在 tasktracker 上，Map 任务把输入分片传给 InputFormat 的 createRecordReder()方法来获得这个分片的 RecordReader，Map 任务用一个 RecordReader 来生成记录的键值对，然后再传递给 Map 函数。FileInputFormat 是所有使用文件作为其数据源的 InputFormat 实现的基类。它提供了两个功能：一个功能是定义哪些文件包含在一个作业的输入中；另一功能是为输入文件生成分片的实现。把分片分割成记录的作业由其子类来完成。FileInputFormat 只分割大文件，这里的大指的是超过 HDFS 块的大小。分片通常与 HDFS 块大小一样。

有些应用程序（如读 XML 文件）可能不希望文件被切分，而是用一个 Mapper 完整处理每一个输入文件。为了避免切分，可以重载 FileInputFormat 的 isSplitable()方法，把返回值设为 false，那么 Map 任务就会只有一个。

TextInputFormat 是默认的 InputFormat。每条记录是一行输入。键是 LongWritable 类型，存储该行在整个文件中的字节偏移量。值是这行的内容，不包括任何行终止符，它是 Text 类型的。

## 【任务实施】

### 一、创建 MapReduce 项目

运行 Eclipse，新建 MapReduce 项目，并添加 XmlTool.jar、Hadoop 和 HBase 的 jar 包，如图 10-4 所示。

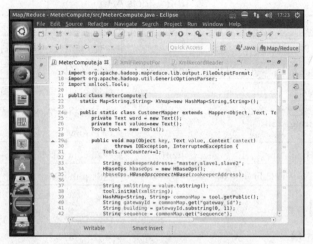

图 10-4　新建 MapReduce 项目

（1）编写 HBaseOps 类

HBaseOps.java 代码如下：

```java
public class HBaseOps {
    static Configuration hbaseConf;
    static HBaseAdmin hbaseAdmin;
    //HBase 连接初始化
    static public HBaseAdmin HBaseOpsconnectHBase(String zookeeperAddress) throws IOException {
        Configuration conf = new Configuration();
        conf.set("hbase.zookeeper.quorum", zookeeperAddress);
        hbaseConf = HBaseConfiguration.create(conf);
        hbaseAdmin = new HBaseAdmin(hbaseConf);
        return hbaseAdmin;
    }
    //关闭 HBase 连接
    static public void disconnect() throws IOException {
        hbaseAdmin.close();
    }
    //插入一行记录
    public void insertRecord(String tableName, String rowkey, String family,
            String qualifier, String value) throws IOException {
        HTable table = new HTable(hbaseConf, tableName);
        Put put = new Put(rowkey.getBytes());
```

```java
        put.add(family.getBytes(), qualifier.getBytes(), value.getBytes());
        table.put(put);
        System.out.println("插入行成功");
    }
    //删除一行记录
    public void deleteRecord(String tableName, String rowkey)
            throws IOException {
        HTable table = new HTable(hbaseConf, tableName);
        Delete del = new Delete(rowkey.getBytes());
        table.delete(del);
        System.out.println("删除行成功");
    }
    //获取一行记录
    public Result getOneRecord(String tableName, String rowkey)
            throws IOException {
        HTable table = new HTable(hbaseConf, tableName);
        Get get = new Get(rowkey.getBytes());
        Result rs = table.get(get);
        return rs;
    }
    //获取所有记录
    public List<Result> getAllRecord(String tableName) throws IOException {
        HTable table = new HTable(hbaseConf, tableName);
        Scan scan = new Scan();
        ResultScanner scanner = table.getScanner(scan);
        List<Result> list = new ArrayList<Result>();
        for (Result r : scanner) {
            list.add(r);
        }
        scanner.close();
        return list;
    }
}
```

（2）重写 Job 的 InputFormatClass 类

① XmlFileInputFormat.java 代码如下：

```java
public class XmlFileInputFormat extends FileInputFormat<Text, Text> {
    @Override
    public RecordReader<Text, Text> createRecordReader(InputSplit arg0,
            TaskAttemptContext arg1) throws IOException, InterruptedException {
        // 重写 RecordReader
        RecordReader<Text, Text> recordReader = new XmlRecordReader();
```

```java
        return recordReader;
    }
    @Override
    protected boolean isSplitable(JobContext context, Path filename) {
        //避免文件拆分
        return false;
    }
}
```

② XmlRecordReader.java 代码如下:

```java
public class XmlRecordReader extends RecordReader<Text,Text> {
    private FileSplit fileSplit;
    private JobContext jobContext;
    private Text currentKey = new Text();
    private Text currentValue = new Text();
    private boolean finishConverting = false;
    @Override
    public void close() throws IOException {
    }
    @Override
    public Text getCurrentKey() throws IOException, InterruptedException {
        return currentKey;
    }
    @Override
    public Text getCurrentValue() throws IOException, InterruptedException {
        return currentValue;
    }
    @Override
    public float getProgress() throws IOException, InterruptedException {
        float progress = 0;
        if(finishConverting){
            progress = 1;
        }
        return progress;
    }
    //初始化,获取文件名并赋值给 currentKey
    @Override
    public void initialize(InputSplit arg0, TaskAttemptContext arg1)
            throws IOException, InterruptedException {
        this.fileSplit = (FileSplit) arg0;
        this.jobContext = arg1;
        String filename = fileSplit.getPath().getName();
```

```java
        this.currentKey = new Text(filename);
    }
    //读取整个 XML 文件内容,并赋值给 currentValue
    @Override
    public boolean nextKeyValue() throws IOException, InterruptedException {
      if(!finishConverting) {
          int len = (int)fileSplit.getLength();
          Path file = fileSplit.getPath();
          FileSystem fs = file.getFileSystem(jobContext.getConfiguration());
          FSDataInputStream in = fs.open(file);
          BufferedReader br = new BufferedReader(new InputStreamReader (in,"gbk"));
          String total="";
          String line="";
          while((line= br.readLine())!= null){
              total =total+line+"\n";
          }
          br.close();
          in.close();
          fs.close();
          currentValue = new Text(total);
          finishConverting = true;
          return true;
      }
      return false;
    }
}
```

(3)编写 MapReduce 程序

MeterCompute.java 代码如下:
```java
public class MeterCompute {
  // KVmap 用于保存不同批次(Job)的最后一组<Key,Value>键值对
  static Map<String,String> KVmap=new HashMap<String,String>();
  public static class CustomerMapper extends Mapper<Object, Text, Text, Text> {
    private Text word = new Text();
    private Text values=new Text();
    //生成 Tools 实例,用于读取 XML 文件内容
    Tools tool = new Tools();
    public void map(Object key, Text value, Context context)
        throws IOException, InterruptedException {
      Tools.runCounter+=1;
      //zookeeperAddress 地址与 Hadoop 集群环境一致
      String zookeeperAddress= "master,slave1,slave2";
      //生成 hbaseOps 实例
```

```java
            HBaseOps hbaseOps = new HBaseOps();
            hbaseOps.HBaseOpsconnectHBase(zookeeperAddress);
            String xmlString = value.toString();
            //调用 Tools 类，xmlString 是整个 XML 文件内容
            tool.initXml(xmlString);
            //commonMap 包含采集器编号、楼栋编号、采集器收集数据的时间
            HashMap<String, String> commonMap = tool.getPublic();
            //gatewayID 表示采集器编号
            String gatewayId = commonMap.get("gateway_id");
            //building 表示楼栋编号
            String building = gatewayId.substring(0, 11);
            //sequence 表示采集器收集数据的时间，也就是读取电表数据的时间
            String sequence = commonMap.get("sequence");
            //contentMap 包含 Key<电表编号>、Value<电表功能号、对应的值>
            HashMap<String, HashMap<String, String>> contentMap = tool.getContent();
            Iterator<Entry<String, HashMap<String, String>>> contentIter = contentMap
                    .entrySet().iterator();
            while (contentIter.hasNext()) {
              Entry<String, HashMap<String, String>> entry = (Entry<String, HashMap<String, String>>) contentIter
                        .next();
              //meterId 表示电表编号
              String meterId = (String) entry.getKey();
              //funId1 表示电表功能编号
              String funId1 = entry.getValue().get("id_1").toString();
              ……
              String funId6 = entry.getValue().get("id_6").toString();
              //funVal1 表示电表功能编号对应的值
              String funVal1 = entry.getValue().get("text_1").toString();
              ……
              String funVal6 = entry.getValue().get("text_6").toString();
              //rowkey 表示采集器编号_电表编号，作为 HBase 的行健
              String rowkey = gatewayId + "_" + meterId ;
              //插入公共部分记录（楼栋编号、采集器编号、采集时间、电表编号）
              hbaseOps.insertRecord("BaseInfo",rowkey,"building","building", building);
              hbaseOps.insertRecord("BaseInfo",rowkey,"gatewayId","gatewayId", gatewayId);
              hbaseOps.insertRecord("BaseInfo",rowkey,"sequence","sequence", sequence);
              hbaseOps.insertRecord("BaseInfo",rowkey,"meterId","meterId",meterId);
              //插入电表功能编号
              hbaseOps.insertRecord("BaseInfo",rowkey,"funId","funId1",funId1);
              ……
              hbaseOps.insertRecord("BaseInfo",rowkey,"funId","funId6",funId6);
```

```java
        //插入电表功能编号对应的功能值
        hbaseOps.insertRecord("BaseInfo",rowkey,"funVal","funVal1",funVal1);
        ……
        hbaseOps.insertRecord("BaseInfo",rowkey,"funVal","funVal6",funVal6);
        //传送给 reducer 的 key：采集器_电表编号
        String reducerKey = gatewayId + "_" + meterId;
        //传送给 reducer 的 value：采集时间_用电量
        String reducerValue = sequence + "_" + funVal1;
        word.set(reducerKey);
        values.set(reducerValue);
        context.write(word, values);
    }
    //关闭 HBase 连接
    hbaseOps.disconnect();
    }
}

public static class CustomerReducer extends Reducer<Text, Text, Text, Text> {
    //private Text result = new Text();
    public void reduce(Text key, Iterable<Text> values,
        Context context) throws IOException, InterruptedException {
        List<String> valueList = new ArrayList<String>();
        //对 map()传递过来的 value<采集时间_用电量>，转为列表 valueList
        for (Text value : values) {
            valueList.add(value.toString());
        }
        //如果 KVmap<Key,Value>不为空，那么 Value 加入列表 valueList
        if (KVmap.containsKey(key.toString())) {
            valueList.add(KVmap.get(key.toString()));
        }
        //列表 valueList 排序（采集时间）
        Collections.sort(valueList);
        //KVmap 保存 Key 最新的值，以便参与下次 Job 计算
        KVmap.put(key.toString(), valueList.get(valueList.size()-1));
        //tlist 存放采集时间
        List<String> tlist = new ArrayList<String>();
        //vlist 存放用电量
        List<String> vlist = new ArrayList<String>();
        for (int i=0;i<valueList.size();i++) {
            String[] mid=valueList.get(i).split("_");
            tlist.add(mid[0]);
            vlist.add(mid[1]);
        }
```

```java
      String zookeeperAddress= "master,slave1,slave2";
      HBaseOps hbaseOps = new HBaseOps();
      hbaseOps.HBaseOpsconnectHBase(zookeeperAddress);
      //rowkey 表示 HBase 的行键盘
      String rowkey = key.toString();
      //kkk[0]表示采集器编号, kkk[1]表示电表编号
      String[] kkk = key.toString().split("_");
      //building 表示楼栋编号
      String building = kkk[0].substring(0, 11);
      //计算用电量变化值,并存入 HBase
      for (int i=0;i<vlist.size()-1;i++) {
        String t1=tlist.get(i);
        String t2=tlist.get(i+1);
        //ch 表示两个时间段的用电量变化值
        int ch=Integer.parseInt(vlist.get(i+1))-Integer.parseInt(vlist.get(i));
        String chs=String.valueOf(ch);
        //插入记录(楼栋编号、采集器编号、电表编号、起始时间、结束时间、用电量变化)
        hbaseOps.insertRecord("ComputeResult",rowkey,"building","building",building);
        hbaseOps.insertRecord("ComputeResult",rowkey,"gatewayId","gatewayId",kkk[0]);
        hbaseOps.insertRecord("ComputeResult",rowkey,"meterId","meterId",kkk[1]);
        hbaseOps.insertRecord("ComputeResult",rowkey,"times","startTime", t1);
        hbaseOps.insertRecord("ComputeResult",rowkey,"times","endTime",t2);
        hbaseOps.insertRecord("ComputeResult",rowkey,"change","change",chs);
      }
      //关闭 HBase 连接
      hbaseOps.disconnect();
    }
  }

  public static void main(String[] args) throws Exception {
    Configuration conf = new Configuration();
    String[] otherArgs = new GenericOptionsParser(conf, args)
      .getRemainingArgs();
    if (otherArgs.length < 2) {
      System.err.println("Usage: wordcount <in> [<in>...] <out>");
      System.exit(2);
    }
    Job job = Job.getInstance(conf, "MeterCompute");
    job.setJarByClass(MeterCompute.class);
    job.setMapperClass(CustomerMapper.class);
```

```
      job.setReducerClass(CustomerReducer.class);
      job.setOutputKeyClass(Text.class);
      job.setOutputValueClass(Text.class);
      //设置job的setInputFormatClass为自定义XmlFileInputFormat.class类
      job.setInputFormatClass(XmlFileInputFormat.class);
      for (int i = 0; i < otherArgs.length - 1; ++i) {
        FileInputFormat.addInputPath(job, new Path(otherArgs[i]));
      }
      FileOutputFormat.setOutputPath(job, new Path(
          otherArgs[otherArgs.length - 1]));
      System.exit(job.waitForCompletion(true) ? 0 : 1);
    }
}
```

## 二、创建 HBase 的表

① 创建 HBase 的 BaseInfo 表，操作如下：

```
hbase(main):001:0> create 'BaseInfo','building','gatewayId','sequence',
'meterId', 'funId', 'funVal'
0 row(s) in 9.1700 seconds
```

BaseInfo 表的列族：楼栋编号、采集器编号、采集时间、电表编号、funId、funVal。

funId 的列：6 个电表的功能项编号（funId1，funId2，…，funId6）

funVal 的列：6 个功能项的读数值（funVal1，funVal2，…，funVal6）

② 创建 HBase 的 ComputeResult 表，操作如下：

```
hbase(main):002:0> create 'ComputeResult','building','gatewayId','meterId',
'times','change'
0 row(s) in 2.3190 seconds

=> Hbase::Table - ComputeResult
```

ComputeResult 表的列族：楼栋编号、采集器编号、电表编号、times、在起始时间和结束时间内的用电量（change）。

Times 的列：起始时间（startTime）、结束时间（endTime）

## 三、配置参数

配置运行参数，如图 10-5 所示，并运行。

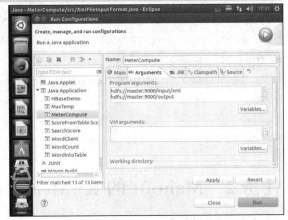

图 10-5　配置运行参数

## 四、查看 HBase 数据库信息

① 查看 BaseInfo 表信息，操作如下：

```
hbase(main):005:0> scan 'BaseInfo'
ROW                    COLUMN+CELL
```

```
    100001A000101_510107D001010001    column=building:building, timestamp=1489311363969,
value=100001A0001
    100001A000101_510107D001010001    column=funId:funId1, timestamp=1489311363984,
value=100001A0001000101
    ……
    100001A000101_510107D001010001    column=funId:funId6, timestamp=1489311364014,
value=100001A0001000106
    100001A000101_510107D001010001    column=funVal:funVal1, timestamp=1489311364017,
value=98
    ……
    100001A000101_510107D001010001    column=funVal:funVal6, timestamp=1489311364032,
value=1083
    ……
24 row(s) in 0.9310 seconds
```

② 查看 ComputeResult 表信息,操作如下:

```
hbase(main):031:0> scan 'ComputeResult'
ROW                          COLUMN+CELL
    100001A000101_510107D001010001    column=building:building, timestamp=
1489311370210, value=100001A0001
    100001A000101_510107D001010001    column=change:change, timestamp=1489311371067,
value=1
    100001A000101_510107D001010001    column=gatewayId:gatewayId, timestamp=
1489311370300,value=100001A000101
    100001A000101_510107D001010001    column=meterId:meterId, timestamp=1489311370427,
value=510107D001010001
    100001A000101_510107D001010001    column=times:endTime, timestamp=1489311370829,
value=2016-11-16-18-15-45
    100001A000101_510107D001010001    column=times:startTime, timestamp=1489311370508,
value=2016-11-16-18-15-35
    100001A000101_510107D001010002    column=building:building, timestamp=1489311371700,
value=100001A0001
    100001A000101_510107D001010002    column=change:change, timestamp=1489311371829,
value=0
    ……
18 row(s) in 0.5400 seconds
```

## 任务 2  Mahout 的 K-Means 计算

### 【任务概述】

Mahout 中有很多经典的数据挖掘算法,其中 K-Means 是比较典型的算法。本任务使用 Mahout 完成 K-Means 计算,测试数据 synthetic_control.data 包括 600 个数据点(行),每个数据点有 60 个属性，将 600 个数据点划分成不同的聚类。

## 【支撑知识】

### 一、Mahout 简介

Mahout 是 Apache Software Foundation（ASF）旗下的一个开源项目，提供一些可扩展的机器学习领域经典算法的实现，旨在帮助开发人员更加方便、快捷地创建智能应用程序。Mahout 包含许多实现，包括聚类、分类、推荐过滤、频繁子项挖掘。此外，通过使用 Apache Hadoop 库，Mahout 可以有效地扩展到云中。

### 二、K-Means 计算

（1）K-Means 聚类算法原理

K-Means 算法接受参数 $K$；然后将事先输入的 $n$ 个数据对象划分为 $K$ 个聚类以便使得所获得的聚类满足：同一聚类中的对象相似度较高；而不同聚类中的对象相似度较小。聚类相似度是利用各聚类中对象的均值所获得一个"中心对象"来进行计算的。

（2）Mahout 下的 K-Means 实现

Mahout 下的 K-Means 算法是在 K-Means 与 Hadoop 中间加上一个中间层来建立 K-Means 与 Hadoop 之间的联系，实现 K-Means 的 MapReduce 的任务分配等工作。具体实现过程如下：

① 参数 input 指定待聚类的所有数据点，clusters 指定初始聚类中心，如果指定参数 $K$，由 org.apache.mahout.clustering.kmeans.RandomSeedGenerator.buildRandom 通过 org.apache.hadoop.fs 直接从 input 指定文件中随机读取 $k$ 个点放入 clusters 中。

② 根据原数据点和上一次迭代（或初始聚类）的聚类中心计算本次迭代的聚类中心，输出到 clusters-N 目录下。该过程由 org.apache.mahout.clustering.kmeans 下的 KMeansMapper\KMeansCombiner\KMeansReducer\KMeansDriver 实现。

- KMeansMapper：在 configure 中初始化 Mapper 时读入上一次迭代产生或初始聚类中心（每个 Mapper 都读入所有的聚类中心）；Map 方法对输入的每个点，计算距离其最近的类，并加入其中输出 Key 为该点所属聚类 ID，Value 为 KMeansInfo 实例，包含点的个数和各分量的累加和。

- KMeansCombiner：本地累加 KMeansMapper 输出的同一聚类 ID 下的点个数和各分量的和。

- KMeansReducer：累加同一聚类 ID 下的点个数和各分量的和，求本次迭代的聚类中心；并根据输入 Delta 判断该聚类是否已收敛：上一次迭代聚类中心与本次迭代聚类中心距离<Delta；输出各聚类中心和其是否收敛标记。

- KMeansDriver：控制迭代过程直至超过最大迭代次数或所有聚类都已收敛，每轮迭代后，KMeansDriver 读取其 clusters-N 目录下的所有聚类，若所有聚类已收敛，则整个 Kmeans 聚类过程收敛了。

## 【任务实施】

### 一、客户端安装 Mahout

① 下载 Mahout 软件包到/home/hadoop 目录下，下载地址如下：

https://mirrors.aliyun.com/apache/mahout/

② 安装 Mahout 软件，操作如下：

```
hadoop@sw-desktop:~$ cd /opt
hadoop@...$ sudo tar xvzf /home/hadoop/apache-mahout-distribution-0.12.2.tar.gz
hadoop@...$ sudo chown -R hadoop:hadoop apache-mahout-distribution-0.12.2
```

③ 修改环境变量，操作如下：

```
hadoop@...$ vi /home/hadoop/.profile
```

添加如下内容：

```
export MAHOUT_HOME=/opt/apache-mahout-distribution-0.12.2
export PATH=$PATH:$MAHOUT_HOME/bin
```

④ 环境变量生效，操作如下：

```
hadoop@...$ source /home/hadoop/.profile
```

⑤ 查看 Mahout 帮助，操作如下：

```
hadoop@sw-desktop:/opt$ mahout --help
MAHOUT_LOCAL is not set; adding HADOOP_CONF_DIR to classpath.
Running on hadoop, using /opt/hadoop-2.7.3/bin/hadoop and HADOOP_CONF_DIR=/opt/hadoop-2.7.3/etc/hadoop
MAHOUT-JOB: /opt/apache-mahout-distribution-0.12.2/mahout-examples-0.12.2-job.jar
Unknown program '--help' chosen.
Valid program names are:
  arff.vector: : Generate Vectors from an ARFF file or directory
  baumwelch: : Baum-Welch algorithm for unsupervised HMM training
  canopy: : Canopy clustering
  cat: : Print a file or resource as the logistic regression models would see it
  cleansvd: : Cleanup and verification of SVD output
  clusterdump: : Dump cluster output to text
  clusterpp: : Groups Clustering Output In Clusters
  cmdump: : Dump confusion matrix in HTML or text formats
  cvb: : LDA via Collapsed Variation Bayes (0th deriv. approx)
  cvb0_local: : LDA via Collapsed Variation Bayes, in memory locally.
  describe: : Describe the fields and target variable in a data set
  evaluateFactorization: : compute RMSE and MAE of a rating matrix factorization against probes
  fkmeans: : Fuzzy K-means clustering
  hmmpredict: : Generate random sequence of observations by given HMM
  itemsimilarity: : Compute the item-item-similarities for item-based collaborative filtering
  kmeans: : K-means clustering
  lucene.vector: : Generate Vectors from a Lucene index
  matrixdump: : Dump matrix in CSV format
  matrixmult: : Take the product of two matrices
```

parallelALS: : ALS-WR factorization of a rating matrix
qualcluster: : Runs clustering experiments and summarizes results in a CSV
recommendfactorized: : Compute recommendations using the factorization of a rating matrix
recommenditembased: : Compute recommendations using item-based collaborative filtering
regexconverter: : Convert text files on a per line basis based on regular expressions
resplit: : Splits a set of SequenceFiles into a number of equal splits
rowid: : Map SequenceFile<Text,VectorWritable> to {SequenceFile<IntWritable, VectorWritable>, SequenceFile<IntWritable,Text>}
rowsimilarity: : Compute the pairwise similarities of the rows of a matrix
runAdaptiveLogistic: : Score new production data using a probably trained and validated AdaptivelogisticRegression model
runlogistic: : Run a logistic regression model against CSV data
seq2encoded: : Encoded Sparse Vector generation from Text sequence files
seq2sparse: : Sparse Vector generation from Text sequence files
seqdirectory: : Generate sequence files (of Text) from a directory
seqdumper: : Generic Sequence File dumper
seqmailarchives: : Creates SequenceFile from a directory containing gzipped mail archives
seqwiki: : Wikipedia xml dump to sequence file
spectralkmeans: : Spectral k-means clustering
split: : Split Input data into test and train sets
splitDataset: : split a rating dataset into training and probe parts
ssvd: : Stochastic SVD
streamingkmeans: : Streaming k-means clustering
svd: : Lanczos Singular Value Decomposition
testnb: : Test the Vector-based Bayes classifier
trainAdaptiveLogistic: : Train an AdaptivelogisticRegression model
trainlogistic: : Train a logistic regression using stochastic gradient descent
trainnb: : Train the Vector-based Bayes classifier
transpose: : Take the transpose of a matrix
validateAdaptiveLogistic: : Validate an AdaptivelogisticRegression model against hold-out data set
vecdist: : Compute the distances between a set of Vectors (or Cluster or Canopy, they must fit in memory) and a list of Vectors
vectordump: : Dump vectors from a sequence file to text
viterbi: : Viterbi decoding of hidden states from given output states sequence

## 二、K-Means 计算

① 下载测试数据到/home/hadoop 目录下，下载地址如下：

http://archive.ics.uci.edu/ml/databases/synthetic_control/synthetic_control.data

② 上传数据，操作如下：

```
hadoop@...$ hdfs dfs -mkdir -p /user/hadoop/testdata
hadoop@...$ hdfs dfs -put /home/hadoop/synthetic_control.data /user/hadoop/testdata
```

③ 运行 K-Means 计算，操作如下：

```
hadoop@...$ mahout org.apache.mahout.clustering.syntheticcontrol.kmeans.Job \
--numClusters 3 --t1 0.5 --t2 10 --maxIter 10 \
-i /user/hadoop/testdata/synthetic_control.data \
-o /user/hadoop/output -ow


……
   1.0 : [distance=2223.618730647693]: [24.691,28.323,25.75,32.889,30.297,32.642,27.154,29.053,30.635,35.075,27.497,35.392,30.997,31.898,31.567,26.617,25.706,26.077,30.565,34.839,27.622,30.479,30.082,25.034,34.137,33.03,35.789,29.711,33.794,32.502,31.826,28.748,31.36,30.977,28.964,29.017,27.047,21.903,15.372,24.714,26.378,25.758,16.965,25.801,25.941,17.484,25.715,25.783,25.252,16.545,17.18,26.245,24.21,15.276,20.186,25.481,20.278,26.209,24.502,26.251]
   1.0 : [distance=2771.69984757083846]: [25.535,35.615,29.184,35.006,34.519,25.275,30.613,34.722,24.983,32.041,30.265,28.395,28.964,32.036,33.502,24.275,26.928,26.076,24.719,30.423,32.728,32.984,25.009,33.649,25.913,35.456,27.444,32.726,34.1,29.262,31.463,26.964,24.343,31.768,21.981,26.878,17.664,23.232,23.508,17.183,18.586,18.864,22.923,19.235,25.574,24.67,15.367,18.541,18.896,18.75,27.067,24.772,23.357,23.92,16.747,26.083,18.237,24.436,18.673,24.763]
17/03/20 08:30:42 INFO ClusterDumper: Wrote 3 clusters
17/03/20 08:30:42 INFO MahoutDriver: Program took 171834 ms (Minutes: 2.8639)
```

④ 查看运行结果目录，操作如下：

```
hadoop@...$ hdfs dfs -ls /user/hadoop/output
Found 10 items
-rw-r--r--   3 hadoop supergroup    194 2017-03-20 08:30 /user/hadoop/output/_policy
drwxr-xr-x   - hadoop supergroup      0 2017-03-20 08:30 /user/hadoop/output/clusteredPoints
drwxr-xr-x   - hadoop supergroup      0 2017-03-20 08:28 /user/hadoop/output/clusters-0
drwxr-xr-x   - hadoop supergroup      0 2017-03-20 08:28 /user/hadoop/output/clusters-1
drwxr-xr-x   - hadoop supergroup      0 2017-03-20 08:29 /user/hadoop/output/clusters-2
drwxr-xr-x   - hadoop supergroup      0 2017-03-20 08:29 /user/hadoop/output/clusters-3
```

    drwxr-xr-x   - hadoop supergroup    0 2017-03-20 08:29 /user/hadoop/output/clusters-4

    drwxr-xr-x   - hadoop supergroup    0 2017-03-20 08:30 /user/hadoop/output/clusters-5-final

    drwxr-xr-x   - hadoop supergroup    0 2017-03-20 08:28 /user/hadoop/output/data

    drwxr-xr-x   - hadoop supergroup    0 2017-03-20 08:28 /user/hadoop/output/random-seeds

⑤ 聚类结果生成序列文件，操作如下：

    hadoop@...$ mahout seqdumper -i /user/hadoop/output/clusteredPoints -o clusteredPoints

    MAHOUT_LOCAL is not set; adding HADOOP_CONF_DIR to classpath.
    Running on hadoop, using /opt/hadoop-2.7.3/bin/hadoop and HADOOP_CONF_DIR=/opt/hadoop-2.7.3/etc/hadoop
    MAHOUT-JOB: /opt/apache-mahout-distribution-0.12.2/mahout-examples-0.12.2-job.jar
    17/03/20 08:42:02 INFO AbstractJob: Command line arguments: {--endPhase=[2147483647], --input=[output/clusteredPoints], --output=[clustered], --startPhase=[0], --tempDir=[temp]}
    17/03/20 08:42:03 INFO MahoutDriver: Program took 1332 ms (Minutes: 0.0222)

⑥ 查看聚类结果，操作如下：

    hadoop@sw-desktop:~$ cat clusteredPoints
    ……
    Key: 538: Value: wt: 1.0 distance: 1642.536860250293 vec: [35.899,26.672,34.191,35.827,25.101,24.856,25.814,30.63,34.212,32.587,31.032,34.304,24.555,35.87,30.683,29.058,28.637,29.855,32.037,32.979,26.118,26.107,25.096,22.703,17.698,16.281,18.186,24.016,24.553,21.452,15.836,21.311,20.879,22.559,21.694,25.856,20.533,21.542,25.766,26.018,20.82,24.959,18.959,23.346,16.068,22.836,21.939,25.722,19.671,26.299,21.879,16.002,15.288,16.946,17.534,16.846,16.546,15.927,18.084,17.475]

    Key: 538: Value: wt: 1.0 distance: 1679.466786890618 vec: [24.538,24.28,28.281,27.132,26.662,32.11,32.81,30.483,35.859,25.387,31.301,25.429,26.866,30.852,24.478,25.665,25.296,30.263,29.657,25.295,25.022,35.264,26.109,9.6,12.675,16.575,19.76,13.349,18.137,7.993,16.751,16.341,15.349,9.476,9.943,16.609,12.331,8.645,19.457,10.836,10.349,9.726,14.575,18.959,15.822,17.364,11.915,13.762,12.402,19.628,19.644,11.524,15.419,12.67,13.116,8.235,12.042,19.31,12.999,17.46]

    Key: 538: Value: wt: 1.0 distance: 2049.45415174948 vec: [34.335,30.938,31.953,31.146,24.519,24.393,27.696,29.874,26.767,33.089,31.371,26.233,26.383,35.661,32.663,27.685,29.277,31.761,34.65,24.94,33.434,26.849,28.714,26.581,34.825,34.026,8.823,12.634,12.694,6.279,13.644,16.651,18.078,7.975,9.274,9.208,12.879,12.729,6.976,17.832,13.33,6.326,12.131,11.842,16.716,10.425,9.

```
445,14.4,15.696,11.028,10.608,15.19,9.076,17.909,9.846,15.013,13.913,11.743
,11.699,10.152]
Count: 600
```

## 任务3  决策树和随机森林的分类预测

### 【任务概述】

使用 Scala/Python 对鸢尾花（Iris）数据进行决策树分类模型训练和样本分类预测，绘制决策树图；使用随机森林进行模型训练和样本分类预测。

### 【任务分析】

#### 一、Iris 数据分析

① 鸢尾花（Iris）数据下载地址如下：

https://archive.ics.uci.edu/ml/machine-learning-databases/iris/iris.data

② 基本数据格式如下：

5.1,3.5,1.4,0.2,Iris-setosa

4.9,3.0,1.4,0.2,Iris-setosa

4.7,3.2,1.3,0.2,Iris-setosa

……

5.9,3.0,5.1,1.8,Iris-virginica

鸢尾花（Iris）的基本信息如下。

花的属性：萼片宽度（Sepal Width）、萼片长度（Sepal Length）、花瓣宽度（Petal Width）、花瓣长度（Petal Length）。

花的种类：山鸢尾（Setosa）、杂色鸢尾（Versicolor）、维吉尼亚鸢尾（Virginica）。

#### 二、主要函数

（1）DecisionTreeClassifier()函数

```
class sklearn.tree.DecisionTreeClassifier(criterion='gini', splitter='best',
max_depth=None, min_samples_split=2, min_samples_leaf=1, min_weight_fraction_leaf=0.0,
max_features=None, random_state=None, max_leaf_nodes=None, min_impurity_decrease=0.0,
min_impurity_split=None, class_weight=None, presort=False)
```

其中：

criterion：string 可选（默认为"gini"），衡量分类的质量。支持的标准有："gini"代表 gini impurity(不纯度)，"entropy"代表 information gain（信息增益）。

splitter:string 可选（默认为"best"），用于每个节点上选择的分类策略。支持的策略有"best"和"random"两种。

max_depth:int or None,可选（默认为"None"），表示树的最大深度。

min_samples_split:int,float 可选（默认为2），划分内部节点所需的最小样本数。

min_samples_leaf:int,float 可选（默认为1），一个叶节点所需要的最小样本数。

min_weight_fraction_leaf:float 可选（默认为 0），叶子节点所需要的最小权值。

max_features:int,float,string or None 可选（默认为 None）寻找最佳分类时考虑的特征数：

如果是 int，那么每次分类都要考虑 max_features 的特征值；

如果是 float，那么 max_features 是一个百分率并且分类时需要考虑的特征数是 int(max_features * n_features)；

如果是 "auto"，那么 max_features=sqrt(n_features)；

如果是 "sqrt"，那么 max_features=sqrt(n_features)；

如果是 "log2"，那么 max_features=log2(n_features)；

如果是 None，那么 max_features=n_features。

（2）RandomForestRegressor()函数

RandomForestRegressor(n_estimators=10, criterion='mse', max_depth=None, min_samples_split=2,min_samples_leaf=1, min_weight_fraction_leaf=0.0, max_features='auto', max_leaf_nodes=None, min_impurity_decrease=0.0, min_impurity_split=None, bootstrap=True, oob_score=False, n_jobs=1, random_state=None, verbose=0, warm_start=False)

其中：

n_estimators：int 可选（默认为 10），决策树的个数越多越好，但性能会越差，至少 100 左右可以达到可接受的性能和误差率。

criterion:string 可选（默认为 "mse"），测量拆分质量，支持的标准是均方误差 "MSE"。

bootstrap：boolean 可选（默认为 True），是否有放回的采样。

oob_score：boolean 可选（默认为 False），在不可见数据中是否使用 out-of-bag 样本估计 R^2。

n_jobs：int 可选（默认为 1），并行 job 个数。如果是-1，那么 CPU 有多少 core，就启动多少 job。

warm_start：boolean 可选（默认为 False），当设置为 True 时，重新使用先前的结果并增加更多的估算，否则使用新的森林。

（3）预测形式

predict_proba(x)：给出带有概率值的结果。每个点在所有 label 的概率和为 1。

predict(x)：直接给出预测结果。内部还是调用的 predict_proba()，根据概率的结果看哪个类型的预测值最高就是哪个类型。

predict_log_proba(x)：和 predict_proba 类似，只是对结果进行 log()处理。

## 【支撑知识】

### 一、决策树

（1）决策树简介

决策树（Decision Tree）是在已知各种情况发生概率的基础上，通过构成决策树来求取净现值的期望值大于等于零的概率，评价项目风险，判断其可行性的决策分析方法，是直观运用概率分析的一种图解法。由于这种决策分支画成图形很像一棵树的枝干，故称为决策树。

在机器学习中，决策树是一种预测模型，代表的是一种对象属性与对象值之间的映射关系，每一个节点代表某个对象，树中的每一个分叉路径代表某个可能的属性值，而每一个叶子节点则对应从根节点到该叶子节点所经历的路径所表示的对象的值。决策树仅有单一输出，如果有多个输出，可以分别建立独立的决策树以处理不同的输出。

决策树算法根据数据的属性采用树状结构建立决策模型，决策树模型常用来解决分类和回归问题。常见的算法包括分类及回归树（Classification And Regression Tree，CART）、ID3（Iterative Dichotomiser 3）、C4.5、Chi-squared Automatic Interaction Detection（CHAID）、Decision Stump、随机森林（Random Forest）、多元自适应回归样条（MARS）及梯度推进机（Gradient Boosting Machine，GBM）。

（2）ID3及基本原理

在信息论中，期望信息越小，信息增益就越大，从而纯度就越高。ID3算法的核心思想就是以信息增益来度量属性的选择，选择分裂后信息增益最大的属性进行分裂。该算法采用自顶向下的贪婪搜索遍历可能的决策空间。

在信息增益中，重要性的衡量标准就是特征能够为分类系统带来多少信息，带来的信息越多，该特征越重要。信息熵将其定义为离散随机事件出现的概率，一个系统越是有序，信息熵就越低，反之一个系统越是混乱，它的信息熵就越高。所以，信息熵可以被认为是系统有序化程度的一个度量。ID3算法主要针对属性选择问题，是决策树学习方法中最具影响和最为典型的算法。该方法是用信息增益度选择测试属性。

香农（Shannon）1948年提出信息论理论。事件 $a_i$ 的信息量 $I(a_i)$ 可以按如下方法进行度量：

$$I(a_i) = p(a_i) \log_2 \frac{1}{p(a_i)} = -p(a_i) \log_2 p(a_i)$$

式中，$p(a_i)$表示事件 $a_i$ 发生的概率。

假如有 $n$ 个互不相容的事件 $a_1, a_2, \cdots a_n$，它们中有且仅有一个发生，则其平均的信息量可以按如下方法进行度量：

$$I(a_1, a_2, \cdots a_n) = \sum_{i=1}^{n} I(a_i) = -\sum_{i=1}^{n} p(a_i) \log_2 p(a_2)$$

在决策树分类中，假设 $S$ 是训练样本集合，$|S|$ 是训练样本数，样本划分为 $n$ 个不同的类 $C_1, C_2, \cdots C_n$，这些类的大小分别标记为 $|C_1|, |C_2|, \cdots |C_n|$，则任意样本 $S$ 属于类 $C_i$ 的概率为

$$P(S_i) = \frac{|C_i|}{|S|}$$

其总的信息熵就可以表示为

$$\text{Entropy}(S) = I(c_1, c_2, \cdots, c_n) = -\sum_{i=1}^{n} \frac{|C_i|}{|S|} \log_2 \frac{|C_i|}{|S|}$$

设一个属性 $A$ 具有 $k$ 个不同的值 $\{a_1, a_2, \cdots a_k\}$，利用属性 $A$ 将训练样本集合 $S$ 划分为 $k$ 个子集 $\{S_1, S_2, \cdots S_k\}$。若选择属性 $A$ 为测试属性，则这些子集就是从集合 $S$ 的节点生长出来的新叶节点。根据属性 $A$ 划分样本的信息熵为

$$\text{Entropy}(S,A) = \sum \left(\frac{|S_v|}{|S|}\right) \times \text{Entropy}(S_v)$$

式中，$\sum$ 是属性 $A$ 的所有可能的值，$S_v$ 是属性 $A$ 有 $v$ 值的 $S$ 子集，$|S_v|$ 是 $S_v$ 中元素的个数，

|S|是S中元数的个数。

属性A划分样本集S后所得到的信息增益（Gain）为
$$Gain(S, A) = Entropy(S) - Entropy(S, A)$$

显然，$Entropy(S, A)$越小，$Gain(S, A)$的值就越大，说明测试属性对分类提供的信息越多，选择A之后对分类的不确定程度越小。属性A的k个不同的值对样本集S的k个子集或分支，通过递归调用，生成其他属性作为节点的子节点和分支来生成整棵决策树。

（3）ID3算法流程

ID3算法的具体流程如下：

① 对当前样本集合计算所有属性的信息增益。

② 选择信息增益最大的属性作为测试属性，把测试属性取值相同的样本划为同一个子样本集。

③ 若子样本集的类别属性只含有单个属性，则分支为叶子节点，判断其属性并标上相应的符号，然后返回调用处；否则，对子样本集递归调用本算法。

二、随机森林

（1）随机森林简介

随机森林（Random Forest，RF）指的是利用多棵树对样本进行训练并预测的一种分类器。该分类器最早由Leo Breiman和Adele Cutler提出，并被注册成了商标。

随机森林就是通过集成学习的思想将多棵树集成的一种算法，它的基本单元是决策树。每棵决策树都是一个分类器，那么对于一个输入样本，n棵树会有n个分类结果。而随机森林集成了所有的分类投票结果，将投票次数最多的类别指定为最终的输出，这就是一种最简单的bagging思想。

作为新兴的、高度灵活的一种机器学习算法，随机森林拥有广泛的应用前景，从市场营销到医疗保健保险，既可以用来做市场营销模拟的建模，统计客户来源、保留和流失，也可用来预测疾病的风险和病患者的易感性。

（2）随机森林的特点

随机森林是一种很灵活实用的方法，它具有如下特点：

① 在当前算法中，具有极高的准确率。

② 能有效运行在大数据集上。

③ 能够处理具有高维特征的输入样本，不需要降维。

④ 能评估各特征值在分类问题的重要性。

⑤ 在建立森林过程中，对泛化误差使用的是无偏估计。

⑥ 对于默认值问题也能够获得很好的结果。

⑦ 在不平衡数据集中，能够均衡误差。

⑧ 生成的森林可以保存在其他数据中使用。

实际上，随机森林的特点不仅仅只有这8点，它就相当于机器学习领域的多面手。

（3）随机森林工作流程

随机森林就是要构建一个森林，森林中的每棵树都是一棵决策树，其中每一棵决策树都是相互独立无关的。当有新的输入样本进入，就让每一棵决策树分别进行判断，看看这个样本应该属于哪一类，然后看看哪一类被选择最多，就预测这个样本为那一类。随机森林尤其适合用分布式并行计算来处理，建决策树的过程可以并行实现，其工作流程如下：

① 原始数据中随机做 $n$ 个抽样。

② 每个抽样都训练一棵决策树，作为根节点的样本。

③ 每个样本有 $M$ 个属性，分裂决策树的节点时，随机从 $M$ 中选择 $m$ 个属性（$m<M$），从 $m$ 个属性中选择 1 个属性作为分裂属性。

④ 形成决策树的过程中，每个节点按照步骤③来分裂，直到不能分裂为止。

⑤ 重复步骤①~④，建立随机森林。

（4）随机森林简单实例

根据已有的训练集已经生成了对应的随机森林，随机森林如何利用某一个人的年龄（Age）、性别（Gender）、教育情况（Education）、工作领域（Industry）及居住地（Residence）共 5 个字段来预测他的收入层次。

收入层次分 3 个段：

① 低于$40,000。

② $40,000~$150,000。

③ 高于$150,000。

随机森林中每一棵树都可以看做一棵CART(分类回归树)，这里假设森林中有 5 棵CART 树，总特征个数 $N$=5，取 $m$=1（这里假设每个CART 树对应一个不同的特征）。

CART 1：特征变量 Age

| | 工资收入层次 | 1 | 2 | 3 |
|---|---|---|---|---|
| Age | 低于 18 | 90% | 10% | 0% |
| | 19~27 | 85% | 14% | 1% |
| | 28~40 | 70% | 23% | 7% |
| | 40~55 | 60% | 35% | 5% |
| | 高于 55 | 70% | 25% | 5% |

CART 2：特征变量 Gender

| | 工资收入层次 | 1 | 2 | 3 |
|---|---|---|---|---|
| Gender | Male | 70% | 27% | 3% |
| | Female | 75% | 24% | 1% |

CART 3：特征变量 Education

| | 工资收入层次 | 1 | 2 | 3 |
|---|---|---|---|---|
| Education | <=High School | 85% | 10% | 5% |
| | Diploma | 80% | 14% | 6% |
| | Bachelor | 77% | 23% | 0% |
| | Master | 62% | 35% | 3% |

CART 4：特征变量 Residence

| Residence | 工资收入层次 | 1 | 2 | 3 |
|---|---|---|---|---|
| | Metro | 70% | 20% | 10% |
| | Non-Metro | 65% | 20% | 15% |

CART 5：特征变量 Industry

| Industry | 工资收入层次 | 1 | 2 | 3 |
|---|---|---|---|---|
| | Finance | 65% | 30% | 5% |
| | Manufacturing | 60% | 35% | 5% |
| | Others | 75% | 20% | 5% |

假设某人信息如下：年龄（Age）35岁，男（Male），大学（Diploma），从事制造业（Manufacturing），在大都市（Metro）工作。

根据这 5 棵 CART 树的分类结果，建立收入层次的分布情况如下：

| CART | 层次 | 1 | 2 | 3 |
|---|---|---|---|---|
| Age | 28~40 | 70% | 23% | 7% |
| Gender | Male | 70% | 27% | 3% |
| Education | Diploma | 80% | 14% | 6% |
| Industry | Manufacturing | 60% | 35% | 5% |
| Residence | Metro | 70% | 20% | 10% |
| 最后估值 | | 70% | 24% | 6% |

最后得出结论，这个人的收入层次 70%是一等，大约 24%为二等，6%为三等，所以，此人工资收入归入①段，收入小于$40,000。

## 【任务实施】

### 一、决策树分类

（1）使用 Scala 语言实现

① 分割鸢尾花（Iris）数据。

训练集：tree1.data，占 80%，行：1-40 51-90 101-140
测试集：tree2.data，占 20%，行：41-50 91-100 141-150

② 上传数据，操作如下：

```
hadoop@sw-desktop:~$ hdfs dfs -mkdir /data
hadoop@sw-desktop:~$ hdfs dfs -put tree*.dat /data
```

③ 打开 Scala IDE 创建项目 DecisionTree，并创建 DecisionTree.scala 类，代码如下：

```
import org.apache.log4j.{Level, Logger}
import org.apache.spark.mllib.feature.HashingTF
import org.apache.spark.mllib.linalg.Vectors
import org.apache.spark.mllib.regression.LabeledPoint
import org.apache.spark.sql.SparkSession
```

```scala
import org.apache.spark.SparkContext
import org.apache.spark.mllib.tree
import org.apache.spark.mllib.tree.model.DecisionTreeModel
object DecisionTree {
  def main(args: Array[String]) {
    val spark = SparkSession
      .builder
      .appName("DecisionTree")
      .getOrCreate()
    val sc = spark.sparkContext
    Logger.getRootLogger.setLevel(Level.WARN)
    //训练数据
    val lines1 = spark.read.textFile("/data/tree1.data").rdd
    //测试数据
    val lines2 = spark.read.textFile("/data/tree2.data").rdd
    //转换成向量
    val tree1 = lines1.map { line =>
      val parts = line.split(',')
      LabeledPoint(
        if(parts(4)=="Iris-setosa") 0.toDouble
        else if (parts(4)=="Iris-versicolor") 1.toDouble
        else 2.toDouble ,
        Vectors.dense(
          parts(0).toDouble,
          parts(1).toDouble,
          parts(2).toDouble,
          parts(3).toDouble))
    }
    val tree2 = lines1.map { line =>
      val parts = line.split(',')
      LabeledPoint(
        if(parts(4)=="Iris-setosa") 0.toDouble
        else if (parts(4)=="Iris-versicolor") 1.toDouble
        else 2.toDouble ,
        Vectors.dense(
          parts(0).toDouble,
          parts(1).toDouble,
          parts(2).toDouble,
          parts(3).toDouble))
    }
    //赋值
    val (trainingData, testData) = (tree1, tree2)
```

```scala
    //分类
    val numClasses = 3
    val categoricalFeaturesInfo = Map[Int, Int]()
    val impurity = "gini"
    //树的最大层次
    val maxDepth = 5
    //特征最大装箱数
    val maxBins = 32
    //模型训练
    val model = tree.DecisionTree.trainClassifier(
       trainingData, numClasses, categoricalFeaturesInfo,
       impurity,maxDepth, maxBins)
    //模型预测
    val labelAndPreds = testData.map { point =>
      val prediction = model.predict(point.features)
      (point.label, prediction)
    }
    //树的错误率
    val testErr = labelAndPreds.filter(r => r._1 != r._2)
                  .count.toDouble /testData.count()
    println("Test Error = " + testErr)
    //打印树的判断值
    println("Learned classification tree model:\n" + model.toDebugString)
    //保存 model 和装载 model
    //model.save(sc, "/data/treedtm")
    // val sameModel = DecisionTreeModel.load(sc, "/data/treedtm ")
    spark.stop()
  }
}
```

④ 导出 jar 包,运行结果如下:

```
hadoop@sw-desktop:~/workspace1$ spark-submit --master spark://master:7077
--class DecisionTree DecisionTree.jar
   ……
Test Error = 0.0
Learned classification tree model:
DecisionTreeModel classifier of depth 5 with 17 nodes
  If (feature 2 <= 1.9)
   Predict: 0.0
  Else (feature 2 > 1.9)
   If (feature 3 <= 1.7)
    If (feature 2 <= 4.9)
```

```
      If (feature 3 <= 1.6)
       Predict: 1.0
      Else (feature 3 > 1.6)
       Predict: 2.0
     Else (feature 2 > 4.9)
      If (feature 3 <= 1.5)
       Predict: 2.0
      Else (feature 3 > 1.5)
       If (feature 0 <= 6.7)
        Predict: 1.0
       Else (feature 0 > 6.7)
        Predict: 2.0
   Else (feature 3 > 1.7)
    If (feature 2 <= 4.8)
     If (feature 0 <= 5.9)
      Predict: 1.0
     Else (feature 0 > 5.9)
      Predict: 2.0
    Else (feature 2 > 4.8)
     Predict: 2.0
```

（2）使用 Python 语言实现

① DecisionTree.py 代码如下：

```
#coding=utf-8
import numpy as np
import matplotlib.pyplot as plt
from sklearn.datasets import load_iris
from sklearn.tree import DecisionTreeClassifier
# 载入iris数据
iris = load_iris()
# 150 个样本，每个样本有 4 个特征
# print iris.data.shape
# 分割数据集 构造训练集/测试集，120/30
# 80%训练   1-40  51-90   101-140
# 20%预测   41-50  91-100  141-150
# 训练集
train_data = np.concatenate((iris.data[0:40, :], iris.data[50:90, :], iris.data[100:140, :]), axis = 0)
# 训练集样本类别
train_target = np.concatenate((iris.target[0:40], iris.target[50:90], iris.target[100:140]), axis = 0)
# 测试集
```

```
    test_data = np.concatenate((iris.data[40:50, :], iris.data[90:100, :],
iris.data[140:150, :]), axis = 0)
    # 测试集样本类别
    test_target = np.concatenate((iris.target[40:50], iris.target[90:100],
iris.target[140:150]), axis = 0)
    # 训练决策树
    clf = DecisionTreeClassifier(criterion='gini',max_depth=5,max_leaf_nodes=32)
    clf.fit(train_data, train_target)
    print clf
    # 预测结果
    predict_target = clf.predict(test_data)
    print predict_target
    # 预测结果与真实结果比对
    print sum(predict_target == test_target)
    # 输出准确率、召回率、F值
    from sklearn import metrics
    print(metrics.classification_report(test_target, predict_target))
    print(metrics.confusion_matrix(test_target, predict_target))
    # 可视化训练好的决策树
    from sklearn.externals.six import StringIO
    from sklearn.tree import export_graphviz
    with open("iris.dot", 'w') as f:
        f = export_graphviz(clf, out_file=f)
```

② 运行 pyspark，操作如下：

```
hadoop@sw-desktop:~$ pyspark
Python 2.7.12 (default, Nov 19 2016, 06:48:10)
……
Welcome to
      ____              __
     / __/__  ___ _____/ /__
    _\ \/ _ \/ _ `/ __/  '_/
   /__ / .__/\_,_/_/ /_/\_\   version 2.1.0
      /_/

Using Python version 2.7.12 (default, Nov 19 2016 06:48:10)
SparkSession available as 'spark'.
>>>
```

③ 在 pyspark 中执行代码，结果如下：

```
DecisionTreeClassifier(class_weight=None, criterion='gini', max_depth=5,
            max_features=None, max_leaf_nodes=32, min_impurity_split=1e-07,
```

```
            min_samples_leaf=1, min_samples_split=2,
            min_weight_fraction_leaf=0.0, presort=False, random_state=None,
            splitter='best')
[0 0 0 0 0 0 0 0 0 0 1 1 1 1 1 1 1 1 1 1 2 2 2 2 2 2 2 2 2 2]
30
             precision    recall   f1-score   support
          0      1.00       1.00      1.00        10
          1      1.00       1.00      1.00        10
          2      1.00       1.00      1.00        10
avg / total      1.00       1.00      1.00        30


[[10  0  0]
 [ 0 10  0]
 [ 0  0 10]]
```

④ 查看训练好的决策树，如图 10-6 所示。

```
$ sudo apt-get install graphviz
$ dot -Tpng iris.dot -o tree.png    # 生成 png 图片
$ dot -Tpdf iris.dot -o tree.pdf    # 生成 pdf
```

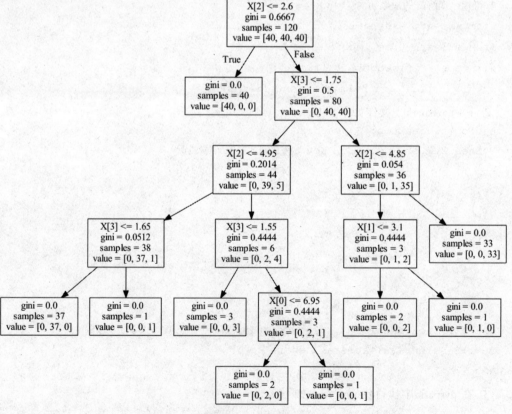

图 10-6　决策树图

## 二、随机森林预测

（1）使用 Scala 语言实现

① RandomForest.scala 代码如下：

```scala
import org.apache.log4j.{Level, Logger}
import org.apache.spark.SparkContext
import org.apache.spark.sql.SparkSession
import org.apache.spark.mllib.evaluation.MulticlassMetrics
import org.apache.spark.mllib.linalg.Vectors
import org.apache.spark.mllib.regression.LabeledPoint
import org.apache.spark.mllib.tree
import org.apache.spark.mllib.tree.model.RandomForestModel
import org.apache.spark.rdd.RDD
object RandomForest {
   /**
    * @param model 随机森林模型
    * @param data  用于交叉验证的数据集
    * */
   def getMetrics(model:RandomForestModel,data: RDD[LabeledPoint]): MulticlassMetrics = {
      //将交叉验证数据集的每个样本的特征向量交给模型预测，并和原本正确的目标特征组成一个tuple
      val predictionsAndLables = data.map { d =>
         (model.predict(d.features), d.label)
      }
      //将结果交给MulticlassMetrics,其可以以不同的方式计算分配器预测的质量
      new MulticlassMetrics(predictionsAndLables)
   }
   /**
    * 在训练数据集上得到最好的参数组合
    * @param trainData 训练数据集
    * @param cvData 交叉验证数据集
    * */
   def getBestParam(trainData: RDD[LabeledPoint], cvData: RDD[LabeledPoint]): Unit = {
      val evaluations = for (impurity <- Array("gini", "entropy");
                   depth <- Array(1, 20);
                   bins <- Array(10, 300)) yield {
         val model = tree.RandomForest.trainClassifier(trainData, 3,
            Map[Int, Int](), 20, "auto", impurity, depth, bins)
         val metrics = getMetrics(model, cvData)
         ((impurity, depth, bins), metrics.precision)
      }
      evaluations.sortBy(_._2).reverse.foreach(println)
```

```scala
    }
    def main(args: Array[String]) {
      val spark = SparkSession
        .builder
        .appName("RandomForest")
        .getOrCreate()
    val sc = spark.sparkContext
    Logger.getRootLogger.setLevel(Level.WARN)
    //iris 数据
    val rawData = spark.read.textFile("/data/iris.data").rdd
    //转换成向量
    val data = rawData.map{ line =>
      val parts = line.split(',')
      LabeledPoint(
        if(parts(4)=="Iris-setosa") 0.toDouble
        else if (parts(4)=="Iris-versicolor") 1.toDouble
        else 2.toDouble ,
        Vectors.dense(
          parts(0).toDouble,
          parts(1).toDouble,
          parts(2).toDouble,
          parts(3).toDouble))
    }
    //训练集、交叉验证集和测试集各占 80%、10%和 10%
    //10%的交叉验证数据集的作用是确定在训练数据集上训练出来的模型的最好参数
    //测试数据集的作用是评估 CV 数据集的最好参数
    val Array(trainData, cvData, testData)
      = data.randomSplit(Array(0.8, 0.1, 0.1))
    trainData.cache()
    cvData.cache()
    testData.cache()
    //随机森林训练参数设置
    //分类数
    val numClasses =3
    //categoricalFeaturesInfo 为空，意味着所有的特征为连续型变量
    val categoricalFeaturesInfo = Map[Int, Int]()
    //树的个数
    val numTrees = 20
    //特征子集采样策略，auto 表示算法自主选取
    val featureSubsetStrategy = "auto"
    //纯度计算
    val impurity = "gini"
    //树的最大层次
```

```
    val maxDepth = 4
    //特征最大装箱数
    val maxBins = 32
    //训练随机森林分类器,trainClassifier 返回的是 RandomForestModel 对象
    //构建随机森林,分类为 3
    val model = tree.RandomForest.trainClassifier(
      trainData, numClasses, categoricalFeaturesInfo,
      numTrees, featureSubsetStrategy, impurity, maxDepth, maxBins)
    val metrics = getMetrics(model, cvData)
    println("--- confusionMatrix ---")
    //混淆矩阵和模型精确率
    println(metrics.confusionMatrix)
    println("--- precision ---")
    println(metrics.precision)
    println("--- (precision,recall) ---")
    //每个类别对应的精确率与召回率
    (0 until 3).map(target => (metrics.precision(target), metrics.recall(target))).foreach(println)
    //保存模型
    model.save(sc,"/data/RFModel")
    //获取最好的参数组合
    getBestParam(trainData, cvData)
    //读取模型
    val rfModel = RandomForestModel.load(sc,"/data/RFModel")
    //进行预测
    val dataAndPreLable = testData.map { point =>
      val prediction = rfModel.predict(point.features)
      (point.label, prediction)
    }
    dataAndPreLable .take(10)
    //树的错误率
    val testErr = dataAndPreLable.filter(r => r._1 != r._2)
      .count.toDouble / testData.count()
    println("Test Error = " + testErr)
    spark.stop()
  }
}
```

② 上传数据并运行,操作如下:

```
hadoop@sw-desktop:~/workspace1$ hdfs dfs -put iris.dat /data
hadoop@sw-desktop:~/workspace1$ spark-submit --master spark://master:7077 --class RandomForest RandomForest.jar
```

③ 运行结果如下:

```
hadoop@sw-desktop:~/workspace1$ spark-submit --master spark://master:7077
```

```
--class RandomForest RandomForest.jar
    ……
    --- confusionMatrix ---
    7.0  0.0  0.0
    0.0  7.0  1.0
    0.0  1.0  6.0
    --- precision ---
    0.9090909090909091
    --- (precision,recall) ---
    (1.0,1.0)
    (0.875,0.875)
    (0.8571428571428571,0.8571428571428571)
    ……
    ((entropy,20,300),0.9090909090909091)
    ((entropy,20,10),0.9090909090909091)
    ((entropy,1,10),0.9090909090909091)
    ((gini,20,300),0.9090909090909091)
    ((gini,20,10),0.9090909090909091)
    ((gini,1,300),0.9090909090909091)
    ((gini,1,10),0.9090909090909091)
    ((entropy,1,300),0.6363636363636364)
    Test Error = 0.0
```

（2）使用 Python 语言实现

① RandomForest.py 代码如下：

```
#coding=utf-8
from sklearn.tree import DecisionTreeRegressor
from sklearn.ensemble import RandomForestRegressor
import numpy as np
from sklearn.datasets import load_iris
# 载入 iris 数据
iris = load_iris()
print iris['target'].shape
# 随机森林回归
rf = RandomForestRegressor()
# 进行模型的训练
rf.fit(iris.data[:150], iris.target[:150])
# 随机挑选 3 个预测不相同的样本
instance = iris.data[[20, 80, 140]]
print instance
print '样本预测: ', rf.predict(instance)
print iris.target[20], iris.target[80], iris.target[140]
```

```
print iris.target_names[iris.target[20]],iris.target_names[iris.target[80]],
iris.target_names[iris.target[140]]
```

② 在 pyspark 执行代码，结果如下：

```
(150,)
[[ 5.4  3.4  1.7  0.2]
 [ 5.5  2.4  3.8  1.1]
 [ 6.7  3.1  5.6  2.4]]
样本预测： [ 0.  1.  2.]
0 1 2
setosa versicolor virginica
```

③ 拆分每一维特征的贡献值，代码如下：

```
from treeinterpreter import treeinterpreter as ti
prediction, bias, contributions = ti.predict(rf, instance)
print "预测： ", prediction
print "先前训练集偏差： ", bias
print "每一维特征的贡献值： "
for c, feature in zip(contributions[0], iris.feature_names):
    print feature, c
```

④ 各节点安装 Treeinterpreter 模块，操作如下：

```
hadoop@...:~$ sudo pip install treeinterpreter
```

⑤ 运行结果如下：

```
预测： [ 0.  1.  2.]
先前训练集偏差： [ 0.976  0.976  0.976]
每一维特征的贡献值：
sepal length (cm) 0.0
sepal width (cm) 0.0
petal length (cm) -0.400666666667
petal width (cm) -0.575333333333
```

## 任务 4　频繁项集计算与关联分析

### 【任务概述】

使用 Apriori 算法对交易数据库中的数据进行频繁项集及其支持度计算，进行关联分析，计算置信度；使用 FP-Tree 算法计算频繁项集及其支持度，进行关联分析，计算置信度。

### 【支撑知识】

#### 一、关联规则

（1）基本概念

从大规模数据集中寻找物品间的隐含关系，被称为关联分析（Association Analysis）或者关联规则学习（Association Rule Learning）。

关联规则挖掘（Association Rule Mining）是数据挖掘中最活跃的研究方法之一，可以用来发现事情之间的联系，最早是为了发现超市交易数据库中不同的商品之间的关系，如表10-3所示。

表10-3 某超市交易数据库

| 交易号TID | 顾客购买的商品 | 交易号TID | 顾客购买的商品 |
| --- | --- | --- | --- |
| T1 | 面包、奶酪、牛奶、茶 | T6 | 面包、茶 |
| T2 | 面包、奶酪、牛奶 | T7 | 啤酒、牛奶、茶 |
| T3 | 蛋糕、牛奶 | T8 | 面包、茶 |
| T4 | 面包、牛奶、茶 | T9 | 面包、奶酪、牛奶、茶 |
| T5 | 面包、蛋糕、牛奶 | T10 | 面包、牛奶、茶 |

定义1：设 $I = \{ i_1, i_2, \cdots, i_m \}$ 是 $m$ 个不同项目的集合，每个 $i_k$ 称为一个项目。项目的集合 $I$ 称为项集。其元素的个数称为项集的长度，长度为 $k$ 的项集称为 $k$-项集。引例中每个商品就是一个项目，项集为 $I = \{$ 面包，啤酒，蛋糕，奶酪，牛奶，茶 $\}$，$I$ 的长度为6。

定义2：每笔交易 $T$ 是项集 $I$ 的一个子集。对应每一个交易有一个唯一标识交易号，记作TID。交易全体构成了交易数据库 $D$，$|D|$ 等于 $D$ 中交易的个数。引例中包含10笔交易，因此，$|D|=10$。

定义3：对于项集 $X$，设定 $count(X \subseteq T)$ 为交易集 $D$ 中包含 $X$ 的交易的数量，则项集 $X$ 的支持度为 $support(X) = count(X \subseteq T)/|D|$。

例如，$X = \{$ 面包，牛奶 $\}$ 出现在 $T1$、$T2$、$T6$、$T5$、$T9$ 和 $T10$ 中，所以，支持度为0.6。

定义4：最小支持度是项集的最小支持阈值，记为SUPmin，代表了用户关心的关联规则的最低重要性。支持度不小于SUPmin的项集称为频繁项集，长度为 $k$ 的频繁集称为 $k$-频繁项集。如果设定SUPmin为0.3，例如，$\{$ 面包，牛奶 $\}$ 的支持度是0.6，所以，是2-频繁项集。

定义5：关联规则是一个蕴含式：$R: X \Rightarrow Y$。

其中 $X \subset I$，$Y \subset I$，并且 $X \cap Y = \Phi$。表示项集 $X$ 在某一交易中出现，则导致 $Y$ 以某一概率也会出现。用户关心的关联规则，可以用两个标准来衡量：支持度和置信度。

定义6：关联规则 $R$ 的支持度是交易集同时包含 $X$ 和 $Y$ 的交易数与 $|D|$ 之比。即

$$support(X \Rightarrow Y) = count(X \cup Y)/|D|$$

支持度反映了 $X$、$Y$ 同时出现的概率。关联规则的支持度等于频繁集的支持度。

定义7：对于关联规则 $R$，置信度是指包含 $X$ 和 $Y$ 的交易数与包含 $X$ 的交易数之比。即

$$confidence(X \Rightarrow Y) = support(X \Rightarrow Y)/support(X)$$

置信度反映了如果交易中包含 $X$，则交易包含 $Y$ 的概率。一般来说，只有支持度和置信度较高的关联规则才是用户感兴趣的。

例如，$X = \{$ 面包 $\}$ 出现在 $T1$、$T2$、$T4$、$T5$、$T6$、$T8$、$T9$ 和 $T10$ 中，$X \cap Y = \{$ 面包，牛奶 $\}$ 出现在 $T1$、$T2$、$T4$、$T5$、$T9$ 和 $T10$ 中，所以，置信度为 $(X \cap Y)/X = 0.75$，认为购买面包和购买牛奶之间存在关联。

定义8：设定关联规则的最小支持度和最小置信度为SUPmin和CONFmin。规则 $R$ 的支持度和置信度均不小于SUPmin和CONFmin，则称为强关联规则。关联规则挖掘的目的就

是找出强关联规则,从而指导商家的决策。

这 8 个定义包含了关联规则相关的几个重要基本概念,关联规则挖掘主要有两个问题:
- 找出交易数据库中所有大于或等于用户指定的最小支持度的频繁项集。
- 利用频繁项集生成所需要的关联规则,根据用户设定的最小置信度筛选出强关联规则。

(2)挖掘过程

关联规则挖掘过程主要包含两个阶段:第一阶段必须先从项集中找出所有的频繁项集(Frequent Itemsets),第二阶段再由这些频繁项集中产生关联规则(Association Rules)。

## 二、Apriori 算法

(1)Apriori 算法简介

Apriori 算法是一种最有影响的挖掘关联规则频繁项集的算法。其主要思想是找出存在于事务数据集中的最大的频繁项集,再利用得到的最大频繁项集与预先设定的最小置信度阈值生成强关联规则。

Apriori 算法是发现频繁项集的一种方法。Apriori 算法的两个输入参数分别是最小支持度和数据集。该算法首先会生成所有单个元素的项集列表。接着扫描数据集来查看哪些项集满足最小支持度要求,那些不满足最小支持度的集合会被去掉。然后,对剩下来的集合进行组合以生成包含两个元素的项集。接下来,再重新扫描交易记录,去掉不满足最小支持度的项集。该过程重复进行直到所有项集都被去掉。

(2)Apriori 算法的伪代码

```
// 找出 1-频繁项集
L[1] =find_frequent_1-itemsets(D);
For(k=2;L[k-1] !=null;k++){
    //产生候选,并剪枝
    C[k] =apriori_gen(L[k-1]);
    //扫描 D 进行候选计数
    For each 事务 t in D {
        C[t] =subset(C[k],t); //得到 t 的子集
        For each 候选 c 属于 C[t]
            c.count++;
    }
    //返回候选项集中不小于最小支持度的项集
    L[k] ={c 属于 C[k] | c.count>=min_sup}
}
Return L[1..k] = 所有的频繁项集;

//连接(join)
Procedure apriori_gen (L[k-1] :frequent[k-1]-itemsets)
    For each 项集 L1 属于 L[k-1]
        For each 项集 L2 属于 L[k-1]
            If( L1==L2 ) then{
                c = L1 连接 L2      //连接步:产生候选
```

```
                //若k-1项集中已经存在子集c则进行剪枝
                if has_infrequent_subset(c, L[k-1] ) then
                    delete c;  //剪枝步:删除非频繁候选
                else
                    add c to C[k];
            }
        Return C[k];
//剪枝(prune)
Procedure has_infrequent_sub (c:candidate k-itemset; L[k-1] :frequent[k-1]-itemsets)
        For each (k-1)-subset s of c
            If s 不属于 L[k-1] then
                Return true;
        Return false;
```

### 三、FP-Tree 算法

（1）FP-Tree 算法简介

Apriori 算法在产生频繁模式完全集前需要对数据库进行多次扫描，同时产生大量的候选频繁集，这就使 Apriori 算法时间和空间复杂度较大。针对 Apriori 算法的固有缺陷，Jiawei Han 提出了 FP-Growth 算法，它采取如下分治策略：将提供频繁项集的数据库压缩到一棵频繁模式树（FP-Tree），但仍保留项集关联信息。FP-Tree 将事务数据表中的各个事务数据项按照支持度排序后，把每个事务中的数据项按降序依次插入到一棵以 NULL 为根结点的树中，同时在每个结点记录该结点出现的支持度。

FP 的全称是 Frequent Pattern，在算法中使用了一种称为频繁模式树（Frequent Pattern Tree）的数据结构。FP-Tree 是一种特殊的前缀树，由频繁项头表和项前缀树构成。所谓前缀树，是一种存储候选项集的数据结构，树的分支用项名标识，树的节点存储后缀项，路径表示项集。

FP-Growth 算法的工作流程如下：首先构建 FP 树，然后利用它来挖掘频繁项集。为构建 FP 树，需要对原始数据集扫描两遍。第一遍对所有元素项的出现次数进行计数。数据库的第一遍扫描用来统计出现的频率，而第二遍扫描中只考虑那些频繁元素。

（2）FP-Tree 构建

① 扫描事务数据库，找到事务中的元素项，依照出现的次数排序,L = { z:5, x:4, r:3, s:3, t:3, y:3, p:2, h:1, q:2, e:1, m:1, j:1, n:1, o:1, u:1, v:1, w:1 }。

② 扫描数据库，将每笔事务中的元素项过滤，去掉不满足最小支持度的元素项（假设最小支持度计数为 3），然后对元素项进行重排序，ordered frequent itemset = { z:5, x:4, r:3, s:3, t:3, y:3 }，如表 10-4 所示，并开始构建 Tree。

表 10-4 事务数据库

| TID | 事务中的元素项 | 过滤及重排序后的事务 |
| --- | --- | --- |
| 1 | r, z, h, j, p | z, r |
| 2 | z, y, x, w, v, u, t, s | z, x, y, s, t |
| 3 | z | z |
| 4 | r, x, n, o, s | x, s, r |
| 5 | y, r, x, z, q, t, p | z, x, y, r, t |
| 6 | y, z, x, e, q, s, t, m | z, x, y, s, t |

● 构建 FP-Tree，从空集开始，将过滤和重排序后的频繁项集一次添加到树中。如果树中已存在现有元素，则增加现有元素的值；

如果现有元素不存在，则向树添加一个分支，如图 10-7 所示。

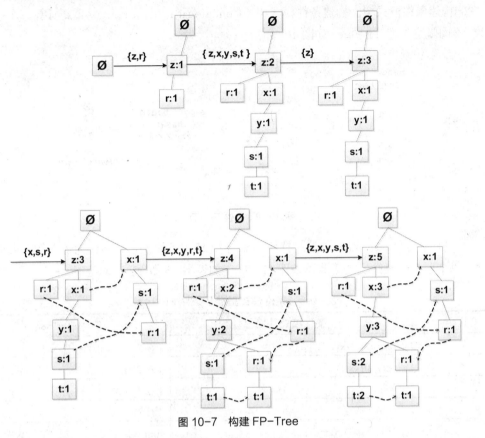

图 10-7　构建 FP-Tree

- 带头指针表的 FP-Tree，头指针表作为一个起始指针来发现相似元素项，每个元素项都构成一条单链表，如图 10-8 所示。

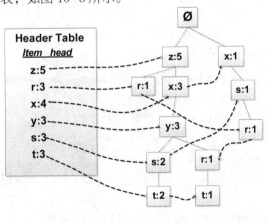

图 10-8　FP-Tree

（3）挖掘频繁项集

① 构建条件模式基准。从头指针表中的每个频繁元素项开始，对每个元素项，获得其对应的条件模式基准（Conditional Pattern Base）。条件模式基准是以所查找元素项为结尾的路径集合。每一条路径其实都是一条前缀路径（Prefix Path）。简而言之，一条前缀路径是介于所

查找元素项与树根节点之间的所有内容。

② 构建条件模式树。构建条件模式树（Conditional Pattern Tree），重新计算每个项目的次数，例如，s-条件 FP-Tree 如图 10-9 所示。

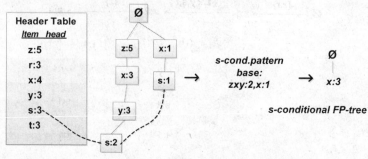

图 10-9　条件 FP-Tree

对于每一个频繁项，都要创建一棵条件 FP 树。使用条件模式基准作为输入数据来构建这些树。每个项目的条件模式基准及条件 FP-Tree 如表 10-5 所示。

表 10-5　条件模式基准及条件 FP-Tree

| Item | 条件模式基准 | 条件 FP-Tree |
| --- | --- | --- |
| t | { (zxys:2), (zxyr:1) } | { (z:3, x3, y3) }\|t |
| s | { (zxy:2), (x:1) } | { (x:3) }\|s |
| y | { (zx:3) } | { (z:3, x:3) }\|y |
| x | { (z:3) } | { (z:3) }\|x |
| r | { (z:1), (xs:1), (zxy:1) } | Empty |
| z | Empty | Empty |

## 【任务实施】

### 一、Apriori 算法

① Apriori 算法代码如下：

```
import org.apache.spark.sql.SparkSession

//返回满足最小支持度的项集的集合(频繁项集)和支持度
def freq(C1:Set[String],dataset:Array[Set[String]],numItems:Int,minSupport:Double) = { if(dataset.filter(d => C1.subsetOf(d)).size/numItems.toDouble >= minSupport) (C1,dataset.filter(d => C1.subsetOf(d)).size/numItems.toDouble) }

def apriori(data: Array[Array[String]], minSupport:Double) = {
    //构建集合表示的数据集 dataset
    val dataset = data.map(_.toSet)

    //数据集 dataset 长度
    var numItems=data.length
```

```scala
    //构建初始候选项集的列表, 即所有候选项集只包含一个元素
    var tmp=sc.parallelize(data)
    val C1 = tmp.flatMap(_.toSet).distinct().collect().map(Set(_))

    //找出1-频繁项集和支持度
    var suppdata =C1.map(freq(_,dataset,numItems,minSupport)).filter(_.!=(()))

    //定义频繁项集
    var L = Array[Array[Set[String]]]()
    val L1 = suppdata.map(_ match{case a:Tuple2[_,_] => a._1 match{ case b:Set[_] => b.asInstanceOf[Set[String]]}})

    L = L :+ L1
    //最初的L1中的每个项集含有一个元素, 新生成的
    //项集应该含有两个元素, 所以, k=2
    var k=2
    while(L(k-2).length>0){
        var Ck = Array[Set[String]]()
        //产生候选项
        for((var1,index) <- L(k-2).zipWithIndex;var2 <- L(k-2).drop(index+1)
            if  var1.take(k-2).equals(var2.take(k-2))){ Ck  =  Ck  :+ (var1|var2) }
        //找出候选项中不小于最小支持度的项集和支持度
        val suppdata_temp = Ck.map(freq(_,dataset,numItems,minSupport)).filter(_.!=(()))
        //符合条件的候选项并入suppdata
        suppdata ++= suppdata_temp
        L = L :+ suppdata_temp.map(_ match{case a:Tuple2[_,_] => a._1 match{ case b:Set[_] => b.asInstanceOf[Set[String]]}})
        k += 1
    }
    L = L.filter(_.nonEmpty)
    (L,suppdata)
}

val spark = SparkSession.builder.appName("Apriori").getOrCreate()
val sc = spark.sparkContext
// 表10-3中的数据
// 数据 { 面包, 啤酒, 蛋糕, 奶酪, 牛奶, 茶 }
// 转为 { a, b, c, d, e, f }
var data = Array(Array("a","d","e","f"),Array("a","d","e"),Array("c","e"),
```

```
Array("a","e","f"),Array("a","c","e"),Array("a","f"),Array("b","e","f"),
Array("a","f"),Array("a","d","e","f"),Array("a","e","f"))

//最小支持度
var minSupport = 0.3
val apr = apriori(data, minSupport)
val L=apr._1
val S=apr._2
println(L.deep.mkString("\n"))
println(S.deep.mkString("\n"))
```

Apriori 运行结果如下:

```
//频繁项集 L:
Array(Set(d), Set(e), Set(a), Set(f))
Array(Set(d, e), Set(d, a), Set(e, a), Set(e, f), Set(a, f))
Array(Set(d, e, a), Set(e, a, f))
//频繁项集及支持度 S:
(Set(d),0.3)
(Set(e),0.8)
(Set(a),0.8)
(Set(f),0.7)
(Set(d, e),0.3)
(Set(d, a),0.3)
(Set(e, a),0.6)
(Set(e, f),0.5)
(Set(a, f),0.6)
(Set(d, e, a),0.3)
(Set(e, a, f),0.4)
```

② 关联规则代码如下:

```
import org.apache.spark.mllib.fpm.AssociationRules
import org.apache.spark.mllib.fpm.FPGrowth.FreqItemset

val freqItemsets = sc.parallelize(Seq(
  new FreqItemset(Array("a"),8L),
  new FreqItemset(Array("d"),3L),
  new FreqItemset(Array("e"),8L),
  new FreqItemset(Array("f"),7L),
  new FreqItemset(Array("a","d"),3L),
  new FreqItemset(Array("a","e"),6L),
  new FreqItemset(Array("a","f"),6L),
  new FreqItemset(Array("d","e"),3L),
```

```
  new FreqItemset(Array("e","f"),5L),
  new FreqItemset(Array("a","d","e"),3L),
  new FreqItemset(Array("a","e","f"),4L)
))
val ar =new AssociationRules()
  .setMinConfidence(0.5)
val results = ar.run(freqItemsets)

results.collect().foreach { rule =>
  println("["+ rule.antecedent.mkString(",")
    +"=>"
    + rule.consequent.mkString(",")+"],"+ rule.confidence)
}
```

AssociationRules 运行结果如下:

```
[d,e=>a],1.0
[a,f=>e],0.6666666666666666
[a,d=>e],1.0
[a=>e],0.75
[a=>f],0.75
[d=>a],1.0
[d=>e],1.0
[e=>a],0.75
[e=>f],0.625
[e,f=>a],0.8
[f=>a],0.8571428571428571
[f=>e],0.7142857142857143
[a,e=>d],0.5
[a,e=>f],0.6666666666666666
```

例如，[a=>e],0.75 表示 a 发生能够推出 e 发生的置信度为 0.75，即买面包同时也买牛奶的置信度为 0.75。

### 二、FP-Tree 算法

① FP-Tree 代码如下：

```
import org.apache.spark.mllib.fpm.FPGrowth
import org.apache.spark.rdd.RDD
//表10-3中的数据
var data = Array(Array("a","d","e","f"),Array("a","d","e"),Array("c","e"),
Array("a","e","f"),Array("a","c","e"),Array("a","f"),Array("b","e","f"),
Array("a","f"),Array("a","d","e","f"),Array("a","e","f"))
val transactions:RDD[Array[String]] = sc.parallelize(data).cache()
```

```
val fpg =new FPGrowth()
  .setMinSupport(0.3)
  .setNumPartitions(10)
val model = fpg.run(transactions)

model.freqItemsets.collect().foreach { itemset =>
  println(itemset.items.mkString("[",",","]")+", "+ itemset.freq)
}
```

```
val minConfidence =0.5
model.generateAssociationRules(minConfidence).collect().foreach { rule =>
  println(
    rule.antecedent.mkString("[",",","]")
      +" => "+ rule.consequent .mkString("[",",","]")
      +", "+ rule.confidence)
}
```

② 运行结果如下。

● 生成的频繁项集：

```
[e], 8
[a], 8
[a,e], 6
[f], 7
[f,a], 6
[f,a,e], 4
[f,e], 5
[d], 3
[d,a], 3
[d,a,e], 3
[d,e], 3
```

● 生成的关联规则：

```
[d,a] => [e], 1.0
[f] => [a], 0.8571428571428571
[f] => [e], 0.7142857142857143
[a] => [e], 0.75
[a] => [f], 0.75
[a,e] => [f], 0.6666666666666666
[a,e] => [d], 0.5
[f,a] => [e], 0.6666666666666666
[f,e] => [a], 0.8
```

```
[d,e] => [a], 1.0
[d]   => [a], 1.0
[d]   => [e], 1.0
[e]   => [a], 0.75
[e]   => [f], 0.625
```

## 【同步训练】

### 一、简答题

（1）什么是决策树？什么是随机森林？

（2）支持度如何定义？置信度如何定义？

（3）Apriori 算法的主要思想是什么？

（4）简单描述 FP-Growth 算法的工作流程。

### 二、编程题

（1）编写 MapReduce 程序，查找每个楼栋的采集器每 10 秒采集到的最大瞬间电流并写入 HBase。

（2）训练一个决策树分类器，输入身高和体重，判断该人是胖还是瘦。基本数据有 10 个样本，每个样本有 2 个属性，分别表示身高和体重，fat 表示胖，thin 表示瘦，如下所示：

1.550 thin     1.560 fat     1.640 thin     1.660 fat     1.760 thin
1.780 fat     1.860 thin     1.890 fat     1.970 thin     1.980 fat

（3）使用 Apriori 算法，计算数据集 dataSet = [[1, 3, 4], [2, 3, 5], [1, 2, 3, 5], [2, 5]]的频繁项集及其支持度。

（4）使用 FP-Tree 算法，计算表 10-4 中商品的频繁项集及其支持度，以及挖掘其关联规则。

# 参 考 文 献

[1] 王鹏，李俊杰，谢志明，等. 云计算和大数据技术——概念 应用与实战[M]. 北京：人民邮电出版社，2016.

[2] 林子雨. 大数据技术原理与应用——概念、存储、处理、分析与应用[M]. 北京：人民邮电出版社，2015.

[3] 李国杰，程学旗. 大数据研究：未来科技及经济社会发展的重大战略领域——大数据的研究现状与科学思考[J]. 中国科学院院刊，2012(6)：647-657.

[4] 程学旗，靳小龙，王元卓，等. 大数据系统和分析技术综述[J]. 软件学报，2014，(09)：1889-1908.

[5] 彭宇，庞景月，刘大同，等. 大数据：内涵、技术体系与展望[J]. 电子测量与仪器学报，2015，(04)：469-482.

[6] 李学龙，龚海刚. 大数据系统综述[J]. 中国科学:信息科学，2015，(01)：1-44.

[7] 中国大数据技术与产业大发展白皮书[R]. 中国计算机学会，2013.

[8] Tom White. Hadoop 权威指南（第 2 版）[M]. 周敏奇，王晓玲，金澈清，等，译. 北京：清华大学出版社，2011.

[9] Lars George. HBase 权威指南[M]. 代志远，刘佳，蒋杰，译. 北京：人民邮电出版社，2013.

[10] 李俊杰，石慧，谢志明，等. 云计算和大数据技术实战[M]. 北京：人民邮电出版社，2015.

[11] 张良均，樊哲，赵云龙，等. Hadoop 大数据分析与挖掘实战[M]. 北京：机械工业出版社，2016.

[12] Edward Capriolo，Dean Wampler，Jason Rutherglen. Hive 编程指南[M]. 曹坤，译. 北京：人民邮电出版社，2013.

[13] Alan Gates. Pig 编程指南[M]. 曹坤，译. 北京：人民邮电出版社，2013.

[14] Holden Karau，Andy Konwinski，Patrick Wendell，et al. Spark 快速大数据分析[M]. 王道远，译. 北京：人民邮电出版社，2015.

[15] 张良均，王路，谭立云，等. Python 数据分析与挖掘实战[M]. 北京：机械工业出版社，2016.